T0220931

**FPGA-based Implementation
of Signal Processing Systems**

FPGA-based Implementation of Signal Processing Systems

Second Edition

Roger Woods
Queen's University, Belfast, UK

John McAllister
Queen's University, Belfast, UK

Gaye Lightbody
University of Ulster, UK

Ying Yi
SN Systems – Sony Interactive Entertainment, UK

This edition first published 2017
© 2017 John Wiley & Sons, Ltd

All rights reserved. No part of this publication may be reproduced, stored in a retrieval system, or transmitted, in any form or by any means, electronic, mechanical, photocopying, recording or otherwise, except as permitted by law. Advice on how to obtain permission to reuse material from this title is available at http://www.wiley.com/go/permissions.

The right of Roger Woods, John McAllister, Gaye Lightbody and Ying Yi to be identified as the authors of this work has been asserted in accordance with law.

Registered Offices
John Wiley & Sons, Inc., 111 River Street, Hoboken, NJ 07030, USA
John Wiley & Sons, Ltd., The Atrium, Southern Gate, Chichester, West Sussex, PO19 8SQ, UK

Editorial Office
The Atrium, Southern Gate, Chichester, West Sussex, PO19 8SQ, UK

For details of our global editorial offices, customer services, and more information about Wiley products visit us at www.wiley.com.

Wiley also publishes its books in a variety of electronic formats and by print-on-demand. Some content that appears in standard print versions of this book may not be available in other formats.

Limit of Liability/Disclaimer of Warranty
While the publisher and authors have used their best efforts in preparing this work, they make no representations or warranties with respect to the accuracy or completeness of the contents of this work and specifically disclaim all warranties, including without limitation any implied warranties of merchantability or fitness for a particular purpose. No warranty may be created or extended by sales representatives, written sales materials or promotional statements for this work. The fact that an organization, website, or product is referred to in this work as a citation and/or potential source of further information does not mean that the publisher and authors endorse the information or services the organization, website, or product may provide or recommendations it may make. This work is sold with the understanding that the publisher is not engaged in rendering professional services. The advice and strategies contained herein may not be suitable for your situation. You should consult with a specialist where appropriate. Further, readers should be aware that websites listed in this work may have changed or disappeared between when this work was written and when it is read. Neither the publisher nor authors shall be liable for any loss of profit or any other commercial damages, including but not limited to special, incidental, consequential, or other damages.

Library of Congress Cataloging-in-Publication Data

Names: Woods, Roger, 1963- author. | McAllister, John, 1979- author. |
    Lightbody, Gaye, author. | Yi, Ying (Electrical engineer), author.
Title: FPGA-based implementation of signal processing systems / Roger Woods,
    John McAllister, Gaye Lightbody, Ying Yi.
Description: Second editon. | Hoboken, NJ : John Wiley & Sons Inc., 2017. |
    Revised edition of: FPGA-based implementation of signal processing systems /
    Roger Woods … [et al.]. 2008. | Includes bibliographical references and index.
Identifiers: LCCN 2016051193 | ISBN 9781119077954 (cloth) | ISBN 9781119077978 (epdf) |
    ISBN 9781119077961 (epub)
Subjects: LCSH: Signal processing–Digital techniques. | Digital integrated
    circuits. | Field programmable gate arrays.
Classification: LCC TK5102.5 .F647 2017 | DDC 621.382/2–dc23 LC record available at
    https://lccn.loc.gov/2016051193

Cover Design: Wiley
Cover Image: © filo/Gettyimages;
(Graph) Courtesy of the authors

Set in 10/12pt WarnockPro by Aptara Inc., New Delhi, India

10 9 8 7 6 5 4 3 2 1

The book is dedicated by the main author to his wife, Pauline, for all for her support and care, particularly over the past two years.

The support from staff from the Royal Victoria Hospital and Musgrave Park Hospital is greatly appreciated.

Contents

Preface

DSP and FPGAs

Digital signal processing (DSP) is the cornerstone of many products and services in the digital age. It is used in applications such as high-definition TV, mobile telephony, digital audio, multimedia, digital cameras, radar, sonar detectors, biomedical imaging, global positioning, digital radio, speech recognition, to name but a few! The evolution of DSP solutions has been driven by application requirements which, in turn, have only been possible to realize because of developments in silicon chip technology. Currently, a mix of programmable and dedicated system-on-chip (SoC) solutions are required for these applications and thus this has been a highly active area of research and development over the past four decades.

The result has been the emergence of numerous technologies for DSP implementation, ranging from simple microcontrollers right through to dedicated SoC solutions which form the basis of high-volume products such as smartphones. With the architectural developments that have occurred in field programmable gate arrays (FPGAs) over the years, it is clear that they should be considered as a viable DSP technology. Indeed, developments made by FPGA vendors would support this view of their technology. There are strong commercial pressures driving adoption of FPGA technology across a range of applications and by a number of commercial drivers.

The increasing costs of developing silicon technology implementations have put considerable pressure on the ability to create dedicated SoC systems. In the mobile phone market, volumes are such that dedicated SoC systems are required to meet stringent energy requirements, so application-specific solutions have emerged which vary in their degree of programmability, energy requirements and cost. The need to balance these requirements suggests that many of these technologies will coexist in the immediate future, and indeed many hybrid technologies are starting to emerge. This, of course, creates a considerable interest in using technology that is programmable as this acts to considerably reduce risks in developing new technologies.

Commonly used DSP technologies encompass software programmable solutions such as microcontrollers and DSP microprocessors. With the inclusion of dedicated DSP processing engines, FPGA technology has now emerged as a strong DSP technology. Their key advantage is that they enable users to create system architectures which allow the resources to be best matched to the system processing needs. Whilst memory resources are limited, they have a very high-bandwidth, on-chip capability. Whilst the prefabricated aspect of FPGAs avoids many of the deep problems met when developing

SoC implementations, the creation of an efficient implementation from a DSP system description remains a highly convoluted problem which is a core theme of this book.

Book Coverage

The book looks to address FPGA-based DSP systems, considering implementation at numerous levels.

- **Circuit-level** optimization techniques that allow the underlying FPGA fabric to be used more intelligently are reviewed first. By considering the detailed underlying FPGA platform, it is shown how system requirements can be mapped to provide an area-efficient, faster implementation. This is demonstrated for a number of DSP transforms and fixed coefficient filtering.
- **Architectural** solutions can be created from a signal flow graph (SFG) representation. In effect, this requires the user to exploit the highly regular, highly computative, data-independent nature of DSP systems to produce highly parallel, pipelined FPGA-based circuit architectures. This is demonstrated for filtering and beamforming applications.
- **System** solutions are now a challenge as FPGAs have now become a heterogeneous platform involving multiple hardware and software components and interconnection fabrics. There is a need for a higher-level system modeling language, e.g. dataflow which will facilitate architectural optimizations but also to address system-level considerations such as interconnection and memory.

The book covers these areas of FPGA implementation, but its key differentiating factor is that it concentrates on the second and third areas listed above, namely the creation of circuit architectures and system-level modeling; this is because circuit-level optimization techniques have been covered in greater detail elsewhere. The work is backed up with the authors' experiences in implementing practical real DSP systems and covers numerous examples including an adaptive beamformer based on a QR-based recursive least squares (RLS) filter, finite impulse response (FIR) and infinite impulse response (IIR) filters, a full search motion estimation and a fast Fourier transform (FFT) system for electronic support measures. The book also considers the development of intellectual property (IP) cores as this has become a critical aspect in the creation of DSP systems. One chapter is given over to describing the creation of such IP cores and another to the creation of an adaptive filtering core.

Audience

The book is aimed at working engineers who are interested in using FPGA technology efficiently in signal and data processing applications. The earlier chapters will be of interest to graduates and students completing their studies, taking the readers through a number of simple examples that show the trade-off when mapping DSP systems into FPGA hardware. The middle part of the book contains a number of illustrative, complex DSP system examples that have been implemented using FPGAs and whose performance clearly illustrates the benefit of their use. They provide insights into how to best use the complex FPGA technology to produce solutions optimized for speed, area and power which the authors believe is missing from current literature. The book summarizes over 30 years of learned experience of implementing complex DSP systems undertaken in many cases with commercial partners.

Second Edition Updates

The second edition has been updated and improved in a number of ways. It has been updated to reflect technology evolutions in FPGA technology, to acknowledge developments in programming and synthesis tools, to reflect on algorithms for Big Data applications, and to include improvements to some background chapters. The text has also been updated using relevant examples where appropriate.

Technology update: As FPGAs are linked to silicon technology advances, their architecture continually changes, and this is reflected in Chapter 5. A major change is the inclusion of the ARM® processor core resulting in a shift for FPGAs to a heterogeneous computing platform. Moreover, the increased use of graphical processing units (GPUs) in DSP systems is reflected in Chapter 4.

Programming tools update: Since the first edition was published, there have been a number of innovations in tool developments, particularly in the creation of commercial C-based high-level synthesis (HLS) and open computing language (OpenCL) tools. The material in Chapter 7 has been updated to reflect these changes, and Chapter 10 has been changed to reflect the changes in model-based synthesis tools.

"Big Data" processing: DSP involves processing of data content such as audio, speech, music and video information, but there is now great interest in collating huge data sets from on-line facilities and processing them quickly. As FPGAs have started to gain some traction in this area, a new chapter, Chapter 12, has been added to reflect this development.

Organization

The FPGA is a heterogeneous platform comprising complex resources such as hard and soft processors, dedicated blocks optimized for processing DSP functions and processing elements connected by both programmable and fast, dedicated interconnections. The book focuses on the challenges of implementing DSP systems on such platforms with a concentration on the high-level mapping of DSP algorithms into suitable circuit architectures.

The material is organized into three main sections.

First Section: Basics of DSP, Arithmetic and Technologies

Chapter 2 starts with a DSP primer, covering both FIR and IIR filtering, transforms including the FFT and discrete cosine transform (DCT) and concluding with adaptive filtering algorithms, covering both the least mean squares (LMS) and RLS algorithms. Chapter 3 is dedicated to computer arithmetic and covers number systems, arithmetic functions and alternative number representations such as logarithmic number representations (LNS) and coordinate rotation digital computer (CORDIC). Chapter 4 covers the technologies available to implement DSP algorithms and includes microprocessors, DSP microprocessors, GPUs and SoC architectures, including systolic arrays. In Chapter 5, a detailed description of commercial FPGAs is given with a concentration on the two main vendors, namely Xilinx and Altera, specifically their UltraScale™/Zynq® and

Stratix® 10 FPGA families respectively, but also covering technology offerings from Lattice and MicroSemi.

Second Section: Architectural/System-Level Implementation

This section covers efficient implementation from circuit architecture onto specific FPGA families; creation of circuit architecture from SFG representations; and system-level specification and implementation methodologies from high-level representations. Chapter 6 covers only briefly the efficient implementation of FPGA designs from circuit architecture descriptions as many of these approaches have been published; the text covers distributed arithmetic and reduced coefficient multiplier approaches and shows how these have been applied to fixed coefficient filters and DSP transforms. Chapter 7 covers HLS for FPGA design including new sections to reflect Xilinx's Vivado HLS tool flow and also Altera's OpenCL approach. The process of mapping SFG representations of DSP algorithms onto circuit architectures (the starting point in Chapter 6) is then described in Chapter 8. It shows how dataflow graph (DFG) descriptions can be transformed for varying levels of parallelism and pipelining to create circuit architectures which best match the application requirements, backed up with simple FIR and IIR filtering examples.

One of the ways to perform system design is to create predefined designs termed IP cores which will typically have been optimized using the techniques outlined in Chapter 8. The creation of such IP cores is outlined in Chapter 9 and acts to address the key to design productivity by encouraging "design for reuse." Chapter 10 considers model-based design for heterogeneous FPGA and focuses on dataflow modeling as a suitable design approach for FPGA-based DSP systems. The chapter outlines how it is possible to include pipelined IP cores via the white box concept using two examples, namely a normalized lattice filter (NLF) and a fixed beamformer example.

Third Section: Applications to Big Data, Low Power

The final section of the book, consisting of Chapters 11–13, covers the application of the techniques. Chapter 11 looks at the creation of a soft, highly parameterizable core for RLS filtering, showing how a generic architecture can be created to allow a range of designs to be synthesized with varying performance. Chapter 12 illustrates how FPGAs can be applied to Big Data applications where the challenge is to accelerate some complex processing algorithms. Increasingly FPGAs are seen as a low-power solution, and FPGA power consumption is discussed in Chapter 13. The chapter starts with a discussion on power consumption, highlights the importance of dynamic and static power consumption, and then describes some techniques to reduce power consumption.

Acknowledgments

The authors have been fortunate to receive valuable help, support and suggestions from numerous colleagues, students and friends, including: Michaela Blott, Ivo Bolsens, Gordon Brebner, Bill Carter, Joe Cavallaro, Peter Cheung, John Gray, Wayne Luk, Bob Madahar, Alan Marshall, Paul McCambridge, Satnam Singh, Steve Trimberger and Richard Walke.

The authors' research has been funded from a number of sources, including the Engineering and Physical Sciences Research Council, Xilinx, Ministry of Defence, Qinetiq,

BAE Systems, Selex and Department of Employment and Learning for Northern Ireland.

Several chapters are based on joint work that was carried out with the following colleagues and students: Moslem Amiri, Burak Bardak, Kevin Colgan, Tim Courtney, Scott Fischaber, Jonathan Francey, Tim Harriss, Jean-Paul Heron, Colm Kelly, Bob Madahar, Eoin Malins, Stephen McKeown, Karen Rafferty, Darren Reilly, Lok-Kee Ting, David Trainor, Richard Turner, Fahad M Siddiqui and Richard Walke.

The authors thank Ella Mitchell and Nithya Sechin of John Wiley & Sons and Alex Jackson and Clive Lawson for their personal interest and help and motivation in preparing and assisting in the production of this work.

List of Abbreviations

1D	One-dimensional
2D	Two-dimensional
ABR	Auditory brainstem response
ACC	Accumulator
ADC	Analogue-to-digital converter
AES	Advanced encryption standard
ALM	Adaptive logic module
ALU	Arithmetic logic unit
ALUT	Adaptive lookup table
AMD	Advanced Micro Devices
ANN	Artificial neural network
AoC	Analytics-on-chip
API	Application program interface
APU	Application processing unit
ARM	Advanced RISC machine
ASIC	Application-specific integrated circuit
ASIP	Application-specific instruction processor
AVS	Adaptive voltage scaling
BC	Boundary cell
BCD	Binary coded decimal
BCLA	Block CLA with intra-group, carry ripple
BRAM	Block random access memory
CAPI	Coherent accelerator processor interface
CB	Current block
CCW	Control and communications wrapper
CE	Clock enable
CISC	Complex instruction set computer
CLA	Carry lookahead adder
CLB	Configurable logic block
CNN	Convolutional neural network
CMOS	Complementary metal oxide semiconductor
CORDIC	Coordinate rotation digital computer
CPA	Carry propagation adder
CPU	Central processing unit
CSA	Conditional sum adder

CSDF	Cyclo-static dataflow
CWT	Continuous wavelet transform
DA	Distributed arithmetic
DCT	Discrete cosine transform
DDR	Double data rate
DES	Data Encryption Standard
DFA	Dataflow accelerator
DFG	Dataflow graph
DFT	Discrete Fourier transform
DG	Dependence graph
disRAM	Distributed random access memory
DM	Data memory
DPN	Dataflow process network
DRx	Digital receiver
DSP	Digital signal processing
DST	Discrete sine transform
DTC	Decision tree classification
DVS	Dynamic voltage scaling
DWT	Discrete wavelet transform
E^2PROM	Electrically erasable programmable read-only memory
EBR	Embedded Block RAM
ECC	Error correction code
EEG	Electroencephalogram
EPROM	Electrically programmable read-only memory
E-SGR	Enhanced Squared Givens rotation algorithm
EW	Electronic warfare
FBF	Fixed beamformer
FCCM	FPGA-based custom computing machine
FE	Functional engine
FEC	Forward error correction
FFE	Free-form expression
FFT	Fast Fourier transform
FIFO	First-in, first-out
FIR	Finite impulse response
FPGA	Field programmable gate array
FPL	Field programmable logic
FPU	Floating-point unit
FSM	Finite state machine
FSME	Full search motion estimation
GFLOPS	Giga floating-point operations per second
GMAC	Giga multiply-accumulates
GMACS	Giga multiply-accumulate per second
GOPS	Giga operations per second
GPUPU	General-purpose graphical processing unit
GPU	Graphical processing unit
GRNN	General regression neural network
GSPS	Gigasamples per second

HAL	Hardware abstraction layer
HDL	Hardware description language
HKMG	High-K metal gate
HLS	High-level synthesis
I2C	Inter-Integrated circuit
I/O	Input/output
IC	Internal cell
ID	Instruction decode
IDE	Integrated design environment
IDFT	Inverse discrete Fourier transform
IEEE	Institute of Electrical and Electronic Engineers
IF	Instruction fetch
IFD	Instruction fetch and decode
IFFT	Inverse fast Fourier transform
IIR	Infinite impulse response
IM	Instruction memory
IoT	Internet of things
IP	Intellectual property
IR	Instruction register
ITRS	International Technology Roadmap for Semiconductors
JPEG	Joint Photographic Experts Group
KCM	Constant-coefficient multiplication
KM	Kernel memory
KPN	Kahn process network
LAB	Logic array blocks
LDCM	Logic delay measurement circuit
LDPC	Low-density parity-check
LLVM	Low-level virtual machine
LMS	Least mean squares
LNS	Logarithmic number representations
LPDDR	Low-power double data rate
LS	Least squares
lsb	Least significant bit
LTI	Linear time-invariant
LUT	Lookup table
MA	Memory access
MAC	Multiply-accumulate
MAD	Minimum absolute difference
MADF	Multidimensional arrayed dataflow
MD	Multiplicand
ME	Motion estimation
MIL-STD	Military standard
MIMD	Multiple instruction, multiple data
MISD	Multiple instruction, single data
MLAB	Memory LAB
MMU	Memory management unit
MoC	Model of computation

MPE	Media processing engine
MPEG	Motion Picture Experts Group
MPSoC	Multi-processing SoC
MR	Multiplier
MR-DFG	Multi-rate dataflow graph
msb	Most significant bit
msd	Most significant digit
MSDF	Multidimensional synchronous dataflow
MSI	Medium-scale integration
MSPS	Megasamples per second
NaN	Not a Number
NLF	Normalized lattice filter
NRE	Non-recurring engineering
OCM	On-chip memory
OFDM	Orthogonal frequency division multiplexing
OFDMA	Orthogonal frequency division multiple access
OLAP	On-line analytical processing
OpenCL	Open computing language
OpenMP	Open multi-processing
ORCC	Open RVC-CAL Compiler
PAL	Programmable Array Logic
PB	Parameter bank
PC	Program counter
PCB	Printed circuit board
PCI	Peripheral component interconnect
PD	Pattern detect
PE	Processing element
PL	Programmable logic
PLB	Programmable logic block
PLD	Programmable logic device
PLL	Phase locked loop
PPT	Programmable power technology
PS	Processing system
QAM	Quadrature amplitude modulation
QR-RLS	QR recursive least squares
RAM	Random access memory
RAN	Radio access network
RCLA	Block CLA with inter-block ripple
RCM	Reduced coefficient multiplier
RF	Register file
RISC	Reduced instruction set computer
RLS	Recursive least squares
RNS	Residue number representations
ROM	Read-only memory
RT	Radiation tolerant
RTL	Register transfer level
RVC	Reconfigurable video coding

SBNR	Signed binary number representation
SCU	Snoop control unit
SD	Signed digits
SDF	Synchronous dataflow
SDK	Software development kit
SDNR	Signed digit number representation
SDP	Simple dual-port
SERDES	Serializer/deserializer
SEU	Single event upset
SFG	Signal flow graph
SGR	Squared Givens rotation
SIMD	Single instruction, multiple data
SISD	Single instruction, single data
SMP	Shared-memory multi-processors
SNR	Signal-to-noise ratio
SoC	System-on-chip
SOCMINT	Social media intelligence
SoPC	System on programmable chip
SPI	Serial peripheral interface
SQL	Structured query language
SR-DFG	Single-rate dataflow graph
SRAM	Static random access memory
SRL	Shift register lookup table
SSD	Shifted signed digits
SVM	Support vector machine
SW	Search window
TCP	Transmission Control Protocol
TFLOPS	Tera floating-point operations per second
TOA	Time of arrival
TR	Throughout rate
TTL	Transistor-transistor logic
UART	Universal asynchronous receiver/transmitter
ULD	Ultra-low density
UML	Unified modeling language
VHDL	VHSIC hardware description language
VHSIC	Very high-speed integrated circuit
VLIW	Very long instruction word
VLSI	Very large scale integration
WBC	White box component
WDF	Wave digital filter

1

Introduction to Field Programmable Gate Arrays

1.1 Introduction

Electronics continues to make an impact in the twenty-first century and has given birth to the computer industry, mobile telephony and personal digital entertainment and services industries, to name but a few. These markets have been driven by developments in silicon technology as described by Moore's law (Moore 1965), which is represented pictorially in Figure 1.1. This has seen the number of transistors double every 18 months. Moreover, not only has the number of transistors doubled at this rate, but also the costs have decreased, thereby reducing the cost per transistor at every technology advance.

In the 1970s and 1980s, electronic systems were created by aggregating standard components such as microprocessors and memory chips with digital logic components, e.g. dedicated integrated circuits along with dedicated input/output (I/O) components on printed circuit boards (PCBs). As levels of integration grew, manufacturing working PCBs became more complex, largely due to greater component complexity in terms of the increase in the number of transistors and I/O pins. In addition, the development of multi-layer boards with as many as 20 separate layers increased the design complexity. Thus, the probability of incorrectly connecting components grew, particularly as the possibility of successfully designing and testing a working system before production was coming under greater and greater time pressures.

The problem became more challenging as system descriptions evolved during product development. Pressure to create systems to meet evolving standards, or that could change after board construction due to system alterations or changes in the design specification, meant that the concept of having a "fully specified" design, in terms of physical system construction and development on processor software code, was becoming increasingly challenging. Whilst the use of programmable processors such as microcontrollers and microprocessors gave some freedom to the designer to make alterations in order to correct or modify the system after production, this was limited. Changes to the interconnections of the components on the PCB were restricted to I/O connectivity of the processors themselves. Thus the attraction of using programmability interconnection or "glue logic" offered considerable potential, and so the concept of field

FPGA-based Implementation of Signal Processing Systems,
Second Edition. Roger Woods, John McAllister, Gaye Lightbody and Ying Yi.
© 2017 John Wiley & Sons, Ltd. Published 2017 by John Wiley & Sons, Ltd.

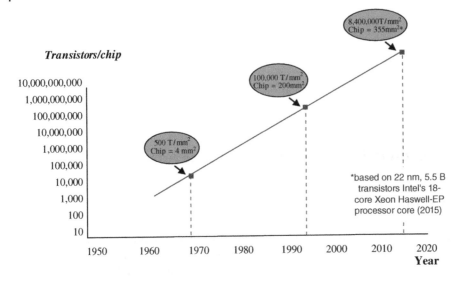

Figure 1.1 Moore's law

programmable logic (FPL), specifically field programmable gate array (FPGA) technology, was born.

From this unassuming start, though, FPGAs have grown into a powerful technology for implementing digital signal processing (DSP) systems. This emergence is due to the integration of increasingly complex computational units into the fabric along with increasing complexity and number of levels in memory. Coupled with a high level of programmable routing, this provides an impressive heterogeneous platform for improved levels of computing. For the first time ever, we have seen evolutions in heterogeneous FPGA-based platforms from Microsoft, Intel and IBM. FPGA technology has had an increasing impact on the creation of DSP systems. Many FPGA-based solutions exist for wireless base station designs, image processing and radar systems; these are, of course, the major focus of this text.

Microsoft has developed acceleration of the web search engine Bing using FPGAs and shows improved ranking throughput in a production search infrastructure. IBM and Xilinx have worked closely together to show that they can accelerate the reading of data from web servers into databases by applying an accelerated Memcache2; this is a general-purpose distributed memory caching system used to speed up dynamic database-driven searches (Blott and Vissers 2014). Intel have developed a multicore die with Altera FPGAs, and their recent purchase of the company (Clark 2015) clearly indicates the emergence of FPGAs as a core component in heterogeneous computing with a clear target for data centers.

1.2 Field Programmable Gate Arrays

The FPGA concept emerged in 1985 with the XC2064^{TM} FPGA family from Xilinx. At the same time, a company called Altera was also developing a programmable device, later to become the EP1200, which was the first high-density programmable logic device (PLD). Altera's technology was manufactured using 3-μm complementary metal oxide semiconductor (CMOS) electrically programmable read-only memory (EPROM) technology and required ultraviolet light to erase the programming, whereas Xilinx's technology was based on conventional static random access memory (SRAM) technology and required an EPROM to store the programming.

The co-founder of Xilinx, Ross Freeman, argued that with continuously improving silicon technology, transistors were going to become cheaper and cheaper and could be used to offer programmability. This approach allowed system design errors which had only been recognized at a late stage of development to be corrected. By using an FPGA to connect the system components, the interconnectivity of the components could be changed as required by simply reprogramming them. Whilst this approach introduced additional delays due to the programmable interconnect, it avoided a costly and time-consuming PCB redesign and considerably reduced the design risks.

At this stage, the FPGA market was populated by a number of vendors, including Xilinx, Altera, Actel, Lattice, Crosspoint, Prizm, Plessey, Toshiba, Motorola, Algotronix and IBM. However, the costs of developing technologies not based on conventional integrated circuit design processes and the need for programming tools saw the demise of many of these vendors and a reduction in the number of FPGA families. SRAM technology has now emerged as the dominant technology largely due to cost, as it does not require a specialist technology. The market is now dominated by Xilinx and Altera, and, more importantly, the FPGA has grown from a simple glue logic component to a complete system on programmable chip (SoPC) comprising on-board physical processors, soft processors, dedicated DSP hardware, memory and high-speed I/O.

The FPGA evolution was neatly described by Steve Trimberger in his FPL2007 plenary talk (see the summary in Table 1.1). The evolution of the FPGA can be divided into three eras. The age of *invention* was when FPGAs started to emerge and were being used as system components typically to provide programmable interconnect giving protection to design evolutions and variations. At this stage, design tools were primitive, but designers were quite happy to extract the best performance by dealing with lookup tables (LUTs) or single transistors.

As highlighted above, there was a rationalization of the technologies in the early 1990s, referred to by Trimberger as the great architectural shakedown. The age of *expansion* was when the FPGA started to approach the problem size and thus design complexity was key. This meant that it was no longer sufficient for FPGA vendors to just produce

Table 1.1 Three ages of FPGAs

Period	Age	Comments
1984–1991	Invention	Technology is limited, FPGAs are much smaller than the application problem size. Design automation is secondary, architecture efficiency is key
1992–1999	Expansion	FPGA size approaches the problem size. Ease of design becomes critical
2000–present	Accumulation	FPGAs are larger than the typical problem size. Logic capacity limited by I/O bandwidth

place and route tools and it became critical that hardware description languages (HDLs) and associated synthesis tools were created. The final *evolution* period was the period of accumulation when FPGAs started to incorporate processors and high-speed interconnection. Of course, this is very relevant now and is described in more detail in Chapter 5 where the recent FPGA offerings are reviewed.

This has meant that the FPGA market has grown from nothing in just over 20 years to become a key player in the IC industry, worth some $3.9 billion in 2014 and expected to be worth around $7.3 billion in 2022 (MarketsandMarkets 2016). It has been driven by the growth in the automotive sector, mobile devices in the consumer electronics sector and the number of data centers.

1.2.1 Rise of Heterogeneous Computing Platforms

Whilst Moore's law is presented here as being the cornerstone for driving FPGA evolution and indeed electronics, it also has been the driving force for computing. However, all is not well with computing's reliance on silicon technology. Whilst the number of transistors continues to double, the scaling of clock speed has not continued at the same rate. This is due to the increase in power consumption, particularly the increase in static power. The issue of the heat dissipation capability of packaging means that computing platform providers such as Intel have limited their processor power to 30 W. This resulted in an adjustment in the prediction for clock rates between 2005 and 2011 (as illustrated in Figure 1.2) as clock rate is a key contributor to power consumption (ITRS 2005).

In 2005, the International Technology Roadmap for Semiconductors (ITRS) predicted that a 100 GHz clock would be achieved in 2020, but this estimation had to be revised first in 2007 and then again in 2011. This has been seen in the current technology where a clock rate of some 30 GHz was expected in 2015 based on the original forecast, but we see that speeds have been restricted to 3–4 GHz. This has meant that the performance per gigahertz has effectively stalled since 2005 and has generated the interest by major

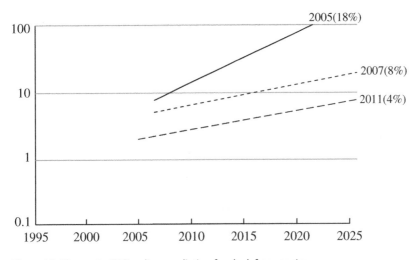

Figure 1.2 Change in ITRS scaling prediction for clock frequencies

computing companies in exploring different architectures that employ FPGA technology (Putnam *et al.* 2014; Blott and Vissers 2014).

1.2.2 Programmability and DSP

On many occasions, the growth indicated by Moore's law has led people to argue that transistors are essentially free and therefore can be exploited, as in the case of programmable hardware, to provide additional flexibility. This could be backed up by the observation that the cost of a transistor has dropped from one-tenth of a cent in the 1980s to one-thousandth of a cent in the 2000s. Thus we have seen the introduction of hardware programmability into electronics in the form of FPGAs.

In order to make a single transistor programmable in an SRAM technology, the programmability is controlled by storing a "1" or a "0" on the gate of the transistor, thereby making it conduct or not. This value is then stored in an SRAM cell which, if it requires six transistors, will will mean that we need seven transistors to achieve one programmable equivalent in FPGA. The reality is that in an overall FPGA implementation, the penalty is nowhere as harsh as this, but it has to be taken into consideration in terms of ultimate system cost.

It is the ability to program the FPGA hardware after fabrication that is the main appeal of the technology; this provides a new level of reassurance in an increasingly competitive market where "right first time" system construction is becoming more difficult to achieve. It would appear that that assessment was vindicated in the late 1990s and early 2000s: when there was a major market downturn, the FPGA market remained fairly constant when other microelectronic technologies were suffering. Of course, the importance of programmability has already been demonstrated by the microprocessor, but this represented a new change in how programmability was performed.

The argument developed in the previous section presents a clear advantage of FPGA technology in overcoming PCB design errors and manufacturing faults. Whilst this might have been true in the early days of FPGA technology, evolution in silicon technology has moved the FPGA from being a programmable interconnection technology to making it into a system component. If the microprocessor or microcontroller was viewed as programmable system component, the current FPGA devices must also be viewed in this vein, giving us a different perspective on system implementation.

In electronic system design, the main attraction of the microprocessor is that it considerably lessens the risk of system development. As the hardware is fixed, all of the design effort can be concentrated on developing the code. This situation has been complemented by the development of efficient software compilers which have largely removed the need for the designer to create assembly language; to some extent, this can even absolve the designer from having a detailed knowledge of the microprocessor architecture (although many practitioners would argue that this is essential to produce good code). This concept has grown in popularity, and embedded microprocessor courses are now essential parts of any electrical/electronic or computer engineering degree course.

A lot of this process has been down to the software developer's ability to exploit an underlying processor architecture, the von Neumann architecture. However, this advantage has also been the limiting factor in its application to the topic of this text, namely DSP. In the von Neumann architecture, operations are processed sequentially, which allows relatively straightforward interpretation of the hardware for programming purposes; however, this severely limits the performance in DSP applications which exhibit high levels of parallelism and have operations that are highly data-independent. This cries out for parallel realization, and whilst DSP microprocessors go some way toward addressing this situation by providing concurrency in the form of parallel hardware and software "pipelining," there is still the concept of one architecture suiting all sizes of the DSP problem.

This limitation is overcome in FPGAs as they allow what can be considered to be a second level of programmability, namely programming of the underlying processor architecture. By creating an architecture that best meets the algorithmic requirements, high levels of performance in terms of area, speed and power can be achieved. This concept is not new as the idea of deriving a system architecture to suit algorithmic requirements has been the cornerstone of application-specific integrated circuit (ASIC) implementations. In high volumes, ASIC implementations have resulted in the most cost-effective, fastest and lowest-energy solutions. However, increasing mask costs and the impact of "right first time" system realization have made the FPGA a much more attractive alternative.

In this sense, FPGAs capture the performance aspects offered by ASIC implementation, but with the advantage of programmability usually associated with programmable processors. Thus, FPGA solutions have emerged which currently offer several hundreds of giga operations per second (GOPS) on a single FPGA for some DSP applications, which is at least an order of magnitude better performance than microprocessors.

1.3 Influence of Programmability

In many texts, Moore's law is used to highlight the evolution of silicon technology, but another interesting viewpoint particularly relevant for FPGA technology is Makimoto's wave, which was first published in the January 1991 edition of *Electronics Weekly*. It is based on an observation by Tsugio Makimoto who noted that technology has shifted between standardization and customization. In the 1960s, 7400 TTL series logic chips were used to create applications; and then in the early 1970s, the custom large-scale integration era emerged where chips were created (or customized) for specific applications such as the calculator. The chips were now increasing in their levels of integration and so the term "medium-scale integration" (MSI) was born. The evolution of the microprocessor in the 1970s saw the swing back towards standardization where one "standard" chip was used for a wide range of applications.

The 1980s then saw the birth of ASICs where designers could overcome the fact that the sequential microprocessor posed severe limitations in DSP applications where higher levels of computations were needed. The DSP processor also emerged, such as the TMS32010, which differed from conventional processors as they were based on the Harvard architecture which had separate program and data memories and separate buses. Even with DSP processors, ASICs offered considerable potential in terms of processing power and, more importantly, power consumption. The development of the FPGA from a "glue component" that allowed other components to be connected together to form a system to become a component or even a system itself led to its increased popularity.

The concept of coupling microprocessors with FPGAs in heterogeneous platforms was very attractive as this represented a completely programmable platform with microprocessors to implement the control-dominated aspects of DSP systems and FPGAs to implement the data-dominated aspects. This concept formed the basis of FPGA-based custom computing machines (FCCMs) which formed the basis for "configurable" or reconfigurable computing (Villasenor and Mangione-Smith 1997). In these systems, users could not only implement computational complex algorithms in hardware, but also use the programmability aspect of the hardware to change the system functionality, allowing the development of "virtual hardware" where hardware could 'virtually" implement systems that are an order of magnitude larger (Brebner 1997).

We would argue that there have been two programmability eras. The first occurred with the emergence of the microprocessor in the 1970s, where engineers could develop programmable solutions based on this fixed hardware. The major challenge at this time was the software environments; developers worked with assembly language, and even when compilers and assemblers emerged for C, best performance was achieved by hand-coding. Libraries started to appear which provided basic common I/O functions, thereby allowing designers to concentrate on the application. These functions are now readily available as core components in commercial compilers and assemblers. The need for high-level languages grew, and now most programming is carried out in high-level programming languages such as C and Java, with an increased use of even higher-level environments such as the unified modeling language (UML).

The second era of programmability was ushered in by FPGAs. Makimoto indicates that field programmability is standardized in manufacture and customized in application. This can be considered to have offered hardware programmability if you think in terms of the first wave as the programmability in the software domain where the hardware remains fixed. This is a key challenge as most computer programming tools work on the fixed hardware platform principle, allowing optimizations to be created as there is clear direction on how to improve performance from an algorithmic representation. With FPGAs, the user is given full freedom to define the architecture which best suits the application. However, this presents a problem in that each solution must be hand-crafted and every hardware designer knows the issues in designing and verifying hardware designs!

Some of the trends in the two eras have similarities. In the early days, schematic capture was used to design early circuits, which was synonymous with assembly-level programming. Hardware description languages such as VHSIC Hardware Description Language (VHDL) and Verilog then started to emerge that could used to produce a higher level of abstraction, with the current aim to have C-based tools such as SystemC and Catapult® from Mentor Graphics as a single software-based programming environment (Very High Speed Integrated Circuit (VHSIC) was a US Department of Defense funded program in the late 1970s and early 1980s with the aim of producing the next generation of integrated circuits). Initially, as with software programming languages, there was mistrust in the quality of the resulting code produced by these approaches.

With the establishment of improved cost-effectiveness, synthesis tools are equivalent to the evolution of efficient software compilers for high-level programming languages, and the evolution of library functions allowed a high degree of confidence to be subsequently established; the use of HDLs is now commonplace for FPGA implementation. Indeed, the emergence of intellectual property (IP) cores mirrored the evolution of libraries such as I/O programming functions for software flows; they allowed common functions to be reused as developers trusted the quality of the resulting implementation produced by such libraries, particularly as pressures to produce more code within the same time-span grew. The early IP cores emerged from basic function libraries into complex signal processing and communications functions such as those available from the FPGA vendors and the various web-based IP repositories.

1.4 Challenges of FPGAs

In the early days, FPGAs were seen as glue logic chips used to plug components together to form complex systems. FPGAs then increasingly came to be seen as complete systems in themselves, as illustrated in Table 1.1. In addition to technology evolution, a number of other considerations accelerated this. For example, the emergence of the FPGA as a DSP platform was accelerated by the application of distributed arithmetic (DA) techniques (Goslin 1995; Meyer-Baese 2001). DA allowed efficient FPGA implementations to be realized using the lookup table or LUT-based/adder constructs of FPGA blocks and allowed considerable performance gains to be gleaned for some DSP transforms such as fixed coefficient filtering and transform functions such as the fast Fourier transform (FFT). Whilst these techniques demonstrated that FPGAs could produce highly effective solutions for DSP applications, the idea of squeezing the last aspect of performance out of the FPGA hardware and, more importantly, spending several person-months creating such innovative designs was now becoming unacceptable.

The increase in complexity due to technology evolution meant that there was a growing gap in the scope offered by current FPGA technology and the designer's ability to develop solutions efficiently using currently available tools. This was similar to the "design productivity gap" (ITRS 1999) identified in the ASIC industry where it was perceived that ASIC design capability was only growing at 25% whereas Moore's law growth was 60%. The problem is not as severe in FPGA implementation as the designer does not have to deal with sub-micrometer design issues. However, a number of key issues exist:

- **Understanding how to map DSP functionality into FPGA**. Some of the aspects are relatively basic in this arena, such as multiply-accumulate (MAC) and delays being mapped onto on-board DSP blocks, registers and RAM components, respectively. However, the understanding of floating-point versus fixed-point, wordlength optimization, algorithmic transformation cost functions for FPGA and impact of routing delay are issues that must be considered at a system level and can be much harder to deal with at this level.
- **Design languages**. Currently hardware description languages such as VHDL and Verilog and their respective synthesis flows are well established. However, users are now looking at FPGAs, with the recent increase in complexity resulting in the integration of both fixed and programmable microprocessor cores as a complete system. Thus, there is increased interest in design representations that more clearly represent system descriptions. Hence there is an increased electronic design automation focus on using C as a design language, but other representations also exist such as those methods based on model of computation (MoC), e.g. synchronous dataflow.

- **Development and use of IP cores**. With the absence of quick and reliable solutions to the design language and synthesis issues, the IP market in SoC implementation has emerged to fill the gap and allow rapid prototyping of hardware. Soft cores are particularly attractive as design functionality can be captured using HDLs and efficiently translated into the FPGA technology of choice in a highly efficient manner by conventional synthesis tools. In addition, processor cores have been developed which allow dedicated functionality to be added. The attraction of these approaches is that they allow application-specific functionality to be quickly created as the platform is largely fixed.

- **Design flow**. Most of the design flow capability is based around developing FPGA functionality from some form of higher-level description, mostly for complex functions. The reality now is that FPGA technology is evolving at such a rate that systems comprising FPGAs and processors are starting to emerge as an SoC platform or indeed, FPGAs as a single SoC platform as they have on-board hard and soft processors, high-speed communications and programmable resource, and this can be viewed as a complete system. Conventionally, software flows have been more advanced for processors and even multiple processors as the architecture is fixed. Whilst tools have developed for hardware platforms such as FPGAs, there is a definite need for software for flows for heterogeneous platforms, i.e. those that involve both processors and FPGAs.

These represent the challenges that this book aims to address and provide the main focus for the work that is presented.

Bibliography

Blott M, Vissers K 2014 Dataflow architectures for 10Gbps line-rate key-value-stores. In *Proc. IEEE Hot Chips*, Palo Alto, CA.

Brebner G 1997 The swappable logic unit. In *Proc. IEEE Symp. on FCCM*, Napa, CA, pp. 77–86.

Clark D 2015 Intel completes acquisition of Altera. *Wall Street J.*, December 28.

Goslin G 1995 Using Xilinx FPGAs to design custom digital signal processing devices. In *Proc. DSPX*, pp. 565–604.

ITRS 1999 International Technology Roadmap for Semiconductors, Semiconductor Industry Association. Downloadable from http://public.itrs.net (accessed February 16, 2016).

ITRS 2005 International Technology Roadmap for Semiconductors: Design. available from http://www.itrs.net/Links/2005ITRS/Design2005.pdf (accessed February 16, 2016).

MarketsandMarkets 2016 FPGA Market, by Architecture (SRAM Based FPGA, Anti-Fuse Based FPGA, and Flash Based FPGA), Configuration (High-End FPGA, Midrange FPGA, and Low-End FPGA), Application, and Geography - Trends & Forecasts to 2022. Report Code: SE 3058, downloadable from marketsandmarkets.com (accessed February 16, 2016).

Meyer-Baese U 2001 *Digital Signal Processing with Field Programmable Gate Arrays* Springer, Berlin.

Moore GE 1965 Cramming more components onto integrated circuits. In *Electronics*.

Available from http://www.cs.utexas.edu/ fussell/courses/cs352h/papers/moore.pdf (accessed February 16, 2016).

Putnam A, Caulfield AM, Chung ES, Chiou D, Constantinides K, Demme J, Esmaeilzadeh H, Fowers J, Gopal GP, Gray J, Haselman M, Hauck S, Heil S, Hormati A, Kim J-Y, Lanka S, Larus J, Peterson E, Pope S, Smith A, Thong J, Xiao PY, Burger D 2014 A reconfigurable fabric for accelerating large-scale datacenter services. In *Proc. IEEE Int. Symp. on Computer Architecture*, pp. 13–24.

Villasenor J, Mangione-Smith WH 1997 Configurable computing. *Scientific American*, 276(6), 54–59.

2

DSP Basics

2.1 Introduction

In the early days of electronics, signals were processed and transmitted in their natural form, typically an analogue signal created from a source signal such as speech, then converted to electrical signals before being transmitted across a suitable transmission medium such as a broadband connection. The appeal of processing signals digitally was recognized quite some time ago for a number of reasons. Digital hardware is generally superior and more reliable than its analogue counterpart, which can be prone to aging and can give uncertain performance in production.

DSP, on the other hand, gives a guaranteed accuracy and essentially perfect reproducibility (Rabiner and Gold 1975). In addition, there is considerable interest in merging the multiple networks that transmit these signals, such as the telephone transmission networks, terrestrial TV networks and computer networks, into a single or multiple digital transmission media. This provides a strong motivation to convert a wide range of information formats into their digital formats.

Microprocessors, DSP microprocessors and FPGAs are a suitable platform for processing such digital signals, but it is vital to understand a number of basic issues with implementing DSP algorithms on, in this case, FPGA platforms. These issues range from understanding both the sampling rates and computation rates of different applications with the aim of understanding how these requirements affect the final FPGA implementation, right through to the number representation chosen for the specific FPGA platform and how these decisions impact the performance. The choice of algorithm and arithmetic requirements can have severe implications for the quality of the final implementation.

As the main concern of this book is the implementation of such systems in FPGA hardware, this chapter aims to give the reader an introduction to DSP algorithms to such a level as to provide grounding for many of the examples that are described later. A number of more extensive introductory texts that explain the background of DSP systems can be found in the literature, ranging from the basic principles (Lynn and Fuerst 1994; Williams 1986) to more comprehensive texts (Rabiner and Gold 1975).

FPGA-based Implementation of Signal Processing Systems,
Second Edition. Roger Woods, John McAllister, Gaye Lightbody and Ying Yi.
© 2017 John Wiley & Sons, Ltd. Published 2017 by John Wiley & Sons, Ltd.

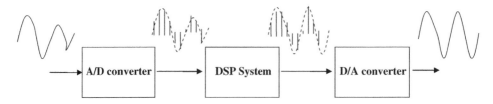

Figure 2.1 Basic DSP system

Section 2.2 gives an introduction to basic DSP concepts that affect hardware implementation. A brief description of common DSP algorithms is then given, starting with a review of transforms, including the FFT, discrete cosine transform (DCT) and the discrete wavelet transform (DWT) in Section 2.3. The chapter then moves on to review filtering in Section 2.4, giving a brief description of finite impulse response (FIR) filters, infinite impulse response (IIR) filters and wave digital filters (WDFs). Section 2.5 is dedicated to adaptive filters and covers both the least mean squares (LMS) and recursive least squares (RLS) algorithms. Concluding comments are given in Section 2.6.

2.2 Definition of DSP Systems

DSP algorithms accept inputs from real-world digitized signals such as voice, audio, video and sensor data (temperature, humidity), and mathematically process them according to the required algorithm's processing needs. A simple diagram of this process is given in Figure 2.1. Given that we are living in a digital age, there is a constantly increasing need to process more data in the fastest way possible.

The digitized signal is obtained as shown in Figure 2.2 where an analogue signal is converted into a pulse of signals and then quantized to a range of values. The input is typically $x(n)$, which is a stream of numbers in digital format, and the output is given as $y(n)$.

Modern DSP applications mainly involve speech, audio, image, video and communications systems, as well as error detection and correction and encryption algorithms. This involves real-time processing of a considerable amount of different types of content at a series of sampling rates ranging from 1 Hz in biomedical applications, right up to tens of megahertz in image processing applications. In a lot of cases, the aim is to process the data to enhance part of the signal (e.g. edge detection in image processing), eliminate interference (e.g. jamming signals in radar applications), or remove erroneous input (e.g. echo or noise cancelation in telephony); other DSP algorithms are essential in capturing, storing and transmitting data, audio, images, and video compression techniques have been used successfully in digital broadcasting and telecommunications.

Figure 2.2 Digitization of analogue signals

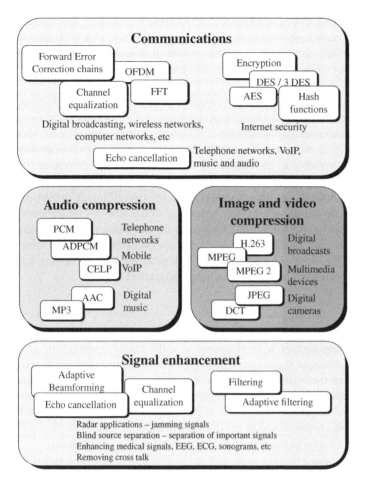

Figure 2.3 Example applications for DSP

Over the years, much of the need for such processing has been standardized; Figure 2.3 shows some of the algorithms required in a range of applications. In communications, the need to provide efficient transmission using orthogonal frequency division multiplexing (OFDM) has emphasized the need for circuits for performing FFTs. In image compression, the evolution initially of the Joint Photographic Experts Group (JPEG) and then the Motion Picture Experts Group (MPEG) led to the development of the JPEG and MPEG standards respectively involving a number of core DSP algorithms, specifically DCT and motion estimation and compensation.

The appeal of processing signals digitally was recognized quite some time ago as digital hardware is generally superior to and more reliable than its analogue counterpart; analogue hardware can be prone to aging and can give uncertain performance in production. DSP, on the other hand, gives a guaranteed accuracy and essentially perfect reproducibility (Rabiner and Gold 1975).

The proliferation of DSP technology has mainly been driven by the availability of increasingly cheap hardware, allowing the system to be easily interfaced to computer technology, and in many cases, to be implemented on the same computers. The need for many of the applications mentioned in Figure 2.3 has driven the need for increasingly complex DSP systems, which in turn has seen the growth of research into developing efficient implementation of some DSP algorithms. This has also driven the need for DSP microprocessors covered in Chapter 4.

A number of different DSP functions can be carried out either in the time domain, such as filtering, or in the frequency domain by performing an FFT (Rabiner and Gold 1975). The DCT forms the central mechanism for JPEG image compression which is also the foundation for the MPEG standards. This DCT algorithm enables the components within the image that are invisible to the naked eye to be identified by converting the spatial image into the frequency domain. They can then be removed using quantization in the MPEG standard without discernible degradation in the overall image quality. By increasing the amount of data removed, greater reduction in file size can be achieved. Wavelet transforms offer both time domain and frequency domain information and have roles not only in applications for image compression, but also in extraction of key information from signals and for noise cancelation. One such example is in extracting key features from medical signals such as the electroencephalogram (EEG).

2.2.1 Sampling

Sampling is an essential process in DSP that allows real-life continuous-time domain signals, in other words analogue signals, to be represented in the digital domain. The process of representation of analogue signals process begins with sampling and is followed by the quantization within the analogue-to-digital converters (ADCs). Therefore, the two most important components in the sampling process are the selection of the samples in time domain and subsequent quantization of the samples within the ADC, which results in quantization noise being added to the digitized analogue signal. The choice of sampling frequency directly affects the size of data processed by the DSP system.

A continuous-time (analogue) signal can be converted into a discrete-time signal by sampling the continuous-time signal at uniformly distributed discrete-time instants. Sampling an analogue signal can be represented by the relation

$$x(n) = x_a(nT), \quad -\infty < n < \infty, \tag{2.1}$$

where $x_a(nT)$ represents the uniformly sampled discrete-time signal. The data stream, $x(n)$, is obtained by sampling the continuous-time signal at the required time interval, given as the sampling instance, T; this is called the sampling period or interval, and its reciprocal is called the sampling rate, F_s.

The question arises as to how precise the digital data need to be to have a meaningful representation of the analogue word. This definition is explained by the Nyquist–Shannon sampling theorem which states that exact reconstruction of a continuous-time baseband signal from its samples is possible if the signal is bandlimited and the sampling frequency is greater than twice the signal bandwidth. The sampling theory was introduced into communication theory by Nyquist (1928) and then into information theory by Shannon (1949).

2.2.2 Sampling Rate

Many computing vendors quote *clock rates*, whereas the rate of computation in DSP systems is given as the *sampling rate*. It is also important to delineate the *throughput*

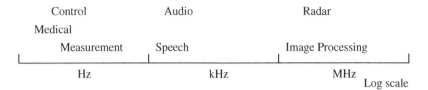

Figure 2.4 Sampling rates for many DSP systems

rate. The sampling rate for many typical DSP systems is given in Figure 2.4 and indicates the rate at which data are fed into and/or from the system. It should not be used to dictate technology choice as, for example, we could have a 128-tap FIR filter requirement for an audio application where the sampling rate may be 44.2 kHz but the throughput rate will be 11.2 megasamples per second (MSPS), as during each sample we need to compute 128 multiplications and 127 additions (255 operations) at the sampling rate.

In simple terms when digitizing an analogue signal, the rate of sampling must be at least twice the maximum frequency f_m (within the signal being digitized) so as to maintain the information and prevent aliasing (Shannon 1949). In other words, the signal needs to be bandlimited, meaning that there is no spectral energy above the maximum frequency f_m. The Nyquist sampling rate f_s is then determined as $2f_m$, usually by human factors (e.g. perception).

A simple example is the sampling of speech, which is standardized at 8 kHz. This sampling rate is sufficient to provide an accurate representation of the spectral components of speech signals, as the spectral energy above 4 kHz, and probably 3 kHz, does not contribute greatly to signal quality. In contrast, digitizing music typically requires a sample rate of 44.2 kHz to cover the spectral range of 22.1 kHz as it is acknowledged that this is more than sufficient to cope with the hearing range of the human ear which typically cannot detect signals above 18 kHz. Moreover, this increase is natural due to the more complex spectral composition of music when compared with speech.

In other applications, the determination of the sampling rate does not just come down to human perception, but involves other aspects. Take, for example, the digitizing of medical signals such as EEGs which are the result of electrical activity within the brain picked up from electrodes in contact with the surface of the skin. In capturing the information, the underlying waveforms can be heavily contaminated by noise. One particular application is a hearing test whereby a stimulus is applied to the subject's ear and the resulting EEG signal is observed at a certain location on the scalp. This test is referred to as the auditory brainstem response (ABR) as it looks for an evoked response from the EEG in the brainstem region of the brain, within 10 ms of the stimulus onset.

The ABR waveform of interest has a frequency range of 100–3000 Hz, therefore bandpass filtering of the EEG signal to this region is performed during the recording process prior to digitization. However, there is a slow response roll-off at the boundaries and unwanted frequencies may still be present. Once digitized the EEG signal may be filtered again, possibly using wavelet denoising to remove the upper and lower contaminating frequencies. The duration of the ABR waveform of interest is 20 ms, 10 ms prior to stimulus and 10 ms afterward. The EEG is sampled at 20 kHz, therefore with a Nyquist frequency of 10 kHz, which exceeds twice the highest frequency component (3 kHz) present in the signal. This equates to 200 samples, before and after the stimulus.

2.3 DSP Transformations

This section gives a brief overview of some of the key DSP transforms mentioned in Chapter 13, including a brief description of applications and their use.

2.3.1 Discrete Fourier Transform

The Fourier transform is the transform of a signal from the time domain representation to the frequency domain representation. In basic terms it breaks a signal up into its constituent frequency components, representing a signal as the sum of a series of sines and cosines.

The Fourier series expansion of a periodic function, $f(t)$, is given by

$$f(t) = \frac{1}{2}a_0 + \sum_{n=1}^{\infty} [a_n \cos(\omega_n t) + b_n \sin(\omega_n t)] \tag{2.2}$$

where, for any non-negative integer n, ω_n is the nth harmonic in radians of $f(t)$ given by

$$\omega_n = n \frac{w\pi}{T}, \tag{2.3}$$

the a_n are the even Fourier coefficients of $f(t)$, given by

$$a_n = \frac{2}{T} \int_{t_1}^{t_2} \cos(\omega_n t) dt, \tag{2.4}$$

and the b_n are the odd Fourier coefficients, given by

$$b_n = \frac{2}{T} \int_{t_1}^{t_2} \sin(\omega_n t) dt \tag{2.5}$$

The discrete Fourier transform (DFT), as the name suggests, is the discrete version of the continuous Fourier transform, applied to sampled signals. The input sequence is finite in duration and hence the DFT only evaluates the frequency components present in this portion of the signal. The inverse DFT will therefore only reconstruct using these frequencies and may not provide a complete reconstruction of the original signal (unless this signal is periodic).

The DFT converts a finite number of equally spaced samples of a function, to a finite list of complex sinusoids, where the transformation is ordered by frequency. This transformation is commonly described as transferring a sampled function from time domain to frequency domain. Given the N-point of equally spaced sampled function $x(n)$ as an input, the N-point DFT is defined by

$$X(k) = \sum_{n=0}^{N-1} x(n) e^{(-j2\pi nk/N)} \quad k = 0, \dots, N-1, \tag{2.6}$$

where n is the time index and k is the frequency index.

The compact version of the DFT can be written using the twiddle factor notation:

$$W_N^{nk} = e^{-2n\pi k/N} = \exp\left(\frac{-2jn\pi k}{N}\right) = \cos\left(\frac{2n\pi k}{N}\right) - j\sin\left(\frac{2jn\pi k}{N}\right). \tag{2.7}$$

Using the twiddle factor notation, equation (2.7) can be written as follows:

$$X(k) = \sum_{n=0}^{N-1} x(n) W_n^{nk}, \quad k = 0, \ldots, N-1. \tag{2.8}$$

The input sequence $x(n)$ can be calculated from $X(k)$ using the inverse discrete Fourier transform (IDFT) given by

$$x(n) = \frac{1}{N} \sum_{n=0}^{N-1} X(k) W_N^{-nk}, \quad n = 0, \ldots, N-1. \tag{2.9}$$

The number of operations performed for one output can be easily calculated as N complex multiplications and $N - 1$ complex additions from equations (2.8) and (2.9). Therefore, the overall conversion process requires N^2 complex multiplications and $N^2 - N$ complex additions. The amount of calculation required for an N-point DFT equation is approximately $2N^2$.

2.3.2 Fast Fourier Transform

In order to reduce the amount of mathematical operations, a family of efficient calculation algorithms called fast Fourier transforms was introduced by Cooley and Tukey (1965). The basic methodology behind the FFT is the computation of large DFTs in small pieces, and their combination with the help of reordering and transposition algorithms. At the end, the combined result gives the same values as with the large sized DFT, but the order of complexity of the main system reduces from N^2 to the order of $N \log(N)$.

The transformed samples are separated by the angle θ and are periodic and mirrored to the left and right of the imaginary and above and below the real axis. This symmetry and periodicity in the coefficients of the transform kernel (W_N) gives rise to a family of FFT algorithms which involves recursively decomposing the algorithm until only two-point DFTs are required. It is computed using the butterfly unit and perfect shuffle network as shown in Figure 2.5:

$$X_k = \sum_{n=0}^{N/2-1} x(n) W_N^{nk} + W_N^{Nk/2} \sum_{n=0}^{N/2-1} x(n + N/2) W_N^{nk}. \tag{2.10}$$

The FFT has immense impact in a range of applications. One particular use is in the central computation within OFDM. This spread spectrum digital modulation scheme is used in communication, particularly within wireless technology, and has resulted in vastly improved data rates within the 802.11 standards, to name just one example. The algorithm relies on the orthogonal nature of the frequency components extracted through the FFT, allowing each of these components to act as a sub-carrier. Note that the receiver uses the inverse fast Fourier transform (IFFT) to detect the sub-carriers and reconstruct the transmission. The individual sub-carriers are modulated using a typical low symbol rate modulation scheme such as phase-shift or quadrature amplitude modulation (QAM), depending on the application.

For the Institute of Electrical and Electronic Engineers (IEEE) 802.11 standard, the data rate ranges up to 54 Mbps depending on the environmental conditions and noise, i.e. phase shift modulation is used for the lower data rates when greater noise is present,

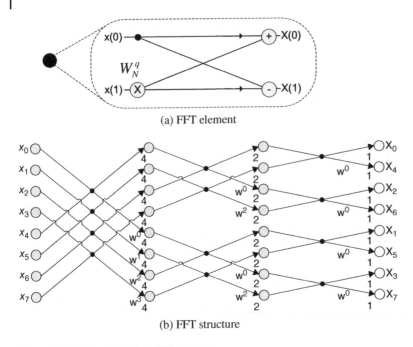

(a) FFT element

(b) FFT structure

Figure 2.5 Eight-point radix-2 FFT structure

and QAM is used in less noisy environments reaching up to 54 Mbps. Figure 2.6 gives an example of the main components within a typical communications chain.

The IEEE 802.11a wireless LAN standard using OFDM is employed in the 5 GHz region of the US ISM band over a channel bandwidth of 125 MHz. From this bandwidth 52 frequencies are used, 48 for data and four for synchronization. The latter point is very important, as the basis on which OFDM works (i.e. orthogonality) relies on the receiver and transmitter being perfectly synchronized.

2.3.3 Discrete Cosine Transform

The DCT is based on the DFT, but uses only real numbers, i.e. the cosine part of the transform, as defined in the equation

$$X(k) = \sum_{n=0}^{N-1} \cos\left[\frac{\pi}{N}\left(n+\frac{1}{2}\right)k\right], \quad k = 0,\ldots,N-1. \tag{2.11}$$

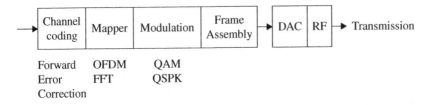

Figure 2.6 Wireless communications transmitter

This two-dimensional (2D) form of the DCT is a vital computational component in the JPEG image compression and also features in MPEG standards:

$$F_{u,v} = \alpha(u)\alpha(v) \sum_{x=0}^{7} \sum_{y=0}^{7} f_{x,y} \cos\left[\frac{\pi}{8}\left(x + \frac{1}{2}\right)u\right] \cos\left[\frac{\pi}{8}\left(y + \frac{1}{2}\right)v\right] \tag{2.12}$$

where u is the horizontal spatial frequency for $0 \le u < 8$, v is the vertical spatial frequency for $0 \le v < 8$, $\alpha(u)$ and $\alpha(v)$ are constants, $f_{x,y}$ is the value of the (x, y) pixel and $F_{u,v}$ is the value of the (u, v) DCT coefficient.

In JPEG image compression, the DCT is performed on the rows and the columns of the image block of 8×8 pixels. The resulting frequency decomposition places the more important lower-frequency components at the top left-hand corner of the matrix, and the frequency of the components increases when moving toward the bottom right-hand part of the matrix.

Once the image has been transformed into numerical values representing the frequency components, the higher frequency components may be removed through the process of quantization as they will have less importance in image quality. Naturally, the greater the amount to be removed the higher the compression ratio; at a certain point, the image quality will begin to deteriorate. This is referred to as lossy compression. The numerical values for the image are read in a zigzag fashion.

2.3.4 Wavelet Transform

A wavelet is a fast-decaying waveform containing oscillations. Wavelet decomposition is a powerful tool for multi-resolution filtering and analysis and is performed by correlating scaled versions of this original wavelet function (i.e. the mother wavelet) against the input signal. This decomposes the signal into frequency bands that maintain a level of temporal information (Mallat 1989). This is particularly useful for frequency analysis of waveforms that are pseudo-stationary where the time-invariant FFT may not provide the complete information.

There are many families of wavelet equations such as the Daubechies, Coiflet and Symmlet (Daubechies 1992). Wavelet decomposition may be performed in a number of ways, namely the continuous wavelet transform (CWT) or DWT which is described in the next section.

Discrete Wavelet Transform

The DWT is performed using a series of filters. At each stage of the DWT, the input signal is passed though a high-pass and a low-pass filter, resulting in the detail and approximation coefficients.

The equation for the low-pass filter is

$$y(n) = (x^*g)(n) = \sum_{-\infty}^{\infty} x(k)g(n-k), \tag{2.13}$$

where g denotes high-pass. By removing half the frequencies at each stage, the signal information can be represented using half the number of coefficients, hence the equations for the low and high filters become

$$y_{\text{low}}(n) = \sum_{-\infty}^{\infty} x(k)g(2n-k) \tag{2.14}$$

and

$$y_{high}(n) = \sum_{-\infty}^{\infty} x(k)h(2n - k),$$ (2.15)

respectively, where h denotes low-pass, and where n has become $2n$, representing the down-sampling process.

Wavelet decomposition is a form of subband filtering and has many uses. By breaking the signal down into the frequency bands, denoising can be performed by eliminating the coefficients representing the highest frequency components and then reconstructing the signal using the remaining coefficients. Naturally, this could also be used for data compression in a similar way to the DCT and has been applied to image compression. Wavelet decomposition is also a powerful transform to use in analysis of medical signals.

2.4 Filters

Digital filtering is achieved by performing mathematical operations on the digitized data; in the analogue domain, filtering is performed with the help of electronic circuits that are formed from various electronic components. In most cases, a digital filter performs operations on the sampled signals with the use stored filter coefficients. With the use of additional components and increased complexity, digital filters could be more expensive than the equivalent analogue filters.

2.4.1 Finite Impulse Response Filter

A simple FIR filter is given by

$$y(n) = \sum_{i=0}^{N-1} a_i x(n - i),$$ (2.16)

where the a_i are the coefficients needed to generate the necessary filtering response such as low-pass or high-pass and N is the number of filter taps contained in the function. The function can be represented using the classical signal flow graph (SFG) representation of Figure 2.7 for $N = 3$ given by

$$y(n) = a_0 x(n) + a_1 x(n - 1) + a_2 x(n - 2).$$ (2.17)

In the classic form, the delay boxes of z^{-1} indicate a digital delay, the branches send the data to several output paths, labeled branches represent a multiplication by the variable shown and the black dots indicate summation functions. However, we find the form

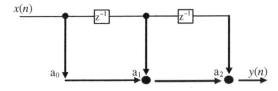

Figure 2.7 Original FIR filter SFG

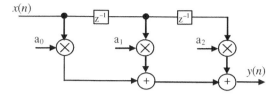

Figure 2.8 FIR filter SFG

given in Figure 2.8 to be easier to understand and use this format throughput the book, as the functionality is more obvious than that given in Figure 2.7.

An FIR filter exhibits some important properties including the following.

- *Superposition*. Superposition holds if a filter produces an output $y(n) + v(n)$ from an input $x(n) + u(n)$, where $y(n)$ is the output produced from input $x(n)$ and $v(n)$ is the output produced from input $u(n)$.
- *Homogeneity*. If a filter produces an output $ay(n)$ from input $ax(n)$ then the filter is said to be homogeneous if the filter produces an output $ay(n)$ from input $ax(n)$.
- *Shift invariance*. A filter is shift invariant if and only if the input of $x(n + k)$ generates an output $y(n + k)$, where $y(n)$ is the output produced by $x(n)$.

If a filter is said to exhibit all these properties then it is said to be a linear time-invariant (LTI) filter. This property allows these filters to be cascaded as shown in Figure 2.9(a) or in a parallel configuration as shown in Figure 2.9(b).

FIR filters have a number of additional advantages, including linear phase, meaning that they delay the input signal but do not distort its phase; they are inherently stable;

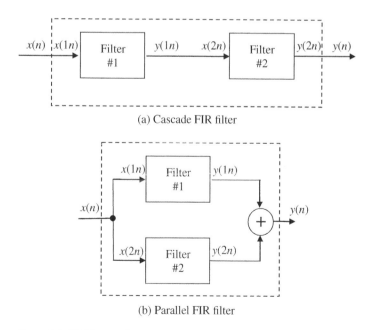

(a) Cascade FIR filter

(b) Parallel FIR filter

Figure 2.9 FIR filter configurations

they are implementable using finite arithmetic; and they have low sensitivity to quantization errors.

Low-Pass FIR Filter

FIR filter implementations are relatively simple to understand as there is a straightforward relationship between the time and frequency domain. A brief summary is now given of digital filter design, but the reader is also referred to some good basic texts (Bourke 1999; Lynn and Fuerst 1994; Williams 1986) which give a much more comprehensive description of filter design. One basic way of developing a digital filter is to start with the desired frequency response, use an inverse filter to get the impulse response, truncate the impulse response and then window the function to remove artifacts (Bourke 1999; Williams 1986). The desired response is shown in Figure 2.10, including the key features that the designer wants to minimize.

Realistically we have to approximate this infinitely long filter with a finite number of coefficients and, given that it needs data from the future, time-shift it so that it does not have negative values. If we can then successfully design the filter and transform it back to the frequency domain we get a ringing in the passband/stopband frequency ranges known as *rippling*, a gradual transition between passband and stopband regions, termed the *transition region*. The ripple is often called the Gibbs phenomenon after Willard Gibbs who identified this effect in 1899, and it is outlined in the FIR filter response in Figure 2.10.

It could be viewed that this is the equivalent of windowing the original frequency plot with a rectangular window; there are other window types, most notably von Hann, Hamming and Kaiser windows (Lynn and Fuerst 1994; Williams 1986) that can be used

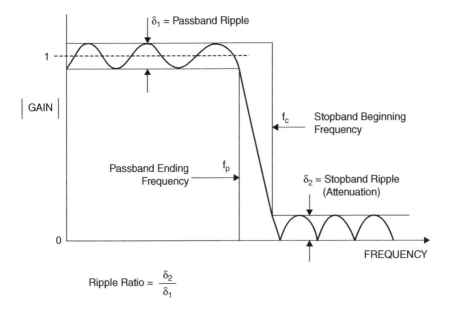

Figure 2.10 Filter specification features

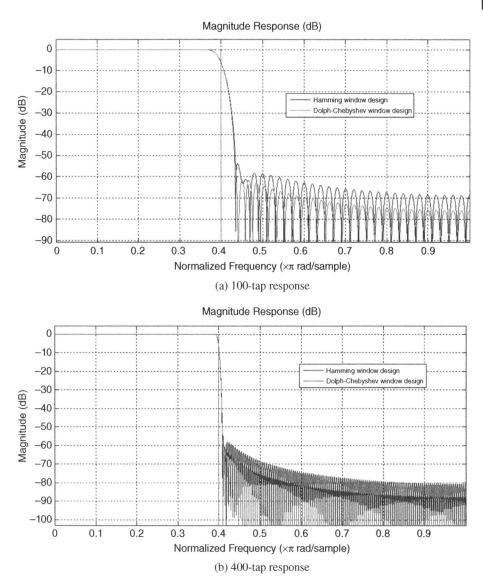

Figure 2.11 Low-pass filter response

to minimize these issues in different ways and to different levels. The result of the design process is the determination of the filter length and coefficient values which best meet the requirements of the filter response.

The number of coefficients has an impact on both the ripple and transition region and is shown for a low-pass filter design, created using the Hamming and Dolph–Chebyshev schemes in MATLAB®. The resulting frequency responses are shown in Figures 2.11 and 2.12 for 100 and 400, taps respectively. The impact of increasing the number of taps in the roll-off between the two bands and the reduction in ripple is clear.

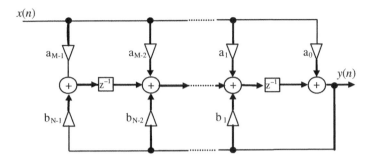

Figure 2.12 Direct form IIR filter

Increasing the number of coefficients clearly allows a better approximation of the filter but at the cost of the increased computation needed to compute the additional taps, and impacts the choice of windowing function.

2.4.2 Infinite Impulse Response filter

The main disadvantage of FIR filters is the large number of taps needed to realize some aspects of the frequency response, namely sharp cut-off resulting in a high computation cost to achieve this performance. This can be overcome by using IIR filters which use previous values of the output as indicated in the equation

$$y(n) = \sum_{i=0}^{N-1} a_i x(n - i) + \sum_{j=1}^{M-1} b_i y(n - j). \tag{2.18}$$

This is best expressed in the transfer function expression

$$H(z) = \frac{\sum_{i=0}^{N-1} a_i x(n - i)}{1 - \sum_{i=1}^{M-1} b_i y(n - j)}, \tag{2.19}$$

and is shown in Figure 2.12.

The design process is different from FIR filters and is usually achieved by exploiting the huge body of analogue filter designs by transforming the s-plane representation of the analogue filter into the z domain. A number of design techniques can be used such as the impulse invariant method, the match z-transform and the bilinear transform. Given an analogue filter with a transfer function, $H_A(s)$, a discrete realization, $H_D(z)$, can be readily deduced by applying a bilinear transform given by

$$H_D(z) = H_A(s)\big|_{s=\frac{2}{T}\left(\frac{z-1}{z+1}\right)} \tag{2.20}$$

This gives a stable digital filter. However, in higher frequencies, distortion or warping is introduced as shown in Figure 2.13. This warping changes the band edges of the digital filter as illustrated in Figure 2.14 and gives a transfer function expression comprising poles and zeros:

$$H(z) = G\frac{(z - \xi_1)(z - \xi_2) \dots (z - \xi_M)}{(z - p_1)(z - p_2) \dots (z - p_N)}. \tag{2.21}$$

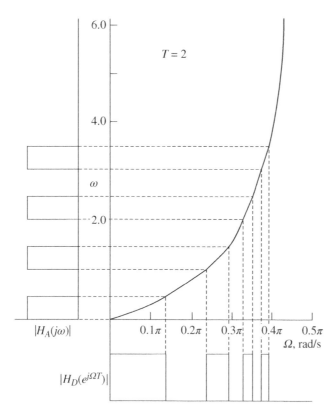

Figure 2.13 IIR filter distortion

The main concern is to maintain stability by ensuring that the poles are located within the unit circle. There is a direct relationship between the location of these zeros and poles and the filter properties. For example, a pole on the unit circle with no zero to annihilate it will produce an infinite gain at a certain frequency (Meyer-Baese 2001).

Due to the feedback loops as shown in Figure 2.12, the structures are very sensitive to quantization errors, a feature which increases as the filter order grows. For this reason, filters are built from second-order IIR filters defined by

$$y(n) = a_0 x(n) + a_1 x(n-1) + a_2 x(n-2) + b_1 y(n-1) + b_2 y(n-2), \qquad (2.22)$$

leading to the structure of Figure 2.15.

2.4.3   Wave Digital Filters

In addition to non-recursive (FIR) and recursive (IIR) filters, a class of filter structures called WDFs is also of considerable interest as they possess a low sensitivity to coefficient variations. This is important in IIR filters as it determines the level of accuracy to which the filter coefficients have to be realized and has a direct correspondence to the dynamic range needed in the filter structure; this affects the internal wordlength sizes and filter performance which will invariably affect throughput rate.

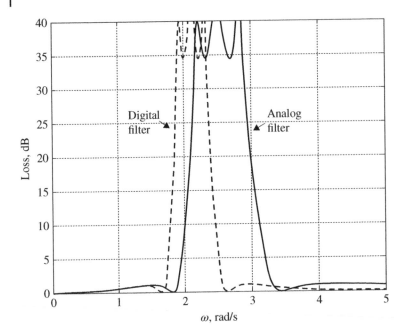

Figure 2.14 Frequency impact of warping

WDFs possess a low sensitivity to attenuation due to their inherent structure, thereby reducing the loss response due to changes in coefficient representation. This is important for many DSP applications for a number of reasons: it allows short coefficient representations to be used which meet the filter specification and which involve only a small hardware cost; structures with low coefficient sensitivities also generate small round-off errors, i.e. errors that result as an effect of limited arithmetic precision within the structure. (Truncation and wordlength errors are discussed in Chapter 3.) As with IIR filters, the starting principle is to generate low-sensitivity digital filters by capturing the low-sensitivity properties of the analogue filter structures.

WDFs represent a class of filters that are modeled on classical analogue filter networks (Fettweis *et al.* 1986; Fettweis and Nossek 1982; Wanhammar 1999) which are typically networks configured in the lattice or ladder structure. For circuits that operate on low frequencies where the circuit dimensions are small relative to the wavelength, the designer can treat the circuit as an interconnection of lumped passive or active components with unique voltages and currents defined at any point in the circuit, on the basis that the phase change between aspects of the circuit will be negligible.

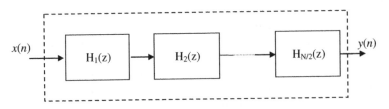

Figure 2.15 Cascade of second-order IIR filter blocks

This allows a number of circuit-level design optimization techniques such as Kirch-hoff's law to be applied. However, for higher-frequency circuits, these assumptions no longer apply and the user is faced with solving Maxwell's equations. To avoid this, the designer can exploit the fact that she is solving the problems only in certain places such as the voltage and current levels at the terminals (Pozar 2005). By exploiting specific types of circuits such as transmission lines which have common electrical propagation times, circuits can then be treated as transmission lines and modeled as distributed components characterized by their length, propagation constant and characteristic impedance.

The process of producing a WDF has been covered by Fettweis *et al.* (1986). The main design technique is to generate filters using transmission line filters and relate these to classical filter structures with lumped circuit elements; this allows the designer to exploit the well-known properties of these structures, termed a *reference filter*. The correspon-dence between the WDF and its reference filter is achieved by mapping the reference structure using a complex frequency variable, ψ, termed *Richard's variable*, allowing the reference structure to be mapped effectively into the ψ domain.

The use of reference structures allows all the inherent passivity and lossless features to be transferred into the digital domain, achieving good filter performance and reducing the coefficient sensitivity, thereby allowing lower wordlengths to be achieved. Fettweis *et al.* (1986) give the simplest and most appropriate choice of ψ as the bilinear transform of the z-variable, given by

$$\psi = \frac{z-1}{z+1} = \tanh(\rho T/2) \tag{2.23}$$

where ρ is the actual complex frequency. This variable has the property that the real frequencies ω correspond to real frequencies ϕ,

$$\phi = \tan(\omega T/2), \quad \rho = j\alpha, \quad \psi = j\phi, \tag{2.24}$$

implying that the real frequencies in the reference domain correspond to real frequen-cies in the digital domain. Other properties described in Fettweis *et al.* (1986) ensure that the filter is causal. The basic principle used for WDF filter design is illustrated in Figure 2.16, taken from Wanhammar (1999). The lumped element filter is shown in Fig-ure 2.16(a) where the various passive components, $L_2 s$, $\frac{1}{C_3 s}$ and $L_4 s$, map to $R_2 \psi$, $\frac{R_3}{\psi}$ and $R_4 \psi$ respectively in the analogous filter given in Figure 2.16(b). Equation (2.23) is then used to map the equivalent transmission line circuit to give the ϕ domain filter in Figure 2.16(c).

WDF Building Blocks

As indicated in Figure 2.16(c), the basic WDF configuration is based upon the various one-, two- and multi-port elements. Figure 2.17 gives a basic description of the two-port element. The network can be described by incident, A, and reflected, B, waves which are related to the port currents, I_1 and I_2, port voltages, V_1 and V_2, and port resistances, R_1 and R_2, by (Fettweis *et al.* 1986)

$$A_1 \cong V_1 + R_1 I_1, \tag{2.25}$$

$$B_2 \cong V_2 + R_2 I_2. \tag{2.26}$$

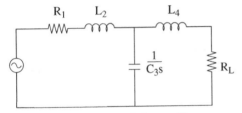

(a) Reference lumped element filter

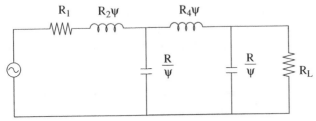

(b) ψ domain filter structure

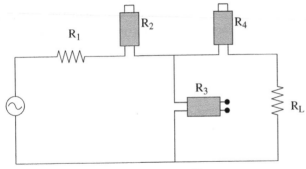

(c) Resulting two-port filter

Figure 2.16 WDF configuration

The transfer function, S_{21}, is given by

$$S_{21} = KB_2/A_1,$$ (2.27)

where

$$K = \sqrt{R_1/R_2}.$$ (2.28)

In a seminal paper, Fettweis *et al.* (1986) show that the loss, α, can be related to the circuit parameters, namely the inductance or capacitance and frequency, ω, such that

Figure 2.17 WDF building blocks

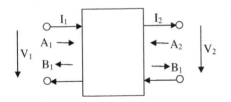

Figure 2.18 Adaptive filter system

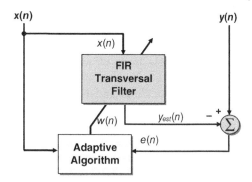

the loss is $\omega = \omega_0$, indicating that for a well-designed filter, the sensitivity of the attenuation is small through its passband, thus giving the earlier stated advantages of lower coefficient wordlengths.

As indicated in Figure 2.17, the basic building blocks for the reference filters are a number of these common two-port and three-port elements or adapters. Some of these are given in Figure 2.18, showing how they are created using multipliers and adders.

2.5 Adaptive Filtering

The basic function of a filter is to remove unwanted signals from those of interest. Obtaining the best design usually requires *a priori* knowledge of certain statistical parameters (such as the mean and correlation functions) within the useful signal. With this information, an optimal filter can be designed which minimizes the unwanted signals according to some statistical criterion.

One popular measure involves the minimization of the mean square of the error signal, where the error is the difference between the desired response and the actual response of the filter. This minimization leads to a cost function with a uniquely defined optimum design for stationary inputs known as a Wiener filter (Widrow and Hoff 1960). However, it is only optimum when the statistical characteristics of the input data match the *a priori* information from which the filter is designed, and is therefore inadequate when the statistics of the incoming signals are unknown or changing (i.e. in a non-stationary environment).

For this situation, a time-varying filter is needed which will allow for these changes. An appropriate solution is an adaptive filter, which is inherently *self-designing* through the use of a *recursive* algorithm to calculate updates for the filter parameters. These then form the taps of the new filter, the output of which is used with new input data to form the updates for the next set of parameters. When the input signals are stationary (Haykin 2002), the algorithm will converge to the optimum solution after a number of iterations, according to the set criterion. If the signals are non-stationary then the algorithm will attempt to track the statistical changes in the input signals, the success of which depends on its inherent convergence rate versus the speed at which statistics of the input signals are changing.

In adaptive filtering, two conflicting algorithms dominate the area, the RLS and the LMS algorithm. The RLS algorithm is a powerful technique derived from the method of least squares (LS). It offers greater convergence rates than its rival LMS algorithm, but this gain is at the cost of increased computational complexity, a factor that has hindered its use in real-time applications.

A considerable body of work has been devoted to algorithms and VLSI archi-tectures for RLS filtering with the aim of reducing this computational complex-ity (Cioffi 1990; Cioffi and Kailath 1984; Döhler 1991; Frantzeskakis and Liu 1994; Gentleman 1973; Gentleman and Kung 1982; Givens 1958; Götze and Schwiegelshohn 1991; Hsieh *et al.* 1993; McWhirter 1983; McWhirter *et al.* 1995; Walke 1997). Much of this work has concentrated on calculating the inverse of the correlation matrix, required to solve for the weights, in a more stable and less computationally intensive manner than straightforward matrix inversion.

The standard RLS algorithm achieves this by recursively calculating updates for the weights using the matrix inversion lemma (Haykin 2002). An alternative and very popu-lar solution performs a set of orthogonal rotations, (e.g. Givens rotations (Givens 1958)) on the incoming data matrix, transforming it into an equivalent upper triangular matrix. The filter parameters can then be calculated by back substitution. This method, known as QR decomposition, is an extension of QR factorization that enables the matrix to be re-triangularized, when new inputs are present, without the need to perform the trian-gularization from scratch. From this beginning, a family of numerically stable and robust RLS algorithms has evolved from a range of QR decomposition methods such as Givens rotations (Givens 1958) and Householder transformations (Cioffi 1990).

2.5.1 Applications of Adaptive Filters

Because of their ability to operate satisfactorily in non-stationary environments, adap-tive filters have become an important part of DSP in applications where the statistics of the incoming signals are unknown or changing. One such application is channel equal-ization (Drewes *et al.* 1998) where the intersymbol interference and noise within a trans-mission channel are removed by modeling the inverse characteristics of the contamina-tion within the channel. Another is adaptive noise cancelation where background noise is eliminated from speech using spatial filtering. In echo cancelation, echoes caused by impedance mismatch are removed from a telephone cable by synthesizing the resound-ing signal and then subtracting it from the original received signal.

The key application for this research is adaptive beamforming (Litva and Lo 1996; Moonen and Proudler 1998; Ward *et al.* 1986). The function of a typical adaptive beam-former is to suppress signals from every direction other than the desired "look direction" by introducing deep nulls in the beam pattern in the direction of the interference. The beamformer output is a weighted combination of signals received by a set of spatially separated antennae, one primary antenna and a number of auxiliary antennae. The pri-mary signal constitutes the input from the main antenna, which has high directivity. The auxiliary signals contain samples of interference threatening to swamp the desired sig-nal. The filter eliminates this interference by removing any signals in common with the primary input signal.

2.5.2 Adaptive Algorithms

There is no distinct technique for determining the optimum adaptive algorithm for a specific application. The choice comes down to a balance between the range of characteristics defining the algorithms, such as rate of convergence (i.e. the rate at which the adaptive algorithm reaches within a tolerance of optimum solution); steady-state error (i.e. the proximity to an optimum solution); ability to track statistical variations in the input data; computational complexity; ability to operate with ill-conditioned input data; and sensitivity to variations in the wordlengths used in the implementation.

Two methods for deriving recursive algorithms for adaptive filters use Wiener filter theory and the LS method, resulting in the LMS and RLS algorithms, respectively. The LMS algorithm offers a very simple yet powerful approach, giving good performance under the right conditions (Haykin 2002). However, its limitations lie with its sensitivity to the condition number of the input data matrix as well as slow convergence rates. In contrast, the RLS algorithm is more elaborate, offering superior convergence rates and reduced sensitivity to ill-conditioned data. On the negative side, the RLS algorithm is substantially more computationally intensive than the LMS equivalent.

Filter coefficients may be in the form of tap weights, reflection coefficients or rotation parameters depending on the filter structure, i.e. transversal, lattice or systolic array, respectively (Haykin 2002). However, in this research both the LMS and RLS algorithms are applied to the basic structure of a transversal filter (Figure 2.18), consisting of a linear combiner which forms a weighted sum of the system inputs, $x(n)$, and then subtracts them from the desired signal, $y(n)$, to produce an error signal, $e(n)$:

$$e(n) = y(n) - \sum_{i=0}^{N-1} w_i x(n). \tag{2.29}$$

In Figure 2.18, $w(n)$ and $w(n + 1)$ are the adaptive and updated adaptive weight vectors respectively, and $y_{est}(n)$ is the estimation of the desired response.

2.5.3 LMS Algorithm

The LMS algorithm is a stochastic gradient algorithm, which uses a fixed step-size parameter to control the updates to the tap weights of a transversal filter as in Figure 2.18 (Widrow and Hoff 1960). The algorithm aims to minimize the mean square error, the error being the difference between $y(n)$ and $y_{est}(n)$. The dependence of the mean square error on the unknown tap weights may be viewed as a multidimensional paraboloid referred to as the error surface, as depicted in Figure 2.19 for a two-tap example (Haykin 2002).

The surface has a uniquely defined minimum point defining the tap weights for the optimum Wiener solution (defined by the Wiener–Hopf equations detailed in the next subsection). However, in the non-stationary environment, this error surface is continuously changing, thus the LMS algorithm needs to be able to track the bottom of the surface.

The LMS algorithm aims to minimize a cost function, $V(w(n))$, at each time step n, by a suitable choice of the weight vector $w(n)$. The strategy is to update the parameter estimate proportional to the instantaneous gradient value, $\frac{dV(w(n))}{dw(n)}$, so that

$$w(n + 1) = w(n) - \mu \frac{dV(w(n))}{dw(n)}, \tag{2.30}$$

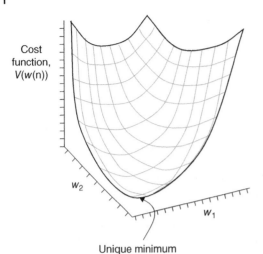

Figure 2.19 Error surface of a two-tap transversal filter

Cost function, $V(w(n))$

w_2

w_1

Unique minimum

where μ is a small positive step size and the minus sign ensures that the parameter estimates descend the error surface. $V(w(n))$ minimizes the mean square error, resulting in the following recursive parameter update equation:

$$w(n+1) = w(n) - \mu x(n)(y(n) - y_{est}(n)). \tag{2.31}$$

The recursive relation for updating the tap weight vector (i.e. equation (2.30)) may be rewritten as

$$w(n+1) = w(n) - \mu x(n)(y(n) - x^T(n)w(n)), \tag{2.32}$$

and represented as filter output

$$y_{est}(n) = w^T(n)x(n), \tag{2.33}$$

estimation error

$$e(n) = y(n) - y_{est}(n), \tag{2.34}$$

and tap weight adaptation

$$w(n+1) = w(n) + \mu x(n)e(n). \tag{2.35}$$

The LMS algorithm requires only $2N+1$ multiplications and $2N$ additions per iteration for an N-tap weight vector. Therefore it has a relatively simple structure and the hardware is directly proportional to the number of weights.

2.5.4 RLS Algorithm

In contrast, RLS is a computationally complex algorithm derived from the method of least squares in which the cost function, $J(n)$, aims to minimize the sum of squared errors, as shown in equation (2.29):

$$J(n) = \sum_{i=0}^{N-1} |e(n-i)|^2. \tag{2.36}$$

Substituting equation (2.29) into equation (2.36) gives

$$J(n) = \sum_{i=0}^{N-1} \left| y(n) - \sum_{i=0}^{N-1} w_k x(n-i) \right|^2 \tag{2.37}$$

Converting from the discrete time domain to a matrix–vector form simplifies the representation of the equations. This is achieved by considering the data values from N samples, so that equation (2.29) becomes

$$e(n) = \begin{bmatrix} e_1 \\ e_2 \\ \vdots \\ e_N \end{bmatrix} = \begin{bmatrix} y_1 \\ y_2 \\ \vdots \\ y_N \end{bmatrix} - \begin{bmatrix} x_1^T \\ x_2^T \\ \vdots \\ x_N^T \end{bmatrix} \begin{bmatrix} W_1 \\ W_2 \\ \vdots \\ W_N \end{bmatrix}, \tag{2.38}$$

which may be expressed as:

$$e(n) = y(n) - X(n)w(n). \tag{2.39}$$

The cost function $J(n)$ may then be represented in matrix form as

$$J(n) = e(n)^T e(n) = (y(n) - X(n)w(n))^T (y(n) - X(n)w(n)). \tag{2.40}$$

This is then multiplied out and simplified to give

$$J(n) = yT(n) - 2y^T(n)X(n)w(n) + w^T(n)X^T(n)X(n)w(n), \tag{2.41}$$

where $X^T(n)$ is the transpose of $X(n)$. To find the optimal weight vector, this expression is differentiated with respect to $w(n)$ and solved to find the weight vector that will drive the derivative to zero. This results in a LS weight vector estimation, w_{LS}, which is derived from the above expression and can be expressed in matrix form as

$$w_{LS}(n) = (X^T(n)X(n))^{-1}X^T y(n) \tag{2.42}$$

These are referred to as the Wiener–Hopf normal equations

$$w_{LS}(n) = \phi(n)^{-1}\theta(n), \tag{2.43}$$

$$\phi(n) = (X^T(n)X(n)), \tag{2.44}$$

$$\theta(n) = X^T(n)y(n), \tag{2.45}$$

where, $\phi(n)$ is the correlation matrix of the input data, $X(n)$, and $\theta(n)$ is the cross-correlation vector of the input data, $X(n)$, with the desired signal vector, $y(n)$. By assuming that the number of observations is larger that the number of weights, a solution can be found since there are more equations than unknowns.

The LS solution given so far is performed on blocks of sampled data inputs. This solution can be implemented recursively, using the RLS algorithm, where the LS weights are updated with each new set of sample inputs. Continuing this adaptation through time would effectively perform the LS algorithm on an infinitely large window of data and would therefore only be suitable for a stationary system. A weighting factor may be included in the LS solution for application in non-stationary environments. This factor assigns greater importance to the more recent input data, effectively creating a moving window of data on which the LS solution is calculated. The forgetting factor, β, is included in the LS cost function (from equation (2.39)) as

$$J(n) = \sum_{i=0}^{N-1} \beta(n-i)e^2(i), \tag{2.46}$$

where $\beta(n-i)$ has the property $0 < \beta(n-i) \leq 1, i = 1, 2, \ldots, N$. One form of the forgetting factor is the exponential forgetting factor:

$$\beta(n-i) = \lambda^{n-i}, \quad i = 1, 2, \ldots, N, \tag{2.47}$$

where λ is a positive constant with a value close to but less than one. Its value is of particular importance as it determines the length of the data window that is used and will affect the performance of the adaptive filter. The inverse of $1 - \lambda$ gives a measure of the "memory" of the algorithm. The general rule is that the longer the memory of the system, the faster the convergence and the smaller the steady-state error. However, the window length is limited by the rate of change in the statistics of the system. Applying the forgetting factor to the Wiener–Hopf normal equations (2.43)–(2.45), the correlation matrix and the cross-correlation matrix become

$$\phi(n) = \sum_{i=0}^{n} \lambda^{n-1} \underline{x}(i)\underline{x}^T(i), \tag{2.48}$$

$$\theta(n) = \sum_{i=0}^{n} \lambda^{n-1} \underline{x}(i)y(i). \tag{2.49}$$

The recursive representations are then expressed as

$$\phi(n) = \left[\sum_{i=1}^{n-1} \lambda^{n-i-1} \underline{x}(i)\underline{x}^T(i) \right] + \underline{x}(n)\underline{x}^T(n), \tag{2.50}$$

or more concisely as

$$\phi(n) = \lambda\phi(n-1) + \underline{x}(n)\underline{x}^T(n). \tag{2.51}$$

Likewise, $\theta(n)$ can be expressed as

$$\theta(n) = \lambda\theta(n-1) + \underline{x}(n)y(n). \tag{2.52}$$

Solving the Wiener–Hopf normal equations to find the LS weight vector requires the evaluation of the inverse of the correlation matrix, as highlighted by the following example matrix–vector expression below:

$$\begin{bmatrix} w_1 \\ w_2 \\ w_3 \end{bmatrix} = \underbrace{\left[\begin{bmatrix} X_{11}X_{12}X_{13} \\ X_{21}X_{22}X_{23} \\ X_{31}X_{32}X_{33} \end{bmatrix}^T \begin{bmatrix} X_{11}X_{12}X_{13} \\ X_{21}X_{22}X_{23} \\ X_{31}X_{32}X_{33} \end{bmatrix}^{-1} \right]} \cdot \underbrace{\left[\begin{bmatrix} X_{11}X_{12}X_{13} \\ X_{21}X_{22}X_{23} \\ X_{31}X_{32}X_{33} \end{bmatrix}^T \begin{bmatrix} y_{11} \\ y_{12} \\ y_{13} \end{bmatrix} \right]}. \tag{2.53}$$

$$\text{correlation matrix} \qquad\qquad \text{cross-correlation matrix}$$

The presence of this matrix inversion creates an implementation hindrance in terms of both numerical stability and computational complexity. Firstly, the algorithm would be subject to numerical problems if the correlation matrix became singular. Also, calculating the inverse for each iteration involves an order of complexity N^3, compared with a complexity of order N for the LMS algorithm, where N is the number of filter taps.

There are two particular methods to solve the LS solution recursively without the direct matrix inversion which reduce this complexity to order N^2. The first technique, referred to as the standard RLS algorithm, recursively updates the weights using the matrix inversion lemma. The alternative and very popular solution performs a set of orthogonal rotations, e.g. Givens rotations (Givens 1958), on the incoming data transforming the square data matrix into an equivalent upper triangular matrix (Gentleman and Kung 1982). The weights can then be calculated by back substitution.

This method, known as QR decomposition (performed using one of a range of orthogonal rotation methods such as Householder transformations or Givens rotations), has been the basis for a family of numerically stable and robust RLS algorithms (Cioffi 1990; Cioffi and Kailath 1984; Döhler 1991; Hsieh *et al.* 1993; Liu *et al.* 1990, 1992; McWhirter 1983; McWhirter *et al.* 1995; Rader *et al.* 1986; Walke 1997). There are versions known as fast RLS algorithms, which manipulate the redundancy within the system to reduce the complexity to the order of N, as mentioned in Section 2.5.5.

Systolic Givens Rotations

The conventional Givens rotation QR algorithm can be mapped onto a highly parallel triangular array (referred to as the QR array (Gentleman and Kung 1982; McWhirter 1983)) built up from two types of cells, a boundary cell (BC) and an internal cell (IC). The systolic array for the conventional Givens RLS algorithm is shown in Figure 2.20. Note that the original version (Gentleman and Kung 1982) did not include the product of cosines formed down the diagonal line of BCs. This modification (McWhirter 1983) is significant as it allows the QR array to both perform the functions for calculating the weights and operate as the filter itself. That is, the error residual (*a posteriori* error) may be found without the need for weight vector extraction. This offers an attractive solution

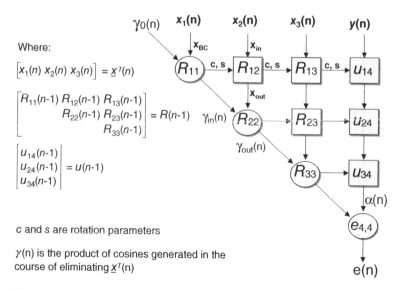

Figure 2.20 Systolic QR array for the RLS algorithm

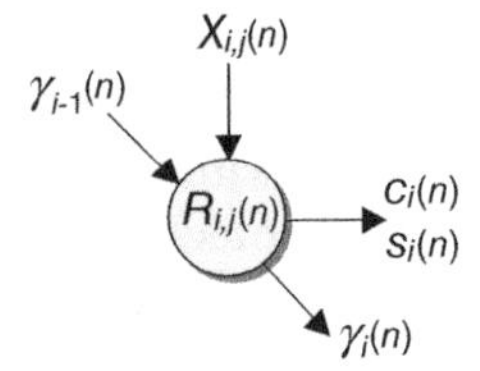

$$R_{i,i}(n) = \{\beta^2 R_{i,i}^2(n-1) + x^2_{i,i}(n)\}^{1/2}$$

$$c_i(n) = \beta \cdot \left(\frac{R_{i,i}(n-1)}{R_{i,i}(n)}\right)$$

$$s_i(n) = \left(\frac{x_{i,i}(n)}{R_{i,i}(n)}\right)$$

$$\gamma_i(n) = c_i(n)\gamma_{i-1}(n)$$

Figure 2.21 BC for QR-RLS algorithm

in applications, such as adaptive beamforming, where the output of interest is the error residual.

All the cells are locally interconnected, which is beneficial, as it is interconnection lengths that have the most influence over the critical paths and power consumption of a circuit. This highly regular structure is referred to as a systolic array. Its processing power comes from the concurrent use of many simple cells rather than the sequential use of a few very powerful cells and is described in detail in the next chapter. The definitions for the BCs and ICs are depicted in Figures 2.21 and 2.22 respectively.

The data vector $x^T(n)$ is input from the top of the array and is progressively eliminated by rotating it within each row of the stored triangular matrix $R(n-1)$ in turn. The rotation parameters c and s are calculated within a BC such that they eliminate the input, $x_{i,i}(n)$. These parameters are then passed unchanged along the row of ICs continuing the rotation. The output values of the ICs, $x_{i+1,j}(n)$, become the input values for the next row. Meanwhile, new inputs are fed into the top of the array, and so the process repeats. In the process, the $R(n)$ and $u(n)$ values are updated to account for the rotation and then stored within the array to be used on the next cycle.

For the RLS algorithm, the implementation of the forgetting factor, λ, and the product of cosines, γ, need to be included within the equations. Therefore the operations of the BCs and ICs have been modified accordingly. A notation has been assigned to the variables within the array. Each R and u term has a subscript, denoted by (i, j), which represents the location of the elements within the R matrix and u vector. A similar notation is assigned to the X input and output variables. The cell descriptions for the updated BCs and ICs are shown in Figures 2.21 and 2.22, respectively. The subscripts are coordinates relating to the position of the cell within the QR array.

2.5.5 Squared Givens Rotations

There are division and square root operations within the cell computation for the standard Givens rotations (Figures 2.21 and 2.22). There has an extensive body of research

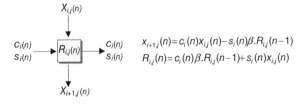

$$x_{i+1,j}(n) = c_i(n)x_{i,j}(n) - s_i(n)\beta.R_{i,j}(n-1)$$

$$R_{i,j}(n) = c_i(n)\beta.R_{i,j}(n-1) + s_i(n)x_{i,j}(n)$$

Figure 2.22 IC for QR-RLS algorithm

into deriving Givens rotation QR algorithms which avoid these complex operations, while reducing the overall number of computations (Cioffi and Kailath 1984; Döhler 1991; Hsieh *et al.* 1993; Walke 1997). One possible QR algorithm is the squared Givens rotation (SGR) (Döhler 1991). Here the Givens algorithm has been manipulated to remove the need for the square root operation within the BC and half the number of multipliers in the ICs.

Studies by Walke (1997) showed that this algorithm provided excellent performance within adaptive beamforming at reasonable wordlengths (even with mantissa wordlengths as short as 12 bits with an increase of 4 bits within the recursive loops). This algorithm turns out to be a suitable choice for the adaptive beamforming design. Figure 2.23 depicts the SFG for the SGR algorithm, and includes the BC and IC descriptions.

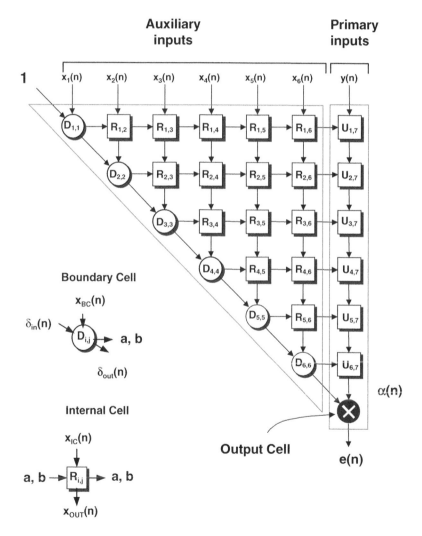

(Note: in the SGR algorithm $D = R^2$)

Figure 2.23 Squared Givens Rotations QR-RLS algorithm

This algorithm still requires the dynamic range of floating-point arithmetic but offers reduced size over fixed-point algorithms, due to the reduced wordlength and operations requirement. It has the added advantage of allowing the use of a multiply-accumulate operation to update R. At little cost in hardware, the wordlength of the accumulator can be increased to improve the accuracy to which R is accumulated, while allowing the overall wordlength to be reduced. This has been referred to as the Enhanced SGR algorithm (E-SGR) (Walke 1997).

However, even with the level of computation reduction achievable by the SGR algorithm, the complexity of the QR cells is still large. In addition, the number of processors within the QR array increases quadratically with the number of inputs, such that for an N-input system, $(N^2 + N)/2$ QR processors are required; furthermore, implementing a processor for each cell could offer data rates far greater than those required by most applications. The following section details the process of deriving an efficient architecture with generic properties for implementing the SGR QR-RLS algorithm.

2.6   Final Comments

The chapter has given a brief review of DSP algorithms with the aim of providing a foundation for the work presented in this book. Some of the examples have been the focus of direct implementation using FPGA technology with the aim of giving enhanced performance in terms of the samples produced per second or a reduction in power consumption. The main focus has been to provide enough background to understand the examples, rather than an exhaustive primer for DSP.

In particular, the material has concentrated on the design of FIR and IIR filtering as this is a topic for speed optimization, particularly the material in Chapter 8, and the design of RLS filters, which is the main topic of Chapter 11 and considers the development of an IP core for an RLS filter solved by QR decomposition. These chapters represent a core aspect of the material in this book.

Bibliography

Bourke P 1999 Fourier method of designing digital filters. Available at http://paulbourke.net/miscellaneous/filter/ (accessed April 2, 2016).

Cioffi J 1990 The fast Householder filters RLS adaptive algorithm RLS adaptive filter. In *Proc. IEEE Int. Conf. on Acoustics, Speech and Signal Processing*, pp. 1619–1621.

Cioffi J, Kailath T 1984 Fast, recursive-least-squares transversal filters for adaptive filtering. *IEEE Trans. on ASSP*, 32(2), 304–337.

Cooley JW, Tukey JW 1965 An algorithm for the machine computation of the complex Fourier series. *Mathematics of Computation*, 19, 297–301.

Daubechies I 1992 *Ten Lectures on Wavelets*. Society for Industrial and Applied Mathematics, Philadelphia.

Döhler R 1991 Squared Givens rotation. *IMA Journal of Numerical Analysis*, 11(1), 1–5.

Drewes C, Hasholzner R, Hammerschmidt JS 1998 On implementation of adaptive equalizers for wireless ATM with an extended QR-decomposition-based RLS-algorithm. In *Proc. IEEE Intl Conf. on Acoustics, Speech and Signal Processing*, pp. 3445–3448.

Fettweis A, Gazsi L, Meerkotter K 1986 Wave digital filter: Theory and practice. *Proc. IEEE*, 74(2), pp. 270–329.

Fettweis A, Nossek J 1982 On adaptors for wave digital filter. *IEEE Trans. on Circuits and Systems*, 29(12), 797–806.

Frantzeskakis EN, Liu KJR 1994 A class of square root and division free algorithms and architectures for QRD-based adaptive signal processing. *IEEE Trans. on Signal Proc.* 42(9), 2455–2469.

Gentleman WM 1973 Least-squares computations by Givens transformations without square roots, *J. Inst. Math. Appl.*, 12, 329–369.

Gentleman W, Kung H 1982 Matrix triangularization by systolic arrays. In *Proc. SPIE*, 298, 19–26.

Givens W 1958 Computation of plane unitary rotations transforming a general matrix to triangular form. *J. Soc. Ind. Appl. Math*, 6, 26–50.

Götze J, Schwiegelshohn U 1991 A square root and division free Givens rotation for solving least squares problems on systolic arrays. *SIAM J. Sci. Stat. Compt.*, 12(4), 800–807.

Haykin S 2002 *Adaptive Filter Theory*. Publishing House of Electronics Industry, Beijing.

Hsieh S, Liu K, Yao K 1993 A unified approach for QRD-based recursive least-squares estimation without square roots. *IEEE Trans. on Signal Processing*, 41(3), 1405–1409.

Litva J, Lo TKY 1996 *Digital Beamforming in Wireless Communications*, Artec House, Norwood, MA.

Liu KJR, Hsieh S, Yao K 1990 Recursive LS filtering using block Householder transformations. In *Proc. of Int. Conf. on Acoustics, Speech and Signal Processing*, pp. 1631–1634.

Liu KJR, Hsieh S, Yao K 1992 Systolic block Householder transformations for RLS algorithm with two-level pipelined implementation. *IEEE Trans. on Signal Processing*, 40(4), 946–958.

Lynn P, Fuerst W 1994 *Introductory Digital Signal Processing with Computer Applications*. John Wiley & Sons, Chicester.

Mallat SG 1989 A theory for multiresolution signal decomposition: the wavelet representation. *IEEE Trans. on Pattern Analysis and Machine Intelligence*, 11(7), 674–693.

McWhirter J 1983 Recursive least-squares minimization using a systolic array. In *Proc. SPIE*, 431, 105–109.

McWhirter JG, Walke RL, Kadlec J 1995 Normalised Givens rotations for recursive least squares processing. In *VLSI Signal Processing, VIII*, pp. 323–332.

Meyer-Baese U 2001 *Digital Signal Processing with Field Programmable Gate Arrays*, Springer, Berlin.

Moonen M, Proudler IK 1998 MDVR beamforming with inverse updating. In *Proc. European Signal Processing Conference*, pp. 193–196.

Nyquist H 1928 Certain topics in telegraph transmission theory. *Trans. American Institute of Electrical Engineers*, 47, 617–644.

Pozar DM 2005 *Microwave Engineering*. John Wiley & Sons, Hoboken, NJ.

Rabiner L, Gold B 1975 *Theory and Application of Digital Signal Processing*. Prentice Hall, New York.

Rader CM, Steinhardt AO 1986 Hyperbolic Householder transformations, definition and applications. In *Proc. of Int. Conf. on Acoustics, Speech and Signal Processing*, pp 2511–2514.

Shannon CE 1949 Communication in the presence of noise. *Proc. Institute of Radio Engineers*, 37(1), 10–21.

Walke R 1997 High sample rate Givens rotations for recursive least squares. PhD thesis, University of Warwick.

Wanhammar L 1999 *DSP Integrated Circuits* Academic Press, San Diego, CA.

Ward CR, Hargrave PJ, McWhirter JG 1986 A novel algorithm and architecture for adaptive digital beamforming. *IEEE Trans. on Antennas and Propagation*, 34(3), 338–346.

Widrow B, Hoff, Jr. ME 1960 Adaptive switching circuits. *IRE WESCON Conv. Rec.*, 4, 96–104.

Williams C 1986 *Designing Digital Filters*. Prentice Hall, New York.

3

Arithmetic Basics

3.1 Introduction

The choice of arithmetic has always been a key aspect for DSP implementation as it not only affects algorithmic performance, but also can impact system performance criteria, specifically area, speed and power consumption. This usually centers around the choice of a floating-point or a fixed-point realization, and in the latter case, a detailed study is needed to determine the minimal wordlength needed to achieve the required performance. However, with FPGA platforms, the choice of arithmetic can have a much wider impact on the performance as the system designer can get a much more direct return from any gains in terms of reducing complexity. This will either reduce the cost or improve the performance in a similar way to an SoC designer.

A key requirement of DSP implementations is the availability of suitable processing elements, specifically adders and multipliers; however, many DSP algorithms (e.g. adaptive filters) also require dedicated hardware for performing division and square roots. The realization of these functions, and indeed the choice of number systems, can have a major impact on hardware implementation quality. For example, it is well known that different DSP application domains (e.g. image processing, radar and speech) can have different levels of bit toggling not only in terms of the number of transitions, but also in the toggling of specific bits (Chandrakasan and Brodersen 1996). More specifically, the signed bit in speech input can toggle quite often, as data oscillates around zero, whereas in image processing the input typically is all positive. In addition, different applications can have different toggling activity in their lower significant bits. This can have a major impact in reducing dynamic power consumption.

For these reasons, it is important that some aspects of computer arithmetic are covered, specifically number representation as well as the implementation choices for some common arithmetic functions, namely adders and multipliers. However, these are not covered in great detail as the reality is that in the case of addition and multiplication, dedicated hardware has been available for some time on FPGA and thus for many applications the lowest-area, fastest- speed and lowest-power implementations will be based on these hardware elements. As division and square root operations are required in some DSP functions, it is deemed important that they are covered here. As dynamic

FPGA-based Implementation of Signal Processing Systems,
Second Edition. Roger Woods, John McAllister, Gaye Lightbody and Ying Yi.
© 2017 John Wiley & Sons, Ltd. Published 2017 by John Wiley & Sons, Ltd.

range is also important, the data representations, namely fixed-point and floating-point, are also seen as critical.

The chapter is organized as follows. In Section 3.2 some basics of computer arithmetic are covered, including the various forms of number representations as well as an introduction to fixed- and floating-point arithmetic. This is followed by a brief introduction to adder and multiplier structures in Section 3.3. Of course, alternative representations also need some consideration so signed digit number representation (SDNR), logarithmic number representations (LNS), residue number representations (RNS), and coordinate rotation digital computer (CORDIC) are considered in Section 3.4. Dividers and circuits for square root are then covered in Sections 3.5 and 3.6, respectively. Some discussion of the choice between fixed-point and floating-point arithmetic for FPGA is given in Section 3.7. In Section 3.8 some conclusions are given and followed by a discussion of some key issues.

3.2 Number Representations

From our early years we are taught to compute in decimal, but the evolution of transistors implies the adoption of binary number systems as a more natural representation for DSP systems. This section starts with a basic treatment of conventional number systems, namely signed magnitude and one's complement, but concentrates on two's complement. Alternative number systems are briefly reviewed later as indicated in the introduction, as they have been applied in some FPGA-based DSP systems.

If x is an $(n + 1)$-bit unsigned number, then the unsigned representation

$$x = \sum_{i=0}^{n} x_i 2^i \tag{3.1}$$

applies, where x_i is the ith binary bit of n and x_0 and x_n are least significant bit (lsb) and most significant bit (msb) respectively. The binary value is converted to decimal by scaling each bit to the relevant significance as shown below, where 1110 is converted to 14:

$$1 \times 2^3 + 1 \times 2^2 + 1 \times 2^1 + 0 \times 2^0 = 14.$$

Decimal to binary conversion is done by successive divisions by 2. For the expression $D = (x_3 x_2 x_1 x_0)_2 = x_3 2^3 + x_2 2^2 + x_1 2^1 + x_0$, we successively divide by 2:

$$
\begin{aligned}
D/2 &= x_3 2^2 + x_2 2^1 + x_1 = Q_1, & \text{remainder} &= x_0 \\
Q_1/2 &= x_3 2^1 + x_2 = Q_2, & \text{remainder} &= x_1 \\
Q_2/2 &= x_3 = Q_3, & \text{remainder} &= x_2 \\
Q_3/2 &= 0 = Q_4, & \text{remainder} &= x_3.
\end{aligned}
$$

So if 14 is converted to binary, this is done as follows:

$$
\begin{aligned}
14/2 &= 7, & \text{remainder} &= 0 = x_0 \\
7/2 &= 3, & \text{remainder} &= 1 = x_1 \\
3/2 &= 1, & \text{remainder} &= 1 = x_2 \\
1/2 &= 0, & \text{remainder} &= 1 = x_3.
\end{aligned}
$$

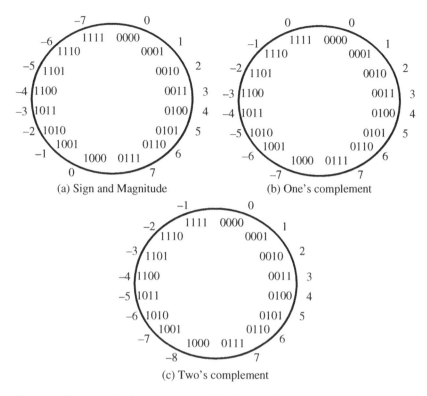

Figure 3.1 Number wheel representation of four-bit numbers

3.2.1 Signed Magnitude

In signed magnitude systems, the $n - 1$ lower significant bits represent the magnitude, and the msb, x_n, bit represents the sign. This is best represented pictorially in Figure 3.1(a), which gives the number wheel representation for a four-bit word. In the signed magnitude notation, the magnitude of the word is decided by the three lower significant bits, and the sign determined by the sign bit, or msb. However, this representation presents a number of problems. First, there are two representations of 0 which must be resolved by any hardware system, particularly if 0 is used to trigger any event (e.g. checking equality of numbers). As equality is normally achieved by checking bit-by-bit, this complicates the hardware. Lastly, operations such as subtraction are more complex, as there is no way to check the sign of the resulting value, without checking the size of the numbers and organizing accordingly. This creates overhead in hardware realization and prohibits the number system's use in practical systems.

3.2.2 One's Complement

In one's-complement systems, the assignment of the bits is done differently. It is based around representing the negative version of the numbers as the inverse or one's complement of the original number. This is achieved by inverting the individual bits, which in practice can easily be achieved through the use of an inverter.

The conversion for an n-bit word is given by

$$\overline{N} = (2^n - 1) - N, \tag{3.2}$$

and the pictorial representation for a four-bit binary word is given in Figure 3.1(b). The problem still exists of two representations of 0. Also, a correction needs to be carried out when performing one's complement subtraction (Omondi 1994). Once again, the need for special treatment of 0 is prohibitive from an implementation perspective.

3.2.3 Two's Complement

In two's-complement systems, the inverse of the number is obtained by inverting the bits of the original word and adding 1. The conversion is given by

$$\overline{N} = 2^n - N, \tag{3.3}$$

and the pictorial representation for a four-bit binary word given in Figure 3.1(c). Whilst this may seem less intuitively obvious than the previous two approaches, it has a number of advantages: there is a single representation for 0, addition and more importantly subtraction can be performed readily in hardware and if the number stays within range, overflow can be ignored in the computation. For these reasons, two's complement has become the dominant number system representation.

This representation therefore efficiently translates into efficient hardware structures for the core arithmetic functions and means that addition and subtraction is easily implemented. As will be seen later, two's complement multiplication is a little more complicated but the single representation of 0 is the differentiator. As will be seen in the next section, the digital circuitry naturally falls out from this.

3.2.4 Binary Coded Decimal

By applying different weighting, a number of other binary codes can be applied as shown in Table 3.1. The following codes are usually called binary coded decimal (BCD). The 2421 is a nine's complement code, i.e. 5 is the inverse of 4, 6 is the inverse of 3, etc. With the Gray code, successive decimal digits differ by exactly one bit. This coding styles

Table 3.1 BCD codes

	BCD		
Decimal	8421	2421	Gray
0	0000	0000	0000
1	0001	0001	0001
2	0010	0010	0011
3	0011	0011	0010
4	0100	0100	0110
5	0101	1011	1110
6	0110	1100	1010
7	0111	1101	1011
8	1000	1110	1001
9	1001	1111	1000

tends to be used in low-power applications where the aim is to reduce the number of transitions (see Chapter 13).

3.2.5 Fixed-Point Representation

Up until now, we have only considered integer representations and not considered the real representations which we will encounter in practical DSP applications. A widely used format for representing and storing numerical binary data is the fixed-point format, where an integer value x represented by $x_{m+n-1}, x_{m+n-2}, \ldots, x_0$ is mapped such that $x_{m+n-1}, x_{m+n-2}, \ldots, x_n$ represents the integer part of the number and the expression, $x_{n-1}, x_{n-2}, \ldots, x_0$ represents the fractional part of the number. This is the interpretation placed on the number system by the user and generally in DSP systems, users represent input data, say $x(n)$, and output data, $y(n)$, as integer values and coefficient word values as fractional so as to maintain the best dynamic range in the internal calculations.

The key issue when choosing a fixed-point representation is to best use the dynamic range in the computation. Scaling can be applied to cover the worst-case scenario, but this will usually result in poor dynamic range. Adjusting to get the best usage of the dynamic range usually means that overflow will occur in some cases and additional circuitry has to be implemented to cope with this condition; this is particularly problematic in two's complement as overflow results in an "overflowed" value of completely different sign to the previous value. This can be avoided by introducing saturation circuitry to preserve the worst-case negative or positive overflow, but this has a nonlinear impact on performance and needs further investigation.

The impact of overflow in two's complement is indicated by the sawtooth representation in Figure 3.2(a). If we consider the four-bit representation represented earlier in Figure 3.1 and look at the addition of 7 (0111) and 1(0001), then we see that this will give 8 (1000) in unsigned binary, but of course this represents -8 in 2's complement which represents the worse possible representation. One approach is to introduce circuitry which will saturate the output to the nearest possible value, i.e. 7 (0111). This is demonstrated in Figure 3.2(b), but the impact is to introduce a nonlinear impact to the DSP operation which needs to be evaluated.

This issue is usually catered for in the high-level modeling stage using tools such as those from MATLAB® or LabVIEW. Typically the designer is able to start with a

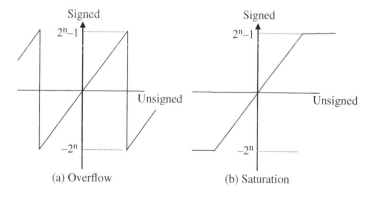

(a) Overflow   (b) Saturation

Figure 3.2 Impact of overflow in two's complement

floating-point representation and then use a fixed-point library to evaluate the performance. At this stage, any impact of overflow can be investigated.

A FPGA-based solution may have timing problems as a result of any additional circuitry introduced for saturation. One conclusion that the reader might draw is that
fixed-point is more trouble than it is worth, but fixed-point implementations are particularly attractive for FPGA implementations (and some DSP microprocessor implementations), as word size translates directly to silicon area. Moreover, a number of optimizations are available that make fixed-point extremely attractive; these are explored in
later chapters.

3.2.6   Floating-Point Representation

Floating-point representations provide a much more extensive means for providing real
number representations and tend to be used extensively in scientific computation applications, but also increasingly in DSP applications. In floating-point, the aim is to represent the real number using a *sign* (S), *exponent* (Exp) and *mantissa* (M), as shown in
Figure 3.3. The most widely used form of floating-point is the IEEE Standard for Binary
Floating-Point Arithmetic (IEEE 754). This specifies four formats:

- single precision (32-bit);
- double precision (64-bit);
- single extended precision;
- double extended precision.

The single-precision format is a 32-bit representation where 1 bit is used for S, 8 bits
for Exp and 23 bits for M. This is illustrated in Figure 3.3(a) and allows the representation of the number x where x is created by $2^{Exp-127} \times M$ as the exponent is represented
as unsigned, giving a single-extended number of approximately $\pm 10^{38.52}$. The double-
precision format is a simple extension of the concept to 64 bits, allowing a range of
$\pm 10^{308.25}$, and is illustrated in Figure 3.3(b); the main difference between this and single
precision being the offset added to the exponent and the addition of zero padding for
the mantissa.

The following simple example shows how a real number, -1082.5674 is converted
into IEEE 754 floating-point format. It can be determined that $S = 1$ as the number is
negative. The number (1082) is converted to binary by successive division (see earlier),

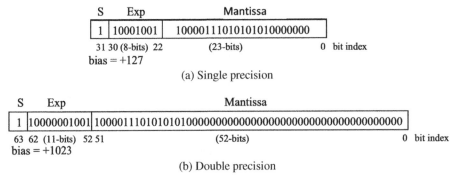

(a) Single precision

(b) Double precision

Figure 3.3 Floating-point representations

Table 3.2 Truth table for a one-bit adder

	Inputs			Outputs	
	A	B	C_{i-1}	S_i	C_i
$C_i = A$ or B	0	0	0	0	0
	0	0	1	1	0
	0	1	0	1	0
$C_i = C_{i-1}$	0	1	1	0	1
	1	0	0	1	0
	1	0	1	0	1
$C_i = A$ or B	1	1	0	0	1
	1	1	1	1	1

giving 10000111010. The fractional part (0.65625) is computed in the same way as above, giving 10101. The parts are combined to give the value 10000111010.10101. The radix point is moved left, to leave a single 1 on the left, giving $1.000011101010101 \times 2^{10}$. Filling with 0s to get the 23-bit mantissa gives the value 10000111010101010000000. In this value the exponent is 10 and, with the 32-bit IEEE 754 format bias of 127, we have 137 which is given as 10001001 in binary (giving the representation in Figure 3.3(a)).

3.3 Arithmetic Operations

This section looks at the implementation of various arithmetic functions, including addition and multiplication but also division and square root. As the emphasis is on FPGA implementation which comprises on-board adders and multipliers, the book concentrates on using these constructions, particularly fixed-point realizations. A brief description of a floating-point adder is given in the following section.

3.3.1 Adders

Addition of two numbers A and B to produce a result S,

$$S = A + B, \tag{3.4}$$

is a common function in computing systems and central to many DSP systems. Indeed, it is a key operation and also forms the basic of multiplication which is, in effect, a series of shifted additions.

A single-bit addition function is given in Table 3.2, and the resulting implementation in Figure 3.4(a). This form comes directly from solving the one-bit adder truth table leading to

$$S_i = A_i \oplus B_i \oplus C_{i-1}, \tag{3.5}$$
$$C_i = A_i \cdot B_i + A_i \cdot C_{i-1} + B_i \cdot C_{i-1}, \tag{3.6}$$

and the logic gate implementation of Figure 3.4(a).

By manipulating the expression for C_i, we can generate the alternative expression

$$C_i = (A_i \oplus B_i)C_{i-1} + A_i \cdot C_{i-1}. \tag{3.7}$$

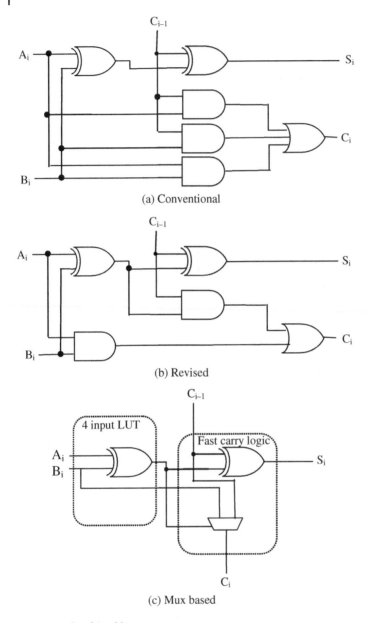

Figure 3.4 One-bit adder structure

This has the advantage of sharing the expression $A_i \oplus B_i$ between both the S_i and C_i expressions, saving one gate but as Figure 3.4(b) illustrates, at the cost of an increased gate delay.

The truth table can also be interpreted as follows: when $A_i = B_i$, then $C_i = B_i$ and $S_i = C_{i-1}$; and when $A_i = \overline{B_i}$, then $C_i = C_{i-1}$ and $S_i = \overline{C_{i-1}}$. This implies a multiplexer for the generation of the carry and, by cleverly using $A_i \oplus B_i$ (already generated in order

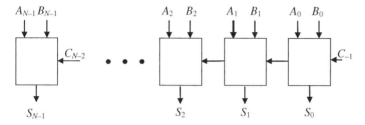

Figure 3.5 *n*-bit adder structure

to develop S_i), very little additional cost is required. This is the preferred construction for FPGA vendors as indicated by the partition of the adder cell in Figure 3.4(c). By providing a dedicated EXOR and mux logic, the adder cell can then be built using a LUT to generate the additional EXOR function.

3.3.2 Adders and Subtracters

Of course, this only represents a single-bit addition, but we need to add words rather than just bits. This can be achieved by chaining one-bit adders together to form a word-level adder as shown in Figure 3.5. This represents the simplest but most area-efficient adder structure. However, the main limitation is the time required to compute the word-level addition which is determined by the time for the carry to propagate from the lsb to the msb. As wordlength increases, this becomes prohibitive.

For this reason, there has been a considerable body of detailed investigations in alternative adder structures to improve speed. A wide range of adder structures have been developed including the carry lookahead adder (CLA) and conditional sum adder (CSA), to name but a few (Omondi 1994). In most cases the structures compromise architecture regularity and area efficiency to overcome the carry limitation.

Carry Lookahead Adder
In the CLA adder, the carry expression given in equation (3.7) is unrolled many times, making the final carry dependent only on the initial value. This can be demonstrated by defining a generate function, G_i, as $G_i = A_i \cdot B_i$, and a propagate function, P_i, as $P_i = A_i \oplus B_i$. Thus we can rewrite equations (3.5) and (3.7) as

$$S_i = P_i \oplus C_{i-1}, \tag{3.8}$$
$$C_i = P_i \cdot C_{i-1} + G_i. \tag{3.9}$$

By performing a series of substitutions on equation (3.9), we can get an expression for the carry out of the fourth addition, namely C_3, which only depends on the carry in of the first adder C_{-1}, as follows:

$$C_0 = G_0 + P_0 \cdot C_{-1}, \tag{3.10}$$
$$C_1 = G(1) + P_1 \cdot C_0 = G_1 + P_1 \cdot G_0 + P_1 \cdot P_0 \cdot C_{-1}, \tag{3.11}$$
$$C_2 = G(2) + P_2 \cdot C_1 = G_2 + P_2 \cdot G_1 + P_2 \cdot P_1 \cdot G_0 + P_2 \cdot P_1 \cdot P_0 \cdot C_{-1}, \tag{3.12}$$
$$C_3 = G(3) + P_3 \cdot C_2 = G_3 + P_3 \cdot G_2 + P_3 \cdot P_2 \cdot G_1 + P_3 \cdot P_2 \cdot P_1 \cdot G_0$$
$$+ P_3 \cdot P_2 \cdot P_1 \cdot P_0 \cdot C_{-1}. \tag{3.13}$$

Adder Comparisons

It is clear from the expressions in equations (3.10)–(3.13) that this results in a very expensive adder structure due to this unrolling. If we use the gate model by Omondi (1994), where any gate delay is given by T and gate cost or area is defined in terms of $2n$ (a two-input AND gates is 2, a three-input AND gates is 3, etc.) and EXORs count as double (i.e. $4n$), then this allows us to generate a reasonable technology-independent model of computation. In this case, the critical path is given as $4T$, which is only one more delay than that for the one-bit adder of Figure 3.4(b).

For the CRA adder structure in Figure 3.4(b), it can be seen that the adder complexity is determined by two 2-input AND gates (cost 2×2), i.e. 4, and one 2-input NOR gate, i.e. 2, and then two 2-input EXOR gates, which is 2×4 (remember there is a double cost for EXOR gates). This gives a complexity of 14, which if we multiply up by 16 gives a complexity of 224 as shown in the first line of Table 3.2. The delay of the first adder cell is $3T$, followed by $n - 2$ delays of $2T$ and a final delay of T, this giving an average of $2nT$ delays, i.e. $32T$.

If we unloop the computation of equations (3.10)–(3.13) a total of 15 times, we can get a structure with the same gate delay of $4T$, but with a very large gate cost i.e. 1264, which is impractical. For this reason, a merger of the CLA technique with the ripple carry structure is preferred. This can be achieved either in the form of the block CLA with inter-block ripple (RCLA) which in effect performs a four-bit addition using a CLA and organizes the structure as a CRA (see Figure 3.6(a)), or a block CLA with intra-group, carry ripple (BCLA) which uses the CLA for the computation of the carry and then uses the lower-cost CRA for the reset of the addition (see Figure 3.6(b)).

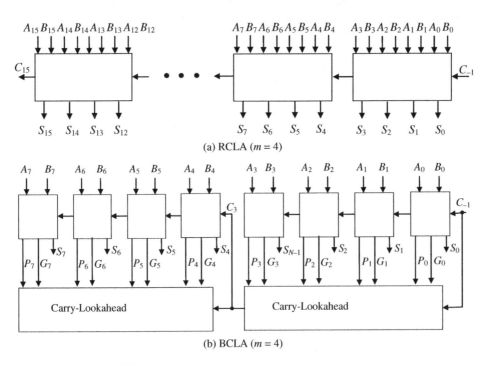

(a) RCLA ($m = 4$)

(b) BCLA ($m = 4$)

Figure 3.6 Alternative CLA structures

These circuits also have varying critical paths as indicated in Table 3.3. A detailed examination of the CLA circuit reveals that it takes $3T$ to produce the first carry, then $2T$ for each of the subsequent stages to produce their carry as the G_i terms will have been produced, and finally $2T$ to create the carry in the final CLA block. This gives a delay for the 16-bit RCLA of $10T$.

The performance represents an estimation of speed against gate cost given as cost unit (CU) and is given by cost multiplied by time divided by 10,000. Of course, this will only come in to play when all circuits meet the timing requirements, which is unusual as it is normally speed that dominates with choice of lowest area then coming as a second measure. However, it is useful in showing some kind of performance measure.

3.3.3 Adder Final Remarks

To a large extent, the variety of different adder structures trade off gate complexity with system regularity, as many of the techniques end up with structures that are much less regular. The aim of much of the research which took place in the 1970s and 1980s was to develop higher-speed structures where transistor switching speed was the dominant feature. However, the analysis in the introduction to this book indicates the key importance of interconnect, and somewhat reduces the impact of using specialist adder structures. Another critical consideration for FPGAs is the importance of being able to scale adder word sizes with application need, and in doing so, offer a linear scale in terms of performance reduction. For this reason, the ripple-carry adder has great appeal in FPGAs and is offered in many of the FPGA structures as a dedicated resource (see Chapter 5).

In papers from the 1990s and early 2000s there was an active debate in terms of adder structure (Hauck *et al.* 2000; Xing and Yu 1971). However, it is clear that, even for adder trees that are commonly used to sum numerous multiplication operations as commonly occurs in DSP applications, the analysis outlined in (Hoe *et al.* 2011) supports the use of the dedicated CRA adder structures on FPGAs.

3.3.4 Multipliers

Multiplication can be simply performed through a series of additions. Consider the example below, which illustrates how the simple multiplication of 5 by 11 is carried out in binary. The usual procedure in computer arithmetic is to align the data in a vertical line and shift right rather than shift left, as shown below. However, rather than perform one single addition at the end to add up all the multiples, each multiple is added to an ongoing product called a partial product. This means that every step in the

Table 3.3 16-bit adder comparisons

Adder type	Time (gate delay)	Cost (CUs)	Performance
CRA (Figure 3.5)	32	224	0.72
Pure CLA	4	1264	0.51
RCLA ($m = 4$) (Figure 3.6(a))	10	336	0.34
BCLA ($m = 4$) (Figure 3.6(b))	14	300	0.42

computation equates to the generation of the multiples using an AND gate and the use of an adder to compute the partial product.

$$
\begin{array}{rl}
5 = 00101 & \text{multiplicand} \\
11 = \underline{01011} & \text{multiplier} \\
00101 & \\
00101 & \\
00000 & \\
00101 & \\
\underline{00000} & \\
55 = 000110111 &
\end{array}
$$

$$
\begin{array}{rll}
5 = 00101 & & \text{multiplicand} \\
11 = \underline{01011} & & \text{multiplier} \\
00000 & & \text{initial partial product} \\
\underline{00101} & & \text{add 1st multiple partial product} \\
00101 & & \\
000101 & & \text{shift right} \\
\underline{00101} & & \text{add 2nd multiple partial product} \\
00101 & & \\
000101 & & \text{shift right} \\
\underline{00000} & & \text{add 3rd multiple partial product} \\
0001111 & & \\
00001111 & & \text{shift right} \\
\underline{00101} & & \text{add 4th multiple partial product} \\
00110111 & & \\
000110111 & & \text{shift right} \\
\underline{00000} & & \text{add 5th multiple partial product} \\
55 = 000110111 & &
\end{array}
$$

A parallel addition can be computed by performing additions at each stage of the multiplication operation. This means that the speed of operation will be defined by the time required to compute the number of additions defined by the multiplicand. However, if the adder structures of Figure 3.5 were to be used, this would result in a very slow multiplier circuit. Use of alternative fast adders structures (some of which were highlighted in Table 3.3) would result in improved performance but this would be a considerable additional cost.

Fast Multipliers

The received wisdom in speeding up multiplications is to either speed up the addition or reduce the number of additions. The latter is achieved by recoding the multiplicand, commonly termed Booth's encoding (discussed shortly). However, increasing the addition speed is achieved by exploiting the carry-save adder structure of Figure 3.7. In conventional addition, the aim is to reduce (or compress) two input numbers into a single output. In multiplication, the aim is to reduce multiple numbers, i.e. multiplicands, down to a single output value. The carry-save adder is a highly efficient structure that allows us to compress three inputs down to two outputs at the cost of a CRA addition but with the speed of the individual cells given in Figure 3.4(b) or (c), namely two or three gate delays.

Thus it is possible to create a carry-save array multiplier as shown in Figure 3.8. An addition is required at each stage, but this is a much faster, smaller CSA addition, allowing a final sum and carry to be quickly generated. A final adder termed a carry

Figure 3.7 Carry-save adder

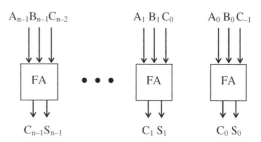

propagation adder (CPA) is then used to compute the final addition using one of the fast addition circuits from earlier.

Even though each addition stage is reduced to two or three gate delays, the speed of the multiplier is then determined by the number of stages. As the word size m grows, the number of stages is then given as $m - 2$. This limitation is overcome in a class of multipliers known as Wallace tree multipliers (Wallace 1964), which allows the addition steps to be performed in parallel. An example is shown in Figure 3.9.

As the function of the carry-save adder is to compress three words to two words, this means that if n is the input wordlength, then after each stage, the words are represented as $3k + l$, where $0 \leq l \leq 2$. This means that the final sum and carry values are produced after $\log_{1.5} n$ rather than $n - 1$ stages as with the carry-save array multiplier, resulting in a much faster implementation.

Figure 3.8 Carry-save array multiplier

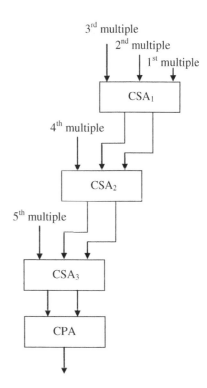

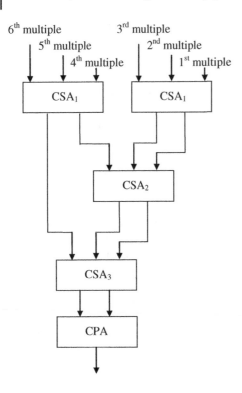

Figure 3.9 Wallace tree multiplier

Booth Encoding

It was indicated earlier that the other way to improve the speed of a multiplier was to reduce the number of additions performed in the multiplications. At first thought, this does not seem obvious as the number of additions is determined by the multiplicand (MD). However, it is possible to encode the binary input in such a way as to reduce the number of additions by two, by exploiting the fact that an adder can easily implement a subtraction.

The scheme is highlighted in Table 3.4 and shows that by examining three bits of the multiplier (MR), namely MR_{i+1}, MR_i and MR_{i-1}, it is is possible to reduce two bit operations down to one operation, either an addition or subtraction. This requires adding to the multiplier the necessary conversion circuitry to detect these sequences. This is

Table 3.4 Modified Booth's algorithm

$MR_{i+1,i}$	MR_{i-1}	Action
00	0	Shift partial product by 2 places
00	1	Add MD and shift partial product by 2 places
01	0	Add MD and shift partial product by 2 places
01	1	Add $2 \times$ MD and shift partial product by 2 places
10	0	Subtract $2 \times$ MD and shift partial product by 2 places
10	1	Subtract MD and shift partial product by 2 places
11	0	Subtract MD and shift partial product by 2 places
11	1	Shift partial product by 2 places

known as the modified Booth's algorithm. The overall result is that the number of additions can be halved.

The common philosophy for fast additions is to combine the Booth encoding scheme with Wallace tree multipliers to produce a faster multiplier implementation. In Yeh and Jen (2000), the authors present an approach for a high-speed Booth encoded parallel multiplier using a new modified Booth encoding scheme to improve performance and a multiple-level conditional-sum adder for the CPA.

3.4 Alternative Number Representations

Over the years, a number of schemes have emerged for either faster or lower-cost implementation of arithmetic processing functions. These have included SDNR (Avizienis 1961), LNS (Muller 2005), RNS (Soderstrand *et al.* 1986) and the CORDIC representation (Voider 1959; Walther 1971). Some of these have been used in FPGA designs specifically for floating-point implementations.

3.4.1 Signed Digit Number Representation

SDNRs were originally developed by Avizienis (1961) as a means to break carry propagation chains in arithmetic operations. In SDNR, each digit is associated with a sign, positive or negative. Typically, the digits are represented in balanced form and drawn from a range $-k$ to $(b-1)-k$, where b is the number base and typically $k = \left\lfloor \frac{b}{2} \right\rfloor$. For balanced ternary which best matches conventional binary, this gives a digit set for x where $x \in (-1, 0, 1)$, or strictly speaking $(1, 0, \bar{1})$ where $\bar{1}$ represents -1. This is known as signed binary number representation (SBNR), and the digits are typically encoded by two bits, namely a sign bit, x_s, and a magnitude bit, x_m, as shown in Table 3.5. Avizienis (1961) was able to demonstrate how such a number system could be used for performing parallel addition without the need for carry propagation (shown for an SBNR adder in Figure 3.10), effectively breaking the carry propagation chain.

A more interesting assignment is the $(+, -)$ scheme where an SBNR digit is encoded as (x^+, x^-), where $x = x^+ + (x^- - 1)$. Alternatively, this can be thought of as $x^- = 0$ implying -1, $x^- = 1$ implying 0, and $x^+ = 0$ implying 0, and $x^+ = 1$ implying 1. The key advantage of this approach is that it provides the ability to construct generalized SBNR adders from conventional adder blocks.

Table 3.5 SDNR encoding

| SDNR digit | SDNR representations | | | |
| | Sig-and-mag | | +/− coding | |
	x_s	x_m	x^+	x^-
x				
0	0	0	0	1
1	0	1	1	0
$\bar{1}$	0	1	0	1
0 or X	1	0	1	0

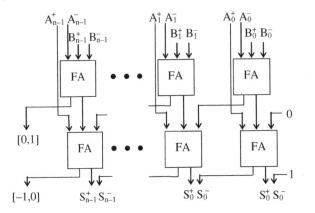

Figure 3.10 SBNR adder

This technique was effectively exploited to allow the design of high- speed circuits for arithmetic processing and digital filtering (Knowles *et al.* 1989) and also for Viterbi decoding (Andrews 1986). In many DSP applications such as filtering, the filter is created with coefficient values such that for fixed-point DSP realizations, the top part of the output word is then required after truncation. If conventional pipelining is used, it will take several cycles for the first useful bit to be generated, seemingly defeating the purpose of using pipelining in the first place.

Using SBNR arithmetic, it is possible to generate the result *msb* or strictly most significant digit (msd) first, thereby allowing the computation to progress much more quickly. In older technologies where speed was at a premium, this was an important differentiator and the work suggested an order-of-magnitude improvement in throughput rate. Of course, the redundant representation had to be converted back to binary, but several techniques were developed to achieve this (Sklansky 1960).

With evolution in silicon processes, SBNR representations are now being overlooked in FPGA design as it requires use of the programmable logic hardware and is relatively inefficient, whereas conventional implementations are able to use the dedicated fast adder logic which will be seen in later chapters. However, its concept is very closely related to binary encoding such as Booth's encoding. There are many fixed-point applications where these number conventions can be applied to reduce the overall hardware cost whilst increasing speed.

3.4.2 Logarithmic Number Systems

The argument for LNS is that it provides a similar range and precision to floating-point but offers advantages in complexity over some floating-point applications. For example, multiplication and division are simplified to fixed-point addition and subtraction, respectively (Haselman *et al.* 2005; Tichy *et al.* 2006).

If we consider that a floating-point number is represented by

$$F = -1^S \times 1.M \times 2^{Exp}, \tag{3.14}$$

then logarithmic numbers can be viewed as a specific case of floating-point numbers where the mantissa is always 1, and the exponent has a fractional part (Koren 2002). The logarithmic equivalent, L, is then described as

$$L = -1^{S_A} \times 2^{Exp_A}, \tag{3.15}$$

where S_A is the sign bit which signifies the sign of the whole number and Exp_A is a two's complement fixed-point number where the negative numbers represent values less than 1.0. In this way, LNS numbers can represent both very large and very small numbers. Typically, logarithmic numbers will have a format where two bits are used for the flag bit (to code for zero, plus/minus infinity, and Not a Number (NaN (Detrey and de Dinechin 2003)), and then k bits and l bits represent the integer and fraction respectively (Haselman *et al.* 2005).

A major advantage of the LNS is that multiplication and division in the linear domain ares replaced by addition or subtraction in the log domain:

$$\log_2\left(\frac{x}{y}\right) = \log_2(x) - \log_2(y). \tag{3.16}$$

However, the operations of addition and subtraction are more complex. In Collange *et al.* (2006), the development of an LNS floating-point library is described and it is shown how it can be applied to some arithmetic functions and graphics applications.

However, LNS has only really established itself in small niche markets, whereas floating-point number systems have become a standard. The main advantage comes from computing a considerable number of operations in the algorithmic domain where the advantages are seen as conversion is problem. Conversions are not exact and error can accumulate for multiple conversions (Haselman *et al.* 2005). Thus whilst there has been some floating-point library developments, FPGA implementations have not been very common.

3.4.3 Residue Number Systems

RNS representations are useful in processing large integer values and therefore have application in computer arithmetic systems, as well as in some DSP applications (see later), where there is a need to perform large integer computations. In RNS, an integer is converted into a number which is an N-tuple of smaller integers called moduli, given by $(m_N, m_{N-1}, \ldots, m_1)$. An integer X is represented in RNS by an N-tuple $(x_N, x_{N-1}, \ldots, x_1)$, where X_i is a non-negative integer, satisfying

$$X = m_i.q_i + x_i, \tag{3.17}$$

where q_i is the largest integer such that $0 \leq q_i \leq (m_i - 1)$ and the value x_i is known as the residue of X modulo m_i. The main advantage of RNS is that additions, subtractions and multiplications are inherently carry-free due to the translation into the format. Unfortunately, other arithmetic operations such as division, comparison and sign detection are very slow and this has hindered the broader application of RNS. For this reason, the work has mostly been applied to DSP operations that involve a lot of multiplications and additions such as FIR filtering (Meyer-Baese *et al.* 1993) and transforms such as the FFT and DCT (Soderstrand *et al.* 1986).

Table 3.6 CORDIC functions

Configuration	Rotation	Vectoring
Linear	$Y = X \times Y$	$Y = X/Y$
Hyperbolic	$X = \cosh(X)$ $Y = \sinh(Y)$	$Z = \text{arctanh}$
Circular	$X = \cos(X)$ $Y = \sin(Y)$	$Z = \text{arctanh}(Y)$ $X = \text{sqr}(X^2 + Y^2)$

Albicocco *et al.* (2014) suggest that in the early days RNS was used to reach the maximum performance in speed, but now it is used primarily to obtain power efficiency and speed–power trade-offs and for reliable systems where redundant RNS are used. It would seem that the number system is suffering the same consequences as SDNRs as dedicated, high-speed computer arithmetic has now emerged in FPGA technology, making a strong case for using conventional arithmetic.

3.4.4 CORDIC

The unified CORDIC algorithm was originally proposed by Voider (1959) and is used in DSP applications for functions such as those shown in Table 3.6. It can operate in one of three configurations (linear, circular and hyperbolic) and in one of two modes (rotation and vectoring) in those configurations. In rotation, the input vector is rotated by a specified angle; in vectoring, the algorithm rotates the input vector to the x-axis while recording the angle of rotation required. This makes it attractive for computing trigonometric operations such as sine and cosine and also for multiplying or dividing numbers.

The following unified algorithm, with three inputs, X, Y and Z, covers the three CORDIC configurations:

$$\begin{aligned} X_{i+1} &= X_i - m \times Y_i \times d_i \times 2^i \\ Y_{i+1} &= Y_i + X_i \times d_i \times 2^i \\ Z_{i+1} &= Z_i - \times e_i. \end{aligned} \tag{3.18}$$

Here m defines the configuration for hyperbolic ($m = -1$), linear ($m = 0$) or circular ($m = 1$), and d_i is the direction of rotation, depending on the mode of operation. For rotation mode $d_i = -1$ if $Z_i < 0$ else $+1$, while in vectoring mode $d_i = +1$ if $Y_i < 0$ else -1. Correspondingly, the value of e^i as the angle of rotation changes depending upon the configuration. The value of e^i is normally implemented as a small lookup table within the FPGA and is defined in Table 3.7 and outlines the pre-calculated values that are typically stored in LUTs, depending upon the configuration.

The reduced computational load experienced in implementing CORDIC operations in performing rotations (Takagi *et al.* 1991) means that it has been used for some DSP

Table 3.7 CORDIC angle of rotation

Configuration	e_i
Linear	2^{-i}
Hyperbolic	$\text{arctanh}(2^{-i})$
Circular	$\arctan(2^{-i})$

applications, particularly those implementing matrix triangularization (Ercegovac and Lang 1990) and RLS adaptive filtering (Ma *et al.* 1992) as this latter application requires rotation operations.

These represent dedicated implementations, however, and the restricted domain of the approaches where a considerable performance gain can be achieved has tended to limit the use of CORDIC. Moreover, given that most FPGA architectures have dedicated hardware based on conventional arithmetic, this somewhat skews the focus towards conventional two's-complement-based processing. For this reason, much of the description and the examples in this text have been restricted to two's complement. However, both main FPGA vendors have CORDIC implementations in their catalogs.

3.5 Division

Division may be thought of as the inverse process of multiplication, but it differs in several aspects that make it a much more complicated function. There are a number of ways of performing division, including recurrence division and division by functional iteration. Algorithms for division and square root have been a major research area in the field of computer arithmetic since the 1950s. The methods can be divided into two main classes, namely digit-by-digit methods and convergence methods. The digit-by-digit methods, also known as direct methods, are somewhat analogous to the pencil-and-paper method of computing quotients and square roots. The results are computed on a digit-by-digit basis, msd first. The convergence methods, which include the Newton–Raphson algorithm and the Taylor series expansion, require the repeated updating of an approximation to the correct result.

3.5.1 Recurrence Division

Digit recurrence algorithms are well-accepted subtractive methods which calculate quotients one digit per iteration. They are analogous to the pencil-and-paper method in that they start with the msbs and work toward the lsbs. The partial remainder is initialized to the dividend, then on each iteration a digit of the quotient is selected according to the partial remainder. The quotient digit is multiplied by the divisor and then subtracted from the partial remainder. If negative, the restoring version of the recurrence divider restores the partial remainder to the previous value, i.e. the results of one subtraction (comparison) determine the next division iteration of the algorithm, which requires the selection of quotient bits from a digit set. Therefore, a choice of quotient bits needs to be made at each iteration by trial and error. This is not the case with multiplication, as the partial products may be generated in parallel and then summed at the end. These factors make division a more complicated algorithm to implement than multiplication and addition.

When dividing two n-bit numbers, this method may require up to $2n + 1$ additions. This can be reduced by employing the non-restoring recurrence algorithm in which the digits of the partial remainder are allowed to take negative and positive values; this reduces the number of additions/subtractions to n. The most popular recurrence division method is an algorithm known as the SRT division algorithm which was named for the three researchers who independently developed it, Sweeney, Robertson and Tocher (Robertson 1958; Tocher 1958).

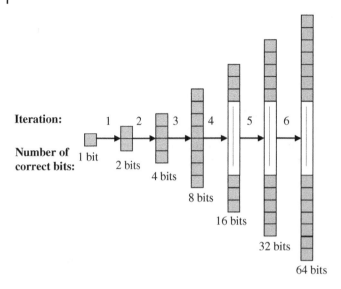

Iteration: 1 2 3 4 5 6

Number of correct bits: 1 bit

2 bits

4 bits

8 bits

16 bits

32 bits

64 bits

Figure 3.11 Quadratic convergence

The recurrence methods offer simple iterations and smaller designs, however, they also suffer from high latencies and converge linearly to the quotient. The number of bits retired at each iteration depends on the radix of the arithmetic being used. Larger radices may reduce the number of iterations required, but will increase the time for each iteration. This is because the complexity of the selection of quotient bits grows exponentially as the radix increases, to the point that LUTs are often required. Therefore, a trade-off is needed between the radix and the complexity; as a result, the radix is usually limited to 2 or 4.

3.5.2 Division by Functional Iteration

The digit recurrence algorithms mentioned in the previous subsection retire a fixed number of bits at each iteration, using only shift and add operations. Functional iterative algorithms employ multiplication as the fundamental operation and produce at least double the number of correct bits with each iteration (Flynn 1970; Ito *et al.* 1995; Obermann and Flynn 1997; Oklobdzija and Ercegovac 1982). This is an important factor as there may be as many as three multiplications in each iteration. However, with the advantage of at least quadratic convergence, a 53-bit quotient can be achieved in six iterations, as illustrated in Figure 3.11.

3.6 Square Root

Methods for performing the square root operation are similar to those for performing division. They fall broadly into the two categories, digit recurrence methods and methods based on convergence techniques. This section gives a brief overview of each.

3.6.1 Digit Recurrence Square Root

Digit recurrence methods can be based on either restoring or non-restoring techniques, both of which operate msd first. The algorithm is subtractive, and after each iteration the resulting bit is set to 0 if a negative value is found, and then the original remainder is 'restored' as the new remainder. If the digit is positive, a 1 is set and the new remainder is used. The "non-restoring" algorithm allows the negative value to persist and then performs a compensation addition operation in the next iteration. The overall process of the square root and division algorithms is very similar, and, as such, there have been a number of implementations of systolic arrays designed to perform both arithmetic functions (Ercegovac and Lang 1991; Heron and Woods 1999).

The performance of the algorithms mentioned has been limited due to the dependence of the iterations and the propagated carries along each row. The full values need to be calculated at each stage to enable a correct comparison and decision to be made. The SRT algorithm is a class of non-restoring digit-by-digit algorithms in which the digit can assume both positive and negative non-zero values. It requires the use of a redundant number scheme (Avizienis 1961), thereby allowing digits to take the values 0, -1 or 1. The most important feature of the SRT method is that the algorithm allows each iteration to be performed without full-precision comparisons at each iteration, thus giving higher performance.

Consider a value R for which the algorithm is trying to find the square root, and S_i the partial square root obtained after i iterations. The scaled remainder, Z_i, at the ith step is

$$Z_i = 2^i \left(R - S_i^2 \right),$$ (3.19)

where $1/4 \leq R < 1$ and hence $1/2 \leq S < 1$. From this, a recurrence relation based on previous remainder calculations can be derived as (McQuillan *et al.* 1993)

$$Z_i = 2^i Z_{i-1} - s_i \left(2S_{i-1} + s_i 2^{-i} \right), \quad i = 2, 3, 4, \ldots,$$ (3.20)

where s_i is the root digit for iteration $i - 1$. Typically, the initial value for Z_0 will be set to R, while the initial estimate of the square root, S_1, is set to 0.5 (due to the initial boundaries placed on R).

There exist higher-radix square root algorithms (Ciminiera and Montuschi 1990; Cortadella and Lang 1994; Lang and Montuschi 1992). However, for most algorithms with a radix greater than 2, there is a need to provide an initial estimate for the square root from a LUT. This relates to the following subsection.

3.6.2 Square Root by Functional Iteration

As with the convergence division in Section 3.6.1, the square root calculation can be performed using functional iteration. It can be additive or multiplicative. If additive, then each iteration is based on addition and will retire the same number of bits with each iteration. In other words, they converge linearly to the solution. One example is the CORDIC implementation for performing the Givens rotations for matrix triangularization (Hamill *et al.* 2000). Multiplicative algorithms offer an interesting alternative as they double the precision of the result with each iteration, that is, they converge quadratically to the result. However, they have the disadvantage of increased computational complexity due to the multiplications within each iterative step.

Similarly to the approaches used in division methods, the square root can be estimated using Newton–Raphson or series convergence algorithms. For the Newton–Raphson method, an iterative algorithm can be found by using

$$x_{i+1} = X_i - \frac{f(x_i)}{f'(x_i)}$$

(3.21)

and choosing $f(x)$ that has a root at the solution. One possible choice is $f(x) = x^2 - b$ which leads to the following iterative algorithm:

$$x_{i+1} = \frac{1}{2}\left(X_i - \frac{b}{x_i}\right).$$

(3.22)

This has the disadvantage of requiring division. An alternative method would be to aim to drive the algorithm toward calculating the reciprocal of the square root, $1/x^2$. For this, $f(x) = 1/x^2 - b$ is used, which leads to the following iterative algorithm:

$$x_{i+1} = \frac{x_i}{2}\left(3 - bx_i^2\right)$$

(3.23)

Once solved, the square root can then be found by multiplying the result by the original value, X, that is, $1/\sqrt{X} \times X = \sqrt{X}$.

Another method for implementing the square root function is to use series convergence, i.e. Goldschmidt's algorithm (Soderquist and Leeser 1995), which produces equations similar to those for division (Even *et al.* 2003). The aim of this algorithm is to compute successive iterations to drive one value to 1 while driving the other value to the desired result. To calculate the square root of a value a, for each iteration:

$$x_{i+1} = x_i \times r_i^2,$$

(3.24)

$$y_{i+1} = y_i \times r_i,$$

(3.25)

where we let $x_0 = y_0 = a$. Then by letting

$$r_i = \frac{3 - y_i}{2},$$

(3.26)

$x \to 1$ and consequently $y_i \to \sqrt{a}$. In other words, with each iteration x is driven closer to 1 while y is driven closer to $\sqrt{a}$. As with the other convergence examples, the algorithm benefits from using an initial estimate of $1/\sqrt{a}$ to pre-scale the initial values of x_0 and y_0.

In all of the examples given for both the division and square root convergence algorithms, vast improvements in performance can be obtained by using a LUT to provide an initial estimate to the desired solution. This is covered in the following subsection.

3.6.3 Initial Approximation Techniques

The number of iterations for convergence algorithms can be vastly reduced by providing an initial approximation to the result read from a LUT. For example, the simplest way of forming the approximation R_0 to the reciprocal of the divisor D is to read an approximation to $1/D$ directly from a LUT. The first m bits of the n-bit input value D are used to address the table entry of p bits holding an approximation to the reciprocal.

Table 3.8 Precision of approximations for example values of g and m

Address bits	Guard bits g	Output bits	Precision
m	0	m	$m + 0.415$ bits
m	1	$m + 1$	$m + 0.678$ bits
m	2	$m + 2$	$m + 0.830$ bits
m	3	$m + 3$	$m + 0.912$ bits

The value held by the table is determined by considering the maximum and minimum errors caused by truncating D from n to m bits.

The time to access a LUT is relatively small so it provides a quick evaluation of the first number of bits to a solution. However, as the size of the input value addressing the LUT increases, the size of the table grows exponentially. For a table addressed by m bits and outputting p bits, the table size will have $2m$ entries of width p bits. Therefore, the size of the LUT soon becomes very large and will have slower access times.

A combination of p and m can be chosen to achieve the required accuracy for the approximation, with the smallest possible table. By denoting the number of bits by which p is larger than m as the number of guard bits g, the total error E_{total} (Sarma and Matula 1993) may be expressed as

$$E_{total} = 2^{m+1} \left(\frac{1}{2^{g+1}} \right). \tag{3.27}$$

Table 3.8 shows the precision of approximations for example values of g and m. These results are useful in determining whether adding a few guard bits might provide sufficient additional accuracy in place of the more costly step in increasing m to $m + 1$ which more than doubles the table size.

Another simple approximation technique is known as read-only memory (ROM) interpolation. Rather than just truncating the value held in memory after the mth bit, the first unseen bit ($m + 1$) is set to 1, and all bits less significant than it are set to 0 (Fowler and Smith 1989). This has the effect of averaging the error. The resulting approximation is then rounded back to the lsb of the table entry by adding a 1 to the bit location just past the output width of the table. The advantage with this technique is its simplicity. However, it would not be practical for large initial approximations as there is no attempt to reduce the table size.

There are techniques for table compression, such as bipartite tables, which use two or more LUTs and then add the output values to determine the approximation (Schulte et al. 1997). To approximate a reciprocal function using bipartite tables, the input operand is divided into three parts as shown in Figure 3.12.

The $n_0 + n_1$ bits provide the address for the first LUT, giving the coefficient a_0 of length p_0 bits. The sections d_0 and d_2, equating to $n_0 + n_2$ bits, provide addresses for the second LUT, giving the second coefficient a_1 of length $p1$ bits. The outputs from the tables are added together to approximate the reciprocal, R_0, using a two-term Taylor series expansion. The objective is to use the first $n_0 + n_1$ msbs to provide the lookup for the first table which holds coefficients based on the values given added with the mid-value of the range of values for d_2. The calculation of the second coefficient is based on the value from sections d_0 and d_2 summed with the mid-value of the range of values for d_1.

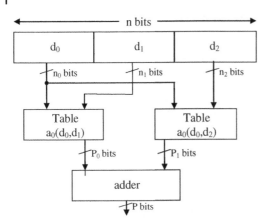

Figure 3.12 Block diagram for bipartite approximation methods

This technique forms a method of averaging so that the errors caused by truncation are reduced. The coefficients for the reciprocal approximation take the form

$$a_0(d_0, d_1) = f(d_0 + d_1 + \delta_2), \tag{3.28}$$
$$a_0(d_0, d_1) = f'(d_0 + \delta_1 + \delta_2)(d_2 - \delta_2), \tag{3.29}$$

where δ_1 and δ_2 are constants exactly halfway between the minimum and maximum values for d_1 and d_2, respectively.

The benefit is that the two small LUTs will have less area than the one large LUT for the same accuracy, even when the size of the addition is considered. Techniques to simplify the bipartite approximation method also exist. One method (Sarma and Matula 1995) eliminates the addition by using each of the two LUTs to store the positive and negative portions of a redundant binary reciprocal value. These are "fused" with slight recoding to round off a couple of low-order bits to obtain the required precision of the least significant bit. With a little extra logic, this recoding can convert the redundant binary values into Booth encoded operands suitable for input into a Booth encoded multiplier.

3.7 Fixed-Point versus Floating-Point

If the natural assumption is that the "most accurate is always best," then there appears to be no choice in determining the number representation, as floating-point will be chosen. Historically, though, the advantage of FPGAs was in highly efficient implementation of fixed-point arithmetic as some of the techniques given in Chapter 7 will demonstrate. However, the situation is changing as FPGA vendors start to make architectural changes which make implementation of floating-point much more attractive, as will be seen in Chapter 5.

The decision is usually made based on the actual application requirements. For example, many applications vary in terms of the data word sizes and the resulting accuracy. Applications can require different input wordlengths, as illustrated in Table 3.9, and can vary in terms of their sensitivity to errors created as a result of limited, internal wordlength. Obviously, smaller input wordlengths will have smaller internal accuracy

Table 3.9 Typical wordlengths

Application	Word sizes (bits)
Control systems	4–10
Speech	8–13
Audio	16–24
Video	8–10

requirements, but the perception of the application will also play a major part in determining the internal wordlength requirements. The eye is tolerant of wordlength limitations in images, particularly if they appear as distortion at high frequencies, whereas the ear is particularly intolerant to distortion and noise at any frequency, but specifically high frequency. Therefore cruder truncation may be possible with some image processing applications, but less so in audio applications.

Table 3.10 gives an estimation of the dynamic range capabilities of some fixed-point representations. It is clear that, depending on the internal computations being performed, many DSP applications can give acceptable signal-to-noise ratio (SNR) with limited wordlengths, say 12–16 bits. Given the performance gain of fixed-point over floating-point in FPGAs, fixed-point realizations have dominated, but the choice will also depend on application input and output wordlengths, required SNR, internal computational complexity and the nature of computation being performed, i.e. whether specialist operations such as matrix inversions or iterative computations are required.

A considerable body of work has been dedicated to reducing the number precision to best match the performance requirements. Constantinides *et al.* (2004) look to derive accurate bit approximations for internal wordlengths by considering the impact on design quality. A floating-point design flow is presented in Fang *et al.* (2002) which takes an algorithmic input and generates floating-point hardware by performing bit width optimization, with a cost function related to hardware, but also to power consumption. This activity is usually performed manually by the designer, using suitable fixed-point libraries in tools such as MATLAB® or LabVIEW, as suggested earlier.

3.7.1 Floating-Point on FPGA

Up until recently, FPGAs were viewed as being poor for floating-point realization. However, the adoption of a dedicated DSP device in each of the main vendors' FPGA families means that floating-point implementation has become much more attractive, particularly if a latency can be tolerated. Table 3.11 gives area and clock speed figures for floating core implementation on a Xilinx Virtex-7 device. The speed is determined by the capabilities of the DSP48E1 core and pipelining within the programmable logic.

Table 3.10 Fixed wordlength dynamic range

Wordlength (bits)	Wordlength range	Dynamic range dB
8	−127 to +127	$20 \log 2^8 \approx 48$
16	−32768 to +32767	$20 \log 2^{16} \approx 96$
24	−8388608 to +8388607	$20 \log 2^{24} \approx 145$

Table 3.11 Xilinx floating-point LogiCORE v7.0 on Virtex-7

Function	DSP48	LUT	Flip-flops	Speed (MHz)
Single range multiplier	2	96	166	462
Double range multiplier	10	237	503	454
Single range accumulator	7	3183	3111	360
Double range accumulator	45	31738	24372	321
Single range divider	0	801	1354	579
Double range divider	0	3280	1982	116.5

The area comparison for floating-point is additionally complicated as the relationship between multiplier and adder area is now changed. In fixed-point, multipliers are generally viewed to be N times bigger than adders, where N is the wordlength. However, in floating-point, the area of floating-point adders is not only comparable to that of floating-point multipliers but, in the case of double point precision, is much larger than that of a multiplier and indeed slower. This corrupts the assumption, at the DSP algorithmic stage, that reducing the number of multiplications in favor of additions is a good optimization.

Calculations in Hemsoth (2012) suggest that the current generation of Xilinx's Virtex-7 FPGAs is about 4.2 times faster than a 16-core microprocessor. This figure is up from a factor of 2.9× as reported in an earlier study in 2010 and suggests that the inclusion of dedicated circuitry is improving the floating-point performance. However, these figures are based on estimated performance and not on a specific application implementation. They indicate 1.33 tera floating-point operations per second (TFLOPS) of single-precision floating-point performance on one device (Vanevenhoven 2011).

Altera have gone one stage further by introducing dedicated hardened circuitry into the DSP blocks to natively support IEEE 754 single-precision floating-point arithmetic (Parker 2012). As all of the complexities of IEEE 754 floating-point are built within the hard logic of the DSP blocks, no programmable logic is consumed and similar clock rates to those for fixed-point designs are achieved. With thousands of floating-point operators built into these hardened DSP blocks, the Altera Arria® 10 FPGAs are rated from 140 giga floating-point operations per second (GFLOPS) to 1.5 TFLOPS across the 20 nm family. This will also be employed in the higher-performance Altera 14 nm Stratix® 10 FPGA family, giving a performance range right up to 10 TFLOPS!

Moreover, the switch to heterogeneous SoC FPGA devices also offers floating-point arithmetic in the form of dedicated ARM processors. As will be seen in subsequent chapters, this presents new mapping possibilities for FPGAs as it is now possible to map the floating-point requirements into the dedicated programmable ARM resources and then employ the fixed-point capabilities of dedicated SoC.

3.8 Conclusions

This chapter has given a brief grounding in computer arithmetic basics and given some idea of the hardware needed to implement basic computer arithmetic functions and some more complex functions such as division and square root. Whilst the chapter outlines the key performance decisions, it is clear that the availability of dedicated adder and multiplier circuitry has made redundant a lot of FPGA-based research into new types of adder/multiplier circuits using different forms of arithmetic.

The chapter has also covered some critical aspects of arithmetic representations and the implications that choice of either fixed- or floating-point arithmetic can have in terms of hardware implementation, particularly given the current FPGA support for floating-point. It clearly demonstrates that FPGA technology is currently very appropriate for fixed-point implementation, but increasingly starting to include floating-point arithmetic capability.

Bibliography

Albicocco P, Cardarilli GC, Nannarelli A, Re M 2014 Twenty years of research on RNS for DSP: Lessons learned and future perspectives. In *Proc. Int. Symp. on Integrated Circuits*, pp. 436–439, doi: 10.1109/ISICIR.2014.7029575.

Andrews M 1986 A systolic SBNR adaptive signal processor. *IEEE Trans. on Circuits and Systems*, 33(2), 230–238.

Avizienis A 1961 Signed-digit number representations for fast parallel arithmetic. *IRE Trans. on Electronic Computers*, 10, 389–400.

Chandrakasan A, Brodersen R 1996 *Low Power Digital Design*. Kluwer, Dordrecht.

Ciminiera L, Montuschi P 1990 Higher radix square rooting. *IEEE Trans. on Computing*, 39(10), 1220–1231.

Collange S, Detrey J, de Dinechin F 2006 Floating point or Ins: choosing the right arithmetic on an application basis. In *Proc. EUROMICRO Conf. on Digital System Design*, pp. 197–203.

Constantinides G, Cheung PYK, Luk W 2004 *Synthesis and Optimization of DSP Algorithms*. Kluwer, Dordrecht.

Cortadella J, Lang T 1994 High-radix division and square-root with speculation. *IEEE Trans. on Computers*, 43(8), 919–931.

Detrey J, de Dinechin FAV 2003 HDL library of LNS operators. In *Proc. 37th IEEE Asilomar Conf. on Signals, Systems and Computers*, 2, pp. 2227–2231.

Ercegovac MD, Lang T 1990 Redundant and on-line cordic: Application to matrix triangularization. *IEEE Trans. on Computers*, 39(6), 725–740.

Ercegovac MD, Lang T 1991 Module to perform multiplication, division, and square root in systolic arrays for matrix computations. *J. of Parallel Distributed Computing*, 11(3), 212–221.

Even G, Seidel PM, Ferguson W 2003 A parametric error analysis of Goldschmidt's division algorithm. In *Proc. IEEE Symp. on Computer Arithmetic*, pp. 165–171.

Fang F, Chen T, Rutenbar RA 2002 Floating-point bit-width optimisation for low-power signal processing applications. In *Proc. of IEEE Int. Conf. on Acoustics, Speech and Signal Processing*, 3, pp. 3208–3211.

Flynn M 1970 On division by functional iteration. *IEEE Trans. on Computers*, 19(8), 702–706.

Fowler DL, Smith JE 1989 High speed implementation of division by reciprocal approximation. In *Proc. IEEE Symp. on Computer Arithmetic*, pp. 60–67.

Hamill R, McCanny J, Walke R 2000 Online CORDIC algorithm and VLSI architecture for implementing QR-array processors. *IEEE Trans. on Signal Processing*, 48(2), 592–598.

Hauck S, Hosler MM, Fry TW 2000 High-performance carry chains for FPGAs. *IEEE Trans. on VLSI Systems*, 8(2), 138–147.

Haselman M, Beauchamp M, Wood A, Hauck S, Underwood K, Hemmert KS 2005 A comparison of floating point and logarithmic number systems for FPGAs. In *Proc. IEEE Symp. on FPGA-based Custom Computing Machines*, pp. 181–190.

Hemsoth N 2012. Latest FPGAs show big gains in floating point performance. *HPCWire*, April 16.

Heron JP, Woods RF 1999 Accelerating run-time reconfiguration on FCCMs. *Proc. IEEE Symp. on FPGA-based Custom Computing Machines*, pp. 260–261.

Hoe DHK, Martinez C, Vundavalli SJ 2011 Design and characterization of parallel prefix adders using FPGAs. In *Proc. IEEE Southeastern Symp. on Syst. Theory*, pp. 168–172.

Ito M, Takagi N, Yajima S 1995 Efficient initial approximation and fast converging methods for division and square root. In *Proc. IEEE Symp. on Computer Arithmetic*, pp. 2–9.

Knowles S, Woods RF, McWhirter JG, McCanny JV 1989 Bit-level systolic architectures for high performance IIR filtering. *J. of VLSI Signal Processing*, 1(1), 9–24.

Koren I 2002 *Computer Arithmetic Algoritmns*, 2nd edition A.K. Peters, Natick, MA.

Lang T, Montuschi P 1992 Higher radix square root with prescaling. *IEEE Trans. on Computers*, 41(8), 996–1009.

Ma J, Deprettere EF, Parhi K 1997 Pipelined cordic based QRD-RLS adaptive filtering using matrix lookahead. In *Proc. IEEE Int. Workshop on Signal Processing Systems*, pp. 131–140.

McQuillan SE, McCanny J, Hamill R 1993 New algorithms and VLSI architectures for SRT division and square root. In *Proc. IEEE Symp. on Computer Arithmetic*, pp. 80–86.

Meyer-Baese U, García A, Taylor F. 2001. Implementation of a communications channelizer using FPGAs and RNS arithmetic. *J. VLSI Signal Processing Systems*, 28(1–2), 115–128.

Muller JM 2005 *Elementary Functions, Algorithms and Implementation*. Birkhäuser, Boston.

Obermann SF, Flynn MJ 1997 Division algorithms and implementations. *IEEE Trans. on Computers*, 46(8), 833–854.

Oklobdzija V, Ercegovac M 1982 On division by functional iteration. *IEEE Trans. on Computers*, 31(1), 70–75.

Omondi AR 1994 *Computer Arithmetic Systems*. Prentice Hall, New York.

Parker M 2012 The industry's first floating-point FPGA. *Altera Backgrounder*. http://bit.ly/2eF01WT

Robertson, J 1958 A new class of division methods. *IRE Trans. on Electronic Computing*, 7, 218–222.

Sarma DD, Matula DW 1993 Measuring the accuracy of ROM reciprocal tables. In *Proc. IEEE Symp. on Computer Arithmetic*, pp. 95–102.

Sarma DD, Matula DW 1995 Faithful bipartite ROM reciprocal tables. In *Proc. IEEE Symp. on Computer Arithmetic*, pp. 17–28.

Schulte MJ, Stine IE, Wires KE 1997 High-speed reciprocal approximations. In *Proc. IEEE Asilomar Conference on Signals, Systems and Computers*, pp. 1183–1187.

Sklansky J 1960 Conditional sum addition logic. *IRE Trans. on Electronic Computers*, 9(6), 226–231.

Soderquist P, Leeser M 1995 An area/performance comparison of subtractive and multiplicative divide/square root implementations. In *Proc. IEEE Symp. on Computer Arithmetic*, pp. 132–139.

Soderstrand MA, Jenkins WK, Jullien GA, Taylor FJ 1986 *Residue Number Systems Arithmetic: Modern Applications in Digital Signal Processing*. IEEE Press, Piscataway, NJ.

Takagi N, Asada T, Yajima S 1991 Redundant CORDIC methods with a constant scale factor for sine and cosine computation. *IEEE Trans. on Computers*, 40(9), 989–995.

Tichy M, Schier J, Gregg D 2006 Efficient floating-point implementation of high-order (N)LMS adaptive filters in FPGA. In Bertels K, Cardoso JMP, Vassiliadis (eds) *Reconfigurable Computing: Architectures and Applications*, Lecture Notes in Computer Science 3985, pp. 311–316. Springer, Berlin.

Tocher K 1958 Techniques of multiplication and division for automatic binary computers. *Quart. J. of Mech. and Applied Mathematics*. 1l(3), 364–384.

Vanevenhoven T 2011, High-level implementation of bit- and cycle-accurate floating-point DSP algorithms with Xilinx FPGAs. *Xilinx White Paper: 7 Series FPGAs WP409 (v1.0)*.

Voider JE 1959 The CORDIC trigonometric computing technique. *IRE Trans. on Electronic Computers*, 8(3), 330–334.

Wallace CS 1964 A suggestion for a fast multiplier. *IEEE Trans. on Electronic Computers*, 13(1), 14–17.

Walther JS 1971 A unified algorithm for elementary functions. In *Proc. Spring Joint Computing Conference*, pp. 379–385.

Xing S, Yu WWh 1998. FPGA adders: performance evaluation and optimal design. *IEEE Design and Test of Computers*, 15(1), 24–29.

Yeh W-C, Jen C-W 2000 High-speed Booth encoded parallel multiplier design, *IEEE Trans. on Computers*, 49(7), 692–701.

4

Technology Review

4.1 Introduction

The technology used for DSP implementation is very strongly linked to the astonishing developments in silicon technology. As was highlighted in the introduction to this book, the availability of a transistor which has continually decreased in cost has been the major driving force in creating new markets and has overseen the development of a number of DSP technologies. Silicon technology has offered an increasingly cheaper platform, and has done so at higher speeds and at a lower power cost. This has inspired a number of core markets, such as computing, mobile telephony and digital TV.

As Chapter 2 clearly indicated, there are numerous advantages for digital systems, specifically guaranteed accuracy, essentially perfect reproducibility and better aging; these developments are seen as key to the continued realization of future systems. The earliest DSP filter circuits were pioneered by Leland B. Jackson and colleagues at Bell Laboratories in the late 1960s and early 1970s (see Jackson 1970). At that time, the main aim was to create silicon chips to perform basic functions such as FIR and IIR filtering. A key aspect was the observation that the binary operation of the transistor was well matched to digital operations required in DSP systems.

From these early days, a number of technologies emerged, ranging from simple microcontrollers which can process systems with sampling rates typically in the moderate kilohertz range, right through to dedicated SoC solutions that give performance in the teraOPS region. The *processor style* architecture has been exploited in various forms, ranging from single- to multicore processor implementations, DSP microprocessors with dedicated hardware to allow specific DSP functionality to be realized efficiently, and reconfigurable processor architectures. Specialized DSP functionality has also been added to conventional central processing units (CPUs) and application-specific instruction processors (ASIPs) that are used for specific markets. All of these are briefly discussed in this chapter with the aim of giving a perspective against which FPGAs should be considered.

The major change between this chapter and its first edition counterpart concerns microprocessors: there has been a major development in architectures, particularly with the evolution of the Intel multicore and Xeon Phi, resulting in a body of work on

FPGA-based Implementation of Signal Processing Systems,
Second Edition. Roger Woods, John McAllister, Gaye Lightbody and Ying Yi.
© 2017 John Wiley & Sons, Ltd. Published 2017 by John Wiley & Sons, Ltd.

parallel programming (Reinders and Jeffers 2014). In addition, multicore DSP architectures have evolved. Finally, graphical processing units (GPUs) are also included as they are now being widely used in many other fields than graphics including DSP. As Chapter 5 is dedicated to the variety of FPGA architectures, the FPGA perspective is only alluded to. Major themes include level of programmability, the programming environment (including tools, compilers and frameworks), the scope for optimization of specifically DSP functionality on the required platform, and the quality of the resulting designs in terms of area, speed, throughput, power and even robustness.

Section 4.2 starts with some remarks on silicon technology scaling and how Dennard scaling has broken down, leading to the evolution of parallelism into conventional DSP platforms. Section 4.3 outlines some further thoughts on architecture and programmability and gives some insights towards the performance limitations of the technologies, and also comments on the importance of programmability. In Section 4.4 the functional requirements of DSP systems are examined, highlighting issues such as computational complexity, parallelism, data independence and arithmetic advantages. The section ends with a brief definition of technology classification and introduces concepts of single instruction, multiple data (SIMD) and multiple instruction, multiple data (MIMD). This is followed by a brief description of microprocessors in Section 4.5 with some more up-to-date description of multicore architectures. DSP processors are then introduced in Section 4.6 along with some multicore examples. Section 4.7 is a new section on GPUs as these devices have started to be used in some DSP applications. For completeness, solutions based on the system-on-chip (SoC) are briefly reviewed in Section 4.8, which includes the development of parallel machines including systolic array architectures. A core development has been the partnering of various technologies, namely ARM processors and DSP microprocessors, and ARM processors incorporated in FPGA fabrics. A number of examples of this evolution are given in Section 4.9. Section 4.10 gives some thoughts on how the various technologies compare and sets the scene for FPGAs in the next chapter.

4.2 Implications of Technology Scaling

Since the first edition of this book, there has been a considerable shift in the direction of evolution of silicon, largely driven by concerns in silicon scaling. Dennard's law builds on Moore's law, relating how the performance of computing is growing exponentially at roughly the same rate as Moore's law. This was driven by the computing and supercomputing industries and therefore, by association, refers to DSP technologies.

The key issue is that Dennard's law is beginning to break down as many computing companies are becoming very concerned by the power consumption of their devices and beginning to limit power consumption for single devices to 130 W (Sutter 2009). This is what Intel has publicly declared. To achieve this power capping requires a slow-down in clock scaling. For example, it was predicted in the 2005 of ITRS Roadmap (ITRS 2005) that clock scaling would continue and we would have expected to have a clock rate of 30 GHz today (ITRS 2011). This was revised in 2007 to 10 GHz, and then in 2011 to 4 GHz, which is what is currently offered by computing chip companies. The implications of this are illustrated in Figure 4.1 reproduced from Sutter (2009).

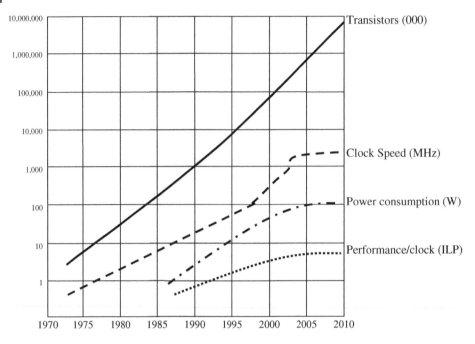

Figure 4.1 Conclusions from Intel CPU scaling

Moore's law has continued unabated, however, in terms of number of devices on a single silicon die, so companies have acted to address the shortfall in clock scaling by shifting towards parallelism and incorporating multiple devices on a single die. As will be seen later, this has implications for the wider adaption of FPGAs because as the guaranteed clock rate ratio of CPUs over FPGAs is considerably reduced, the performance divide widens, making FPGAs an attractive proposition. This would seem to be reflected in the various developments of computing companies in exploiting FPGA technologies largely for data centers where the aim is to be able to reduce the overall power consumption costs, as described in the next chapter.

The authors argue that the main criterion in DSP system implementation is the circuit architecture that is employed to implement the system, i.e. the hardware resources available and how they are interconnected; this has a major part to play in the performance of the resulting DSP system. FPGAs allow this architecture to be created to best match the algorithmic requirements, but this comes at increased design cost. It is interesting to compare the various approaches, and this chapter aims to give an overview of the various technologies available for implementing DSP systems, using relevant examples where applicable, and the technologies are compared and contrasted.

4.3 Architecture and Programmability

In many processor-based systems, design simply represents the creation of the necessary high-level code with some thought given to the underlying technology architecture, in order to optimize code quality and thus improve performance. Crudely speaking, though, performance is sacrificed to provide this level of programmability. Take, for example, the microprocessor architecture based on the von Neumann sequential model where the underlying architecture is fixed and the maximum achievable performance will be determined by efficiently scheduling the algorithmic requirements onto the inherently sequential processing architecture. If the computation under consideration is highly parallel in nature (as is usually the case in DSP), then the resulting performance will be poor.

If we were to take the other extreme and develop an SoC-based architecture that best matches the parallelism of computational complexity of the algorithm (as will be outlined in the final section of this chapter), then the best performance in terms of area, speed and power consumption should be achieved. This requires a number of design activities to ensure that hardware implementation metrics best match the application performance criteria and that the resulting design operates correctly.

To more fully understand this concept of generating a circuit architecture, consider the "state of the art" in 1969. Hardware capability in terms of numbers of transistors was limited and thus highly valued, so the processing in the filters described in Jackson (1970) had to be undertaken in a rather serial fashion. Current FPGA technology provides hundreds of bit parallel multipliers, so the arithmetic style and resulting performance are quite different, implying a very different sort of architecture. The aim is thus to make the best use of the available hardware against the performance criteria of the application. Whilst this approach of developing the hardware to match the performance needs is highly attractive, the architecture development presents a number of problems related to the very process of producing this architecture, namely design time, verification and test of the architecture in all its various modes, and all the issues associated with producing a design that is right first time.

Whilst the implementation of these algorithms on a specific hardware platform can be compared in terms of metrics such as throughput rate, latency, circuit area, energy, and power consumption, one major theme that can also be used to differentiate these technologies is programmability (strictly speaking, ease of programmability). As will become clear in the descriptive material in this section, DSP hardware architectures can vary in their level of programmability. A simple platform with a fixed hardware architecture can then be easily programmed using a high-level software language as, given the fixed nature of the platform, efficient software compilers can be (and indeed have been) developed to create the most efficient realizations. However, as the platform becomes more complex and flexible, the complexity and efficiency of these tools are compromised, as now special instructions have to be introduced to meet this functionality and the problem of parallel processing rears its ugly head.

In this case, the main aim of the compiler is to take source code that may not have been written for the specific hardware architecture, and identify how these special functions might be applied to improve performance. In a crude sense, we suggest that making the circuit architecture programmable achieves the best efficiency in terms of performance, but presents other issues with regard to evolution of the architecture either to meet small changes in applications requirements or relevance to similar applications. This highlights the importance of tools and design environments, described in Chapter 7.

SoC is at the other end of the spectrum from a programmability perspective; in this case, the platform will have been largely developed to meet the needs of the system under consideration or some domain-specific, standardized application. For example,

OFDM access-based systems such as LTE-based mobile phones require specific DSP functionality such as orthogonal frequency division multiple access (OFDMA) which can be met by developing an SoC platform comprising processors and dedicated hardware IP blocks. This is essential to meet the energy requirements for most mobile phone implementations. However, silicon fabrication costs have now pushed SoC implementation into a specialized domain where typically solutions are either for high volume, or have specific domain requirements, e.g. ultra-low power in low- power sensors.

4.4 DSP Functionality Characteristics

DSP operations are characterized as being computationally intensive, exhibiting a high degree of parallelism, possessing data independence, and in some cases having lower arithmetic requirements than other high-performance applications, e.g. scientific computing. It is important to understand these issues more fully in order to judge their impact for mapping DSP algorithms onto hardware platforms such as FPGAs.

4.4.1 Computational Complexity

DSP algorithms can be highly complex. For example, consider the N-tap FIR filter expression given in Chapter 2 and repeated here:

$$y(n) = \sum_{i=0}^{N-1} a_i x(n-i). \tag{4.1}$$

In effect, this computation indicates that a_0 must be multiplied by $x(n)$, followed by the multiplication of a_1 by $x(n-i)$ to which it must be added, and so on. Given that the tap size is N, this means that the computation requires N multiplications followed by $N-1$ additions in order to compute $y(n)$ as shown below:

$$y(n) = a_0 x(n) + a_1 x(n-1) + a_2 x(n-2) + \cdots + a_{N-1} x(n-N+1) \tag{4.2}$$

Given that another computation will start on the arrival of next sample, namely x_{n+1}, this defines the computations required per cycle, namely $2N$ operations (N multiplications and N additions) per sample or two operations per tap. If a processor implementation is targeted, then this requires, say, a loading of the data every cycle, which would need two or three cycles (to load data and coefficients) and storage of the accumulating sum. This could mean an additional three operations per cycle, resulting in six operations per tap or, overall, $6N$ operations per sample. For an audio application with a sampling rate of 44.2 kHz, a 128-tap filter will require 33.9 MSPS, which may seem realistic for some technologies, but when you consider image processing rates of 13.5 MHz, these computational rates quickly explode, resulting in a computation rate of 10 gigasamples per second (GSPS). In addition, this may only be one function within the system and thus represent only a small proportion of the total processing required.

For a processor implementation, the designer will determine if the hardware can meet the throughput requirements by dividing the clock speed of the processor by the number of operations that need to be performed each cycle, as outlined above. This can give a poor return in performance, since if N is large, there will be a large disparity between clock and throughput rates. The clock rate may be fast enough to provide the necessary sampling rate, but it will present problems in system design, both in delivering a very

Figure 4.2 Simple parallel implementation of a FIR filter

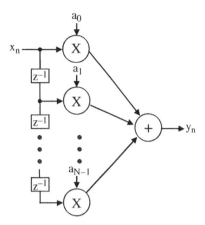

fast clock rate and controlling the power consumption, particularly dynamic power consumption, as this is directly dependent on the clock rate.

4.4.2 Parallelism

The nature of DSP algorithms is such that high levels of parallelism are available. For example, the expression in equation (4.1) can be implemented in a single processor, or a parallel implementation, as shown in Figure 4.1, where each element in the figure becomes a hardware component therefore implying 127 registers for the delay elements, 128 MAC blocks for computing the products, $a_1 x(N - i)$, where $i = 0, 1, 2, \ldots, N - 1$, and their addition which can of course, be pipelined if required.

In this way, we have the hardware complexity to compute an iteration of the algorithm in one sampling period. Obviously, a system with high levels of parallelism and the needed memory storage capability will accommodate this computation in the time necessary. There are other ways to derive the required levels of parallelism to achieve the performance, outlined in Chapter 8.

4.4.3 Data Independence

The data independence property is important as it provides a means for ordering the computation. This can be highly important in reducing the memory and data storage requirements. For example, consider N iterations of the FIR filter computation of equation (4.1), below. It is clear that the $x(n)$ datum is required for all N calculations and there is nothing to stop us performing the calculation in such a way that N computations are performed at the same time for $y(n), y(n + 1), \ldots, y(n + N - 1)$, using the $x(n)$ datum and thus removing any requirement to store it:

$$y(n) = a_0 \underline{x(n)} + a_1 x(n - 1) + a_2 x(n - 2) + \cdots + a_{N-1} x(n - N + 1)$$
$$y(n + 1) = a_0 x(n + 1) + a_1 \underline{x(n)} + a_2 x(n - 1) + \cdots + a_{N-1} x(n - N + 2)$$
$$y(n + 2) = a_0 x(n + 2) + a_1 x(n + 1) + a_2 \underline{x(n)} + \cdots + a_{N-1} x(n + N + 3)$$
$$\vdots = \vdots$$
$$y(n + N - 1) = a_0 x(n + N - 1) + a_1 x(n + 1) + a_2 x(n + N + 1) + \cdots + a_{N-1} \underline{x(n)}.$$

Obviously the requirement is now to store the intermediate accumulator terms. This obviously presents the designer with a number of different ways of performing system optimization, and in this case gives in a variation of schedule in the resulting design. This is just one implication of data independence.

4.4.4 Arithmetic Requirements

In many DSP technologies, the wordlength requirements of the input data are such that the use of internal precision can be considerably reduced. For example, consider the varying wordlengths for the different applications as illustrated at the end of Chapter 3. Typically, the input wordlength will be determined by the precision of the ADC device creating the source material. Depending on the amount and type of computation required (e.g. multiplicative or additive), the internal word growth can be limited, which may mean that a suitable fixed-point realization is sufficient.

The low arithmetic requirement is vital as it means small memory requirements, faster implementations as adder and multiplier speeds are governed by input wordlengths, and smaller area. For this reason, there has been a lot of work to determine maximum wordlengths as discussed in the previous chapter. One of the interesting aspects is that for many processor implementations both external and internal wordlengths will have been predetermined when developing the architecture, but in FPGAs it may be required to carry out detailed analysis to determine the wordlength at different parts of the DSP system (Boland and Constantinides 2013).

All of these characteristics of DSP computation are vital in determining an efficient implementation, and have in some cases driven technology evolution. For example, one the main differences between the early DSP processors and microprocessors was the availability of a dedicated multiplier core. This was viable for DSP processors as they were targeted at DSP applications where multiplication is a core operation, but not for general processing applications, and so multipliers were not added to microprocessors at that time.

4.4.5 Processor Classification

The technology for implementing DSP ranges from microcontrollers right though to single-chip DSP multi-processors, which range from conventional processor architectures with a very long instruction word (VLIW) extension to allow instruction-level parallelism through to dedicated architecture defined for specific application domains. Although there have been other more comprehensive classifications after it, Flynn's classification is the most widely known and used for identifying the instructions and the data as two orthogonal streams in a computer. The taxonomy is summarized in Table 4.1, which includes single instruction, single data (SISD) and multiple instruction, single data (MISD). These descriptions are used widely to describe the various representations for processing elements (PEs).

4.5 Microprocessors

The classical von Neumann microprocessor architecture is shown in Figure 4.3. This sequentially applies a variety of instructions to specified data in turn. The architecture

Table 4.1 Flynn's taxonomy of processors

Class	Description	Examples
SISD	Single instruction stream operating on single data stream	von Neumann processor
SIMD	Several PEs operating in lockstep on individual data streams	VLIW processors
MISD		Few practical examples
MIMD	Several PEs operating independently on separate data streams	Multi-processor

consists of five types of unit: a memory containing data and instructions, an instruction fetch and decode (IFD) instruction, arithmetic logic unit (ALU) and the memory access (MA) unit. These units correspond to the four different stages of processing, which repeat for every instruction executed on the machine:

1. Instruction fetch
2. Instruction decode
3. Execute
4. Memory access.

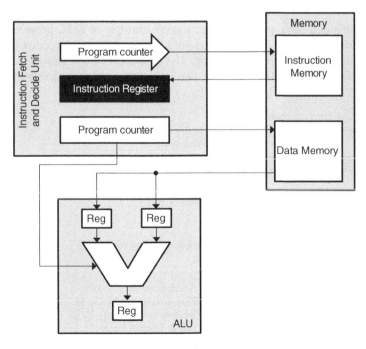

Figure 4.3 Von Neumann processor architecture

During the instruction fetch (IF) stage, the IFD unit loads the instruction at the address in the program counter (PC) into the instruction register (IR). In the second, instruction decode (ID) stage, this instruction is decoded to produce an opcode for the ALU and the addresses of the two data operands, which are loaded into the input registers of the ALU. During the execute stage, the ALU performs the operation specified by the opcode on the input operands to produce the result, which is written back into memory in the MA stage.

In general, these types of SISD machine can be subdivided into two categories, depending on their instruction set style. The complex instruction set computer (CISC) machines have complex instruction formats which can become highly specific for specific operations. This leads to compact code size, but can complicate pipelined execution of these instructions. On the other hand, reduced instruction set computer (RISC) machines have regular, simple instruction formats which may be processed in a regular manner, promoting high throughput via pipelining, but will have increased code size.

The von Neumann processor architecture is designed for general-purpose computing, and is limited for embedded applications, due to its highly sequential nature. This makes this kind of processor architecture suitable for a wide range of applications. However, whilst embedded processors must be flexible, they are often tuned to a particular application and have advanced performance requirements, such as low power consumption or high throughput. A key evolution in this area has been the ARM series of processors which have been primarily developed for embedded applications and, in particular, mobile phone applications. More recently they have expanded to the wider internet of things (IoT) markets and also data servers, specifically microservers (Gillan 2014).

4.5.1 ARM Microprocessor Architecture Family

The ARM family of embedded microprocessors are a good example of RISC processor architectures, exhibiting one of the key trademarks of RISC processor architectures, namely that of instruction execution path pipelining. The pipelines of these processor architectures (as identified for the ARM processor family in Table 4.2) are capable of

Table 4.2 ARM microprocessor family overview

Processor	Instruction sets	Extensions
ARM7TDMI, ARM922T	Thumb	
ARM926EJ-S ARM946E-S ARM966E-S	Improved ARM/Thumb DSP instructions	Jazelle
ARM1136JF-S ARM1176JZF-S ARM11 MPCore	SIMD instructions Unaligned data support	Thumb-2, TrustZonezelle
Cortex-A8/R4/M3/M1	Thumb-2 Thumb-2 Thumb-2	v7A (applications) - NEON v7R (real-time) - H/W divide V7M (microcontroller) - H/W Divide & Thumb-2 only

enabling increased throughput of the unit, but only up to a point. A number of innovations have developed as the processor has developed.

- **ARM7TDMI** has a three-stage pipeline with a single interface to memory.
- **ARM926EJ-S** has a five-stage pipeline with a memory management unit with various
  caches and DSP extensions. The DSP extension is in the form of a single-cycle 32×16-
  bit multiplier and supports instructions that are common in DSP architectures, i.e.
  variations on signed multiply-accumulate, saturated add and subtract, and counting
  leading zeros.
- **ARM1176JZ(F)-S** core has moved to a eight-stage pipeline with improved performance in terms of branch prediction, a vector floating-point unit and intelligent
  energy management.
- **ARM11 MPCore** technology has moved to multicore and has up to four MP11 processors with cache coherency and an interrupt controller.
- **ARM Cortex-A8** has moved to a 14-stage pipeline and has an on-board NEON media
  processor.

With increased pipeline depth comes increased control complexity, a factor which
places a limit on the depth of pipeline which can produce justifiable performance
improvements. After this point, processor architectures must exploit other kinds of parallelism for increased real-time performance. Different techniques and exemplar processor architectures to achieve this are outlined in Section 4.5.2.

The key innovation of the ARM processor is that the company believes that the architecture comprises the instruction set and the programmer's model. Most ARMs implement two instruction sets, namely the 32-bit ARM instruction set and the 16-bit Thumb
instruction set. The Thumb set has been optimized for code density from C code as this
represents a very large proportion of example ARM code. It also gives improved performance from narrow memory which is critical in embedded applications.

The latest ARM cores include a new instruction set, Thumb-2, which provides a mixture of 32-bit and 16-bit instructions and maintains code density with increased flexibility. The Jazelle-DBX cores have been developed to allow the users to include executable
Java bytecode. A number of innovations targeted at DSP have occurred, e.g. as in the
ARM9 family where DSP operations were supported. The evolution in the ARM to support DSP operations has not substantially progressed beyond that highlighted. A more
effective route has been to incorporate the ARM processor with both DSP processors,
which is discussed in Section 4.9.

ARM Programming Route

The most powerful aspect of microprocessors is the mature design flow that allows programming from C/C++ source files. ARM supports an Eclipse-based Integrated Design
Environment (IDE) or IDE-based design flow which provides the user with a C/C++
source editor which helps the designer to spend more time writing code and avoid chas-
ing down syntax errors. The environment will list functions, variables, and declarations,
allows full change history and re-factoring of function names and code segments globally. This provides a short design cycle allowing code to be quickly compiled into ARM
hardware.

It is clear that there have been some architectural developments which make the ARM processor more attractive for DSP applications, but, as indicated earlier, the preferred route has been a joint offering either with DSP processors or FPGAs. ARM also offers an mbed hardware platform (https://mbed.org/) which uses on-line tools.

4.5.2 Parallella Computer

A clear shift in microprocessors has been towards multicores and there have been many examples of multicore structures, including Intel multicore devices and multicore DSP devices (see later). The Parallella platform is an open source, energy-efficient, high-performance, credit-card sized computer which is based on Adapteva's Epiphany multicore chips. The Epiphany chip consists of a scalable array of simple RISC processors programmable in C/C++ connected together with a fast on-chip network within a single shared memory architecture. It comes as either a 16- or 64-core chip. For the 16-core system, 1, 2, 3, 4, 6, 8, 9, 12 or 16 cores can be used simultaneously.

The Epiphany is a 2D scalable array of computing nodes which is connected by a low-latency mesh network-on-chip and has access to shared memory. Figure 4.4 shows the Epiphany architecture and highlights the core components:

- a superscalar, floating-point RISC CPU that can execute two floating-point operations and a 64-bit memory load operation on every clock cycle;
- local memory in each mesh node that provides 32 bytes/cycle of sustained bandwidth and is part of a distributed, shared memory system;
- multicore communication infrastructure in each node that includes a network interface, a multi-channel DMA engine, multicore address decoder, and a network monitor;
- a 2D mesh network that supports on-chip node-to-node communication latencies in nanoseconds, with zero startup overhead.

The Adapteva processor has been encapsulated in a small, high-performance and low-power computer called the Parallella Board. The main processor is a dual-core ARM A9 and also a Zynq FPGA. The board can run a Linux-based operating system. The main memory of the board is contained in an SD card and this also includes the operating files for the system. There is 1 GB of shared memory between the ARM host processor and the Epiphany coprocessor.

The system is programmed in C/C++, but there are a number of extra commands which are specific to the Parallella Board and handle data transfer between the Epiphany and the ARM host processor. It can be programmed with most of the common parallel programming methods, such as SIMD and MIMD, using parallel frameworks like open computing language (OpenCL) or Open Multi-Processing (OpenMP). Two programs need to be written, one for the host processor and one for the Epiphany itself. These programs are then linked when they are compiled.

Programming Route

The Parallela is programmed by the Epiphany software development kit (SDK), known as eSDK, and is based on standard development tools including an optimizing C-compiler, functional simulator, debugger, and multicore IDE. It can directly implement regular

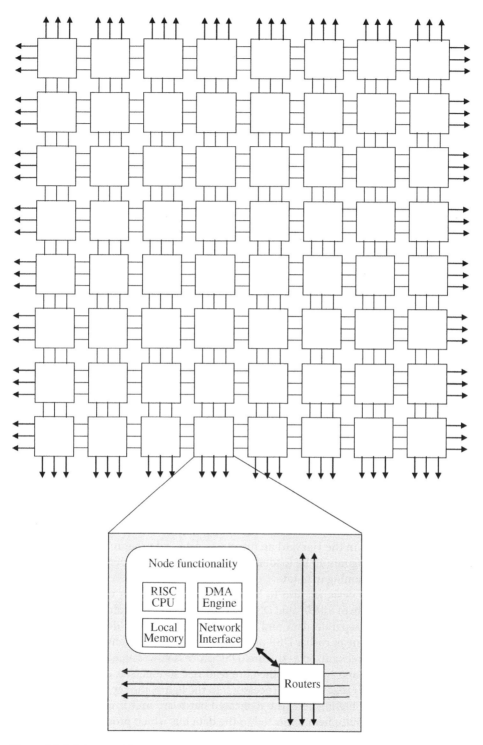

Figure 4.4 Epiphany architecture

ANSI-C and does not require any C-subset, language extensions, or SIMD style programming. The eSDK interfaces with a hardware abstraction layer (HAL) which allows interaction with the hardware through the user application.

The key issue with programming the hardware is to try to create the program in such a way that the functionality will efficiently use the memory of each core which has an internal memory of 32 kB which is split into four separate 8 kB banks. Also, every core also has the ability to quickly access the memory of the other cores which means that streaming will work effectively as data can then be passed in parallel. The cores also have access to the memory shared with the ARM processor, which suggests a means of supplying input data.

Efficiency is then judged by how effectively the computation can be distributed across each core, so it is a case of ensuring an efficient partition. This will ensure that a good usage of the core functionality can be achieved which preserves the highly regular dataflow of the algorithm so that data passing maps to the memory between cores. This may require a spatial appreciation of the architecture to ensure that the functionality efficiency is preserved and a regular matching occurs. This allow the realization to exploit the fast inter-core data transfers and thus avoid multiple accesses in and out of shared memory.

4.6 DSP Processors

As was demonstrated in the previous section, the sequential nature of microprocessor architectures makes them unsuitable for efficient implementation of complex DSP systems. This has spurred the development of dedicated types of processors called DSP microprocessors such as Texas Instrument's TMS32010 which have features that are particularly suited for DSP processing. These features have been encapsulated in the Harvard architecture illustrated in Figure 4.5.

The earlier DSP microprocessors were based on the Harvard architecture. This differs from the von Neumann architecture in terms of memory organization and dedicated DSP functionality. In the von Neumann machine, one memory is used for storing both program code and data, effectively providing a memory bottleneck for DSP implementation as the data independence illustrated in Section 4.4.3 cannot be effectively exploited to provide a speedup. In the Harvard architecture, data and program memory are separate, allowing the program to be loaded into the processor independently of the data which is typically streaming in nature.

DSP processors are designed also to have dedicated processing units, which in the early days took the form of a dedicated DSP block for performing multiply-accumulation quickly. In addition, separate data and program memories and dedicated hardware became the cornerstone of earlier DSP processors. Texas Instrument's TMS32010 DSP (Figure 4.6), which is recognized as the first DSP processor, was an early example of the Harvard architecture and highlights the core features, so it acts as a good example to understand the broad range of DSP processors. In the figure, the separate program and data buses are clearly highlighted. The dedicated hardware unit is clearly indicated in this case as a 16-bit multiplier connected to the data bus which produces a 32-bit output called P32, which can then be be accumulated as indicated by the 32-bit arithmetic

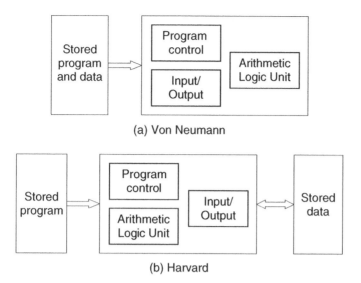

(a) Von Neumann

(b) Harvard

Figure 4.5 Processor architectures

logic unit (ALU) and accumulator (ACC) circuitry which is very effective in computing the function given in equation (4.1).

4.6.1 Evolutions in DSP Microprocessors

A number of modifications have occurred to the original Harvard DSP architecture, which are listed below (Berkeley Design Technology 2000).

- *VLIW.* Modern processor architectures have witnessed an increase in the internal bus wordlengths. This allows a number of operations performed by each instruction in parallel, using multiple processing functional units. If successful, the processor will be able to use this feature to exploit these multiple hardware units; this depends on the computation to be performed and the efficiency of the compiler in utilizing the underlying architecture. This is complicated by the move toward higher-level programming languages which require good optimizing compilers that can efficiently translate the high-level code and eliminate any redundancies introduced by the programmer.
- *Increased number of data buses.* In many recent devices the number of data buses has been increased. The argument is that many DSP operations involve two operands, thus requiring three pleces of information (including the instruction) to be fed from memory. By increasing the number of buses, a speedup is achieved, but this also increases the number of pins on the device. However, some devices gets around this by using a program cache, thereby allowing the instruction bus to double as a data bus when the program is being executed out of the program cache.
- *Pipelining.* Whilst the introduction of VLIW has allowed parallelism, another way to exploit concurrency is to introduce pipelining, both within the processing units in the DSP architecture, and in the execution of the program. The impact of pipelining is to break the processing time into smaller units, thereby allowing several overlapping

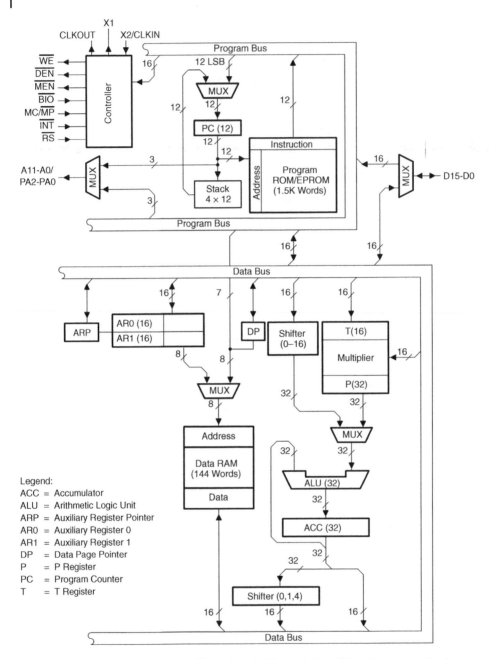

Figure 4.6 TI TMS32010 DSP processor (Reproduced with permission of Texas Instruments Inc.)

computations to take place at once, in the same hardware. However, this comes at the expense of increased latency. Pipelining can also be employed within the processor control unit which controls the program fetch, instruction dispatch and instruction decode operation.

- *Fixed point operations.* Some DSP systems only require fixed-point arithmetic and do not need the full-precision arithmetic offered by some DSP processing units. For this reason, fixed-point and floating-point DSP microprocessors have evolved to match application environments. However, even in fixed-point, some applications do not require the full fixed-point range of some processors, e.g. 32 bits in the TMS320C64xx series processor, and therefore inefficiency exists. For example, for a filter application in image processing applications, the input wordlength may vary between 8 and 16 bits, and coefficients could take 12–16 bits. Thus, the multiplication stage will not require anything larger than a 16×16-bit multiplier. The DSP processors exploit this by organizing the processing unit, e.g. the TMS320C6678, by allowing multiple multiplications to be take place in one time unit, thereby improving the throughput rate. Thus, the processors are not compromised in terms of the internal wordlength used.

These optimizations have evolved over a number of years. and have led to improved performance. However, it is important to consider the operation in order to understand how the architecture performs in some applications.

4.6.2   TMS320C6678 Multicore DSP

The TMS320C6678 multicore fixed- and floating-point DSP microprocessor is based on TI's KeyStone multicore architecture and is illustrated in Figure 4.7. It comprises eight C66x CorePac DSPs, each of which runs at 1.0–1.25 GHz, giving an overall clock rate of up to 10 GHz. This gives 320 giga multiply-accumulates (GMAC) at the clock rate of 1.25 GHz. A key aspect has been the introduction of increasing levels of memory in the form of 32 kB of L1 program cache and 32 kB of L1 data cache with 512 kB of L2 cache per core. The chip also has 4 MB of L2 shared memory. Clearly, with technology evolution and in line with processor developments and the response to the slowdown of Dennard scaling (see Section 4.2), the response has been to create a multicore implementation of previous processors in this case, the C6000, rather than scale the processor.

Another feature of the latest DSP families has been the inclusion of specific functionality, specifically network-on-chip topology in the form of a TeraNet switch fabric which support up to 2 TB of data. There are also dedicated processing engines such as packet and security accelerators to address the networking and security markets in which DSP processors are increasingly being used. The devices also come with an additional features such as 64-bit DDR3 and universal asynchronous receiver/transmitter (UART) interfaces.

The key objective is to be able to exploit the processing capability offered by this multicore platform which depends on both the computation to be performed and the use of optimizing compilers that perform a number of simplifications to improve efficiency. These simplifications include routines to remove all functions that are never called and to simplify functions that return values that are never used, to reorder function declarations and propagate arguments into function bodies (Dahnoun 2000). The compiler also performs a number of optimizations to take advantage of the underlying architecture

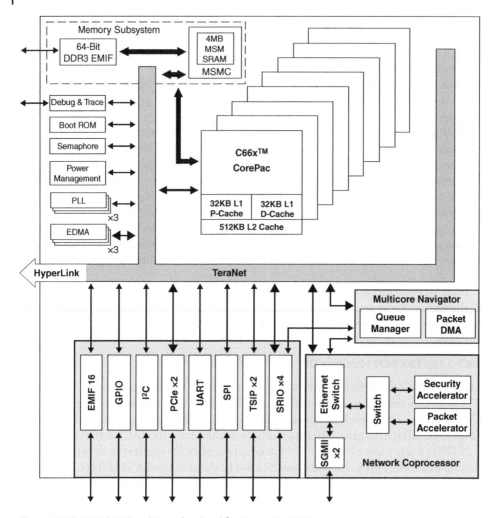

Figure 4.7 TMS320C6678 multicore fixed and floating-point DSP microprocessor

including software pipelining, loop optimizations, loop unrolling and other routines to remove global assignments and expressions (Dahnoun 2000).

4.7 Graphical Processing Units

Another class of processors that has had a major impact is the graphical processing unit (GPU). It was developed particularly to suit applications in image and video processing and as such has resulted in a multi-processor array structure with a memory hierarchy suited to store the initial image data and then portions of the image for highly parallel computing. While originally designed with a fixed pipeline which is highly suitable for graphics, the modern GPU's pipeline is highly programmable and allows for general-purpose computation. For this reason, the technology has now seen wider adoption. The

Figure 4.8 Nvidia GeForce GPU architecture

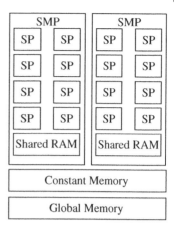

creation of the general-purpose graphical processing unit (GPUPU) presents a powerful computing platform as it comprises a CPU which computes the sequential part of the code and the GPU calculates the computer-intensive part of the code.

There are many GPUs available from different manufacturers, and all offer massively parallel floating-point computation and high-speed, large-capacity RAM. The Nvidia GeForce GPU is illustrated in Figure 4.8 and shows the computing structure. The architecture is generally composed of a number of floating-point streaming processors (SP) optimized for graphics processing, each of which contains a small amount of low-latency shared memory along with a larger bank of SDRAM which is available to all multi-processors. This is contained with the shared-memory multi-processors (SMP). The architecture allows parallel execution of numerous SIMD functions.

Since products from different manufacturers will inevitably differ in their low-level implementation, there are a number of abstract application programming interfaces (APIs) including CUDA and OpenCL. CUDA can used to program the devices and allow the software developer to access the massively parallel SIMD architecture of modern GPUs for general processing tasks. CUDA was developed by Nvidia and is an extension of C to enable programming of GPU devices; it allows easy management of parallelism and handles communications with the host. OpenCL is a broadly supposed open standard, defined by the Khronos Group, that allows programming of both GPUs and CPUs. It is supported by Intel, AMD, Nvidia, and ARM, and is the GPGPU development platform most widely used by developers in both the USA and Asia-Pacific.

The key aspect of GPU is hundreds of cores that can be used for highly parallel implementation of graphics algorithms and high levels of memory. As numerous computations or *threads* can run on each processor engine, the GPU has thousands of threads. Because the architecture is fixed and is now being applied to a wider range of applications, the technology is cheap, certainly compared to DSP microprocessor and FPGAs.

4.7.1 GPU Architecture

GPU implementations work well for streaming applications where large amounts of data will be streamed to the GPU and then processed. Obviously this is the case for image and video processing which is what the hardware was developed for in the first instance. When programming the device, it is paramount to consider the architectural implications of the hardware, to produce efficient code. It could be argued to some some extent that this places much more emphasis on the programmer's ability to get the performance out of the hardware, but this is increasingly the case for multicore technology.

For example, the multi-processors share one off-chip global memory and it is not cached, so it is very important to achieve memory coalescing (Luo *et al.* 2010). Memory coalescing occurs when consecutive threads access consecutive memory locations. In this case, it is important to coalesce several memory transactions into one transaction, as there is a shared memory within each SMP which is common to all the streaming pro-cessors inside the multi-processor. As the shared memory is on chip, it can be accessed within a smaller number of clock cycles, whereas the global memory will be typically an order of magnitude larger.

Thus the memory can be viewed as follows:

- **Global memory** is typically several gigabytes and is available to the GPU processors. It is used for fast caching to the motherboard RAM, as it is used to read and write large amounts of data and is normally associated with blocks of threads.
- **Local shared memory** is smaller (tens of kilobytes) and can be accessed extremely quickly, so it can really speed up computations, since the instruction access cost is much lower compared to global memory. It is usually associated with a block of threads
- **Private thread memory**, as the name suggests, is a very small bank of memory used within each thread for variables and temporary storage during the computation.

By carefully observing the memory structure and ensuring that the computation is inherently parallel enough, it is possible to achieve a speedup. For example, the work on breadth-first search graph operation (Luo *et al.* 2010) shows that a tenfold increase in the number of vertices computed only incurs a fivefold increase in the compute time. GPUs have also been applied to some DSP algorithms such as the low-density parity-check (LDPC) decoder (Wang *et al.* 2011). The challenge has been to use the threads to fully occupy the GPU computation resources when decoding the LDPC codes and organizing the computation in such a way as to minimize the memory access times. Work by Falcao *et al.* (2012) has compared the programming of a GPU and CPU using OpenCL.

4.8 System-on-Chip Solutions

Up to now, the DSP technology offerings have been in the form of some type of prede-fined architectural offering. The major attraction of dedicated ASIC offerings is that the architecture can be developed to specifically match the algorithmic requirements, allow-ing the level of parallelism to be created to ultimately match the performance require-ments. For the earlier 128-tap FIR filter, it is possible to dedicate a multiplier and adder to each multiplication and addition respectively, thereby creating a fully parallel imple-mentation. Moreover, this can then be pipelined in order to speed up the computation, giving if required an N times speedup for an N-tap filter. How the designer can achieve this is described in detail in Chapter 8.

When considering programmability, SoC solutions will have been developed with programmable parts as well as dedicated acceleration blocks. Indeed, the C6000 DSP described in Section 4.6.2 could be considered to be a DSP SoC. Therefore, such an approach would have to be driven by a critical factor such as immense computational performance or power consumption or indeed both, in order to justify the considerable costs and risks in creating a DSP SoC in the first place. This is normally counterbalanced by increasing the levels of programmability, but this causes an increase in test and verification times. Non-recurring engineering (NRE) costs are such that the cost of producing a number of prototypes now typically exceeds the financial resources of most major manufacturers. Thus the argument for using dedicated SoC hardware has to be compelling. Currently, it is mostly only mass market smartphones and other mobile devices which can justify this.

Whilst the sequential model has served well in the sense that it can implement a wide range of algorithms, the real gain from DSP implementation comes from parallelism of the hardware. For this reason, there has been considerable interest in developing parallel hardware solutions evolving from the early days of the transputer. However, it is capturing this level of parallelism that is the key issue. A key architecture which was developed to capture parallelism was the systolic array (Kung and Leiserson 1979; Kung 1988) which forms the starting point for this section.

4.8.1 Systolic Arrays

Systolic array architectures were introduced to address the challenges of very large scale integration (VLSI) design by Kung and Leiserson (1979). In summary, they have the following general features (Kung 1988):

- an array of processors with extensive concurrency;
- small number of processor types;
- control is simple;
- interconnections are local.

Their processing power comes from the concurrent use of many simple cells, rather than the sequential use of a few very powerful cells. They are particularly suitable for parallel algorithms with simple and regular dataflows, such as matrix-based operations. By employing pipelining, the operations in the systolic array can be continually filtered through the array, enabling full efficiency of the processing cells.

A systolic linear array is shown in Figure 4.9. Here, the black circles represent pipeline stages after each processing element. The lines drawn through these pipeline stages are

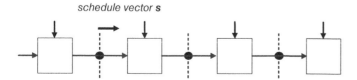

Figure 4.9 Linear systolic array

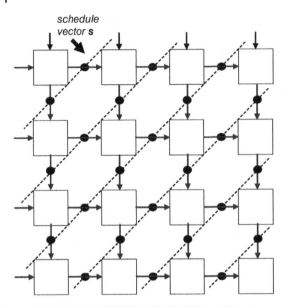

Figure 4.10 Systolic array architecture

the scheduling lines depicting which PEs are operating on the same iteration at the same time; in other words, these calculations are being performed at the same clock cycle. The lines drawn through these pipeline stages are the scheduling lines depicting which PEs are operating on the same iteration at the same time; in other words, these calculations are being performed at the same clock cycle.

Figure 4.10 shows a classical rectangular systolic array each with local interconnections. This type of array is highly suitable for matrix–matrix operations. Each PE receives data only from its nearest neighbor and each processor contains a small memory elements in which intermediate values are stored. The control of the data through the array is by a synchronous clock, which effectively pumps the data through the array; hence the name "systolic" arrays, by analogy with the heart pumping blood around the body. Figure 4.11 depicts the systolic array applied for QR decomposition. The array is built from two types of cells, boundary and internal, all locally interconnected.

The concept of systolic arrays was employed in many DSP applications. McCanny and McWhirter (1987) applied it at the bit level, whereas the original proposer of the technique developed the concept into the iWarp which was an attempt in 1988 by Intel and Carnegie Mellon University to build an entirely parallel computing node in a single microprocessor, complete with memory and communications links (Gross and O'Hallaron 1998). The main issue with this type of development was that it was very application-specific, coping with a range of computationally complex algorithms; instead, the systolic array design concept was applied more successfully to develop a wide range of signal processing chips (Woods *et al.* 2008). Chapter 12 demonstrates how the concept has been successfully applied to the development of an IP core for recursive least squares filtering.

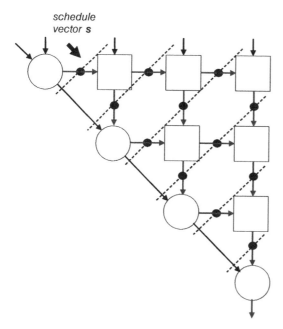

schedule
vector **s**

Figure 4.11 Triangular systolic array architecture

4.9 Heterogeneous Computing Platforms

Since the first edition of this book, there have been many alternative developments due to scaling issues, but rather than evolve existing platforms to cater for DSP processing, the focus has been to derive new forms of multicore platforms:

- **Multicore architectures** such as the Parallela (highlighted in Section 4.5.2) which offers high levels of parallelism, and coprocessing architectures such as the Intel Xeon Phi^{TM} processor which delivers up to 2.3 times higher peak FLOPS and up to 3 times more performance per watt. The technology has been used for 3D image reconstruction in computed tomography which has primarily been accelerated using FPGAs (Hofmann *et al.* 2014). They show how not only parallelization can be applied but also that SIMD vectorization is critical for good performance.
- **DSP/CPU processor architectures** such as KeyStone^{TM} II multicore processors, e.g. the 66AK2Hx platform which comprises a quad-ARM Cortex-A15 MPCore^{TM} processor with up to eight TMS320C66x high-performance DSPs using the KeyStone II multicore architecture. This has been applied for cloud radio access network (RAN) stations; Flanagan (2011) shows how the platform can provide scalability.
- **SoC FPGAs** have evolved by incorporating processors (primarily the ARM processor) on the FPGA fabric. These effectively present a hardware/software system where the FPGA programmable fabric can be used to accelerate the data-intensive computation and the processor can be used for control and interfacing. This is discussed in more detail in the next chapter.

4.10 Conclusions

This chapter has highlighted the variety of technologies used for implementing complex DSP systems. These compare in terms of speed, power consumption and, of course, area, although this is a little difficult to ascertain for processor implementations. The chapter has taken a specific slant on programmability with regard to these technologies and, in particular, has highlighted how the underlying chip architecture can limit the performance. Indeed, the fact that it is possible to develop SoC architectures for ASIC and FPGA technologies is the key feature in achieving the high performance levels. It could be argued that the fact that FPGAs allow circuit architectures and that they are programmable are the dual factors that make them so attractive for some system implementation problems.

Whilst the aim of the chapter has been to present different technologies and, in some cases, compare and contrast them, the reality is that modern DSP systems are now collections of these different platforms. Many companies are now offering complex DSP platforms comprising CPUs, DSP processors and embedded FPGA. Thus the next chapter views FPGAs both as heterogeneous platforms in themselves but also as part of these solutions.

Bibliography

Berkeley Design Technology 2000 Choosing a DSP processor. Available at http://www.bdti.com/MyBDTI/pubs/choose_2000.pdf (accessed November 6, 2016).

Boland DP, Constantinides GA 2013 Word-length optimization beyond straight line code. In *Proc. ACM/SIGDA Int. Symp. on Field Programmable Gate Arrays*, pp. 105–114.

Dahnoun, N 2000 *Digital Signal Processing Implementation Using the TMS320C6000^{TM} DSP Platform*. Prentice Hall, Harlow.

Falcao G, Silva V, Sousa L, Andrade J 2012 Portable LDPC decoding on multicores using OpenCL. *IEEE Signal Processing Mag.*, 29(4), 81–109.

Flanagan T 2011 Creating cloud base stations with TI's KeyStone multicore architecture. Texas Instruments White Paper (accessed June 11, 2015).

Gillan CJ 2014 On the viability of microservers for financial analytics. In *Proc. IEEE Workshop on High Performance Computational Finance*, pp. 29–36.

Gross T, O'Hallaron DR 1998 *iWarp: Anatomy of a Parallel Computing System*. MIT Press, Cambridge, MA.

Hofmann J, Treibig J, Hager G, Welleinet G 2014 Performance engineering for a medical imaging application on the Intel Xeon Phi accelerator. In *Proc. 27th Int. Conf. on Architecture of Computing Systems*, pp. 1–8.

ITRS 2005 International Technology Roadmap for Semiconductors: Design. available from http://www.itrs.net/Links/2005ITRS/Design2005.pdf (accessed February 16, 2016).

International Technology Roadmap for Silicon 2011 *Semiconductor Industry Association*. Available at http://www.itrs.net/ (accessed June 11, 2015).

Jackson LB 1970 Roundoff noise analysis for fixed-point digital filters realized in cascade or parallel form. *IEEE Trans. Audio Electroacoustics* 18, 107–122.

Kung HT, Leiserson CE 1979 Systolic arrays (for VLSI). In *Proc. on Sparse Matrix*, pp. 256–282.

Kung SY 1988 *VLSI Array Processors*. Prentice Hall, Englewood Cliffs, NJ.

Luo L, Wong M, Hwu W-M 2010 An effective GPU implementation of breadth-first search. In *Proc. Design Automation Conf.*, Anaheim, CA, pp. 52–55.

McCanny J, McWhirter JG. 1987 Some systolic array developments in the UK. *IEEE Computer*, 20(7), 51–63.

Reinders J, Jeffers J 2014 *High Performance Parallelism Pearls: Multicore and Many-Core Programming Approaches*. Morgan Kaufmann, Waltham, MA.

Sutter H 2009 The free lunch is over: A fundamental turn toward concurrency in software. Available at http://www.gotw.ca/publications/concurrency-ddj.htm (accessed June 11, 2015).

Wang G, Wu M, Yang S, Cavallaro JR 2011 A massively parallel implementation of QC-LDPC decoder on GPU. In *Proc. IEEE 9th Symp. on Application Specific Processors*, pp. 82–85.

Woods RF, McCanny JV, McWhirter JG. 2008 From bit level systolic arrays to HDTV processor chips. *J. of VLSI Signal Processing* 53(1–2), 35–49.

5

Current FPGA Technologies

5.1 Introduction

The analysis at the end of previous chapter makes it clear that the choice of the specific technology and the resulting design approach directly impacts the performance that will be achieved. For example, the use of simple DSP microcontrollers typically implies a DSP system with relatively low performance requirements such as medical or industrial control systems. The design effort is only that needed to produce efficient C or C++ source code for its implementation, and indeed it may be possible to use the software compilers associated with MATLAB® or LabVIEW that the user may have used as the initial design environment to scope the requirements such as wordlength and system complexity.

This design approach can be applied for the full range of "processor"- style platforms, but it may be required that dedicated handcrafted C code is produced to achieve the necessary performance. This is probably particularly relevant in applications where performance requirements are tight. Also, the hardware may possess dedicated functionality that is not well supported within the high-level tool environment. In these cases, it is clear that the platform will be chosen to meet some superior area, speed and power performance criteria.

Mindspeed's T33xx family of wireless application processors (Mindspeed 2012) have been directly targeted at base stations for mobile services and thus have dedicated functionality such as forward error correction (FEC) and Mindspeed application processor DSP blocks for advanced signal processing and encryption functions. In these cases, the user has to compromise on ease of design, in order to take advantage of the specific architectural feature offered by the technology or use company-specific tool sets to achieve the necessary performance. This notion is taken to the extreme in the SoC arena where the user is faced with creating the circuit architecture to best match the performance requirements.

Working towards this ultimate performance is effectively what an FPGA platform offers. From "glue logic" beginnings, FPGAs have now become an advanced platform for creating high- performance, state-of-the-art systems. Indeed, many high-performance data-processing environments are starting to see the benefits of this technology. The

FPGA-based Implementation of Signal Processing Systems,
Second Edition. Roger Woods, John McAllister, Gaye Lightbody and Ying Yi.
© 2017 John Wiley & Sons, Ltd. Published 2017 by John Wiley & Sons, Ltd.

purpose of this chapter is to give a review of the current FPGA technologies with a focus on how they can be used in creating DSP systems. The chapter acts to stress key features and leaves the detail to the vendors' data sheets. Whilst a number of technologies are available, the focus is on the latest commercial offerings from the two dominant vendors, namely Xilinx and Altera, whilst giving a brief description of the other solutions.

Section 5.2 gives a brief historical perspective on FPGAs, describing how they have emerged from being a fine-grained technology to a complex SoC technology. Section 5.3 describes the Altera Stratix® 10 FPGA family, the most powerful FPGA family that the company offers. The next sections then go on to describe the FPGA technology offerings from Xilinx, specifically the UltraScale™ (Section 5.4) and Zynq® (Section 5.5) FPGA families. The technologies offered by Microsemi and Lattice offer specific features that are very relevant in certain markets. For this reason, Lattice's iCE40isp family of small, low-power, integrated mobile FPGAs is described in Section 5.6, and Microsemi's RTG4, a radiation tolerant (RT) FPGA for signal processing applications, is described in Section 5.7. Section 5.8 attempts to summarize the key features of recent FPGA devices and gives some insights into FPGA-based DSP system design. Some conclusions are given in Section 5.9.

5.2 Toward FPGAs

In the 1970s, logic systems were created by building PCB boards consisting of transistor-transistor logic (TTL) logic chips. However, as functions got larger, the logic size and levels increased and thus compromised the speed of design. Typically, designers used logic minimization techniques, such as those based on Karnaugh maps or Quine–McCluskey minimization, to create a sum of products expression by generating the product terms using AND gates and summing them using an OR gate.

The concept of creating a structure to achieve implementation of this functionality was captured in the early programmable array logic (PAL) device, introduced by Monolithic Memories in 1978. The PAL comprised a programmable AND matrix connected to a fixed OR matrix which allowed sum of products structures to be implemented directly from the minimized expression. The concept of an AND and OR matrix became the key feature of a class of devices known as programmable logic devices (PLDs); a brief classification is given in Table 5.1, the final member of which is of course the ROM.

As illustrated in Figure 5.1, a ROM possesses the same structure only with a fixed AND plane (effectively a decode) and a programmable OR plane. In one sense, an $n \times m$ structure can be viewed as providing the capability of storing four (in general, n^2) two-bit (or m-bit) words, as shown in Figure 5.2. The decoder, which is only required to reduce the number of pins coming into the memory, is used to decode the address input pins,

Table 5.1 PLD types

	AND matrix	OR matrix
ROM	Fixed	Programmable
PLA	Programmable	Fixed
PAL	Programmable	Programmable

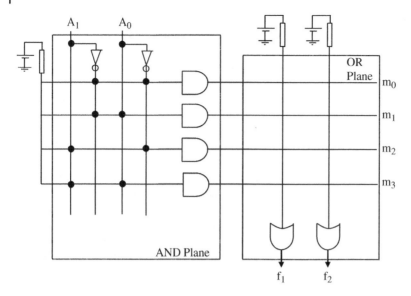

Figure 5.1 PLD architecture

and a storage area or memory array is used to store the data. As the decoder generates the various address lines using AND gates and the outputs are summed using OR gates, this provides the AND–OR configuration needed for Boolean implementation.

In general, as an $n \times 1$-bit ROM could implement any n-input Boolean function, a four-input ROM or LUT became the core component of the very first FPGA, the Xilinx XC2000. The four-input LUT was small enough to achieve efficient utilization of the chip area, but large enough to implement a reasonable range of functions, based on early analysis (Rose *et al.* 1990). If a greater number of inputs is required, then LUTs are cascaded together to provide the implementation, but at a slower speed. This was judged to provide an acceptable trade-off.

The PLD structure had a number of advantages. It clearly matched the process of how the sum of products was created by the logic minimization techniques. The function could then be fitted into one PLD device, or, if not enough product terms were available, then it could be fed back into a second PLD stage. Another major advantage was that the circuit delay is deterministic, either comprising one logic level or two, etc. However,

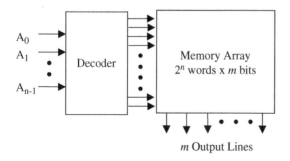

Figure 5.2 Storage view of PLD architecture

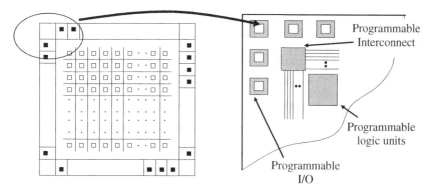

Figure 5.3 Early Manhattan architecture

the real advantage comes in the form of the programmability which reduces the risk in PCB development, allowing possible errors to be fixed by adjusting the PLD logic implementation. But as integration levels grew, the idea of using the PLD as a building block became an attractive FPGA proposition as illustrated by the early Altera MAX® family. As mentioned earlier, Xilinx opted for the LUT approach.

5.2.1 Early FPGA Architectures

The early FPGA offerings were based around the Manhattan-style architecture shown in Figure 5.3 where each individual cell comprised simple logic structures and cells were linked by programmable connections. Thus, the FPGA could be viewed as comprising the following:

- programmable logic units that can be programmed to realize different digital functions;
- programmable interconnect to allow different blocks to be connected together;
- programmable I/O pins.

This was ideal for situations where FPGAs were viewed as glue logic as programmability was then the key to providing redundancy and protection against PCB board manufacture errors; it might even provide a mechanism to correct design faults. However, technology evolution, outlined by Moore's law, now provided scalability for FPGA vendors. During the 1980s, this was exploited by FPGA vendors in scaling their technology in terms of numbers of levels of interconnectivity and number of I/Os. However, it was recognized that this approach had limited scope, as scaling meant that interconnect was becoming a major issue and technology evolution now raised the interesting possibility that dedicated hardware could be included, such as dedicated multipliers and, more recently, ARM™ processors. In addition, the system interconnectivity issue would be alleviated by including dedicated interconnectivity in the form of Serializer/Deserializer (SERDES) and RapidIO. Technology evolution has had a number of implications for FPGA technology:

- **Technology debate** In the early days, three different technologies emerged, namely conventional SRAM, anti-fuse and electrically erasable programmable read-only memory (E²PROM) technologies. The latter two technologies both require special steps to create either the anti-fuse links or the special transistors to provide the E^2PROM transistor. Technological advances favored SRAM technology as it required only standard technology: this became particularly important for Altera and Xilinx, as FPGA fabrication was being outsourced and meant that no specialist technology interaction with the silicon fabrication companies was needed. Indeed, it is worth noticing that silicon manufacturers now see FPGA technologies as the most advanced technology to test their fabrication facilities.

- **Programmable resource functionality** A number of different offerings again exist in terms of the basic logic block building resource used to construct systems. Early offerings (e.g. Algotronix and Crosspoint) offered simple logic functions or multiplexers as the logic resource, but with interconnect playing an increasing role in determining system performance, these devices were doomed. Coarser-grained technologies such as the PLD-type structure (and, more particularly, the LUT), dominated because they are flexible and well understood by computer programmers and engineers. Examining the current FPGA offerings, it is clear that the LUT-based structure now dominates with the only recent evolution an increase in the size of the LUT from a four-input to a five- or six-input version in the Xilinx Virtex/ UltraScale™ technology and to an eight-input version in the Altera Stratix® family.
- **Change in the FPGA market** Growing complexity meant that the FPGA developed from being primarily a glue logic component to being a major component in a complex system. With DSP being a target market, FPGA vendors had to compare their technology offerings in terms of new competitors, primarily the DSP processor developers such as TI and Analog Devices presented in Chapter 4.
- **Tool flow** Initially, FPGAs were not that complex, so up until the mid 1990s, tools were basic. Eventually they had to become more complex, moving toward automatic place and route tools. These still play a major role in vendors' tool flows. With increasing complexity, there has been an identified need for system-level design tools: DSP Builder and SDK for OpenCL from Altera, and Vivado and the SDSoC™ development environment from Xilinx. This may be an increasingly problematic issue as tools tend to lag well behind technology developments; it is a major area of focus in this book.

It has now got to the stage that FPGAs represent system platforms. This is recognized by both major vendors who now describe their technology in these terms. Xilinx describes its Zynq® UltraScale+™ FPGA as a technology comprising heterogeneous multi-processing engines, while Altera describes its Stratix® 10 FPGA family as featuring its third-generation hard processor system. Thus, the technology has moved from the era of programmable cells connected by programmable interconnect, as highlighted at the start of this section, to devices that are complex, programmable SoCs which comprise a number of key components, namely dedicated DSP processor blocks and processor engines.

5.3 Altera Stratix® V and 10 FPGA Family

Altera offers a series of FPGAs covering a range of performance needs and application domains (see Table 5.2). Its leading-edge FPGA is the Stratix® 10. Details were

Table 5.2 Altera FPGA family

Family	Brief description
Stratix®	High-performance FPGA and SoC family with multiple DSP blocks embedded memory, memory interfaces, transceiver blocks which supports partial reconfiguration
Arria	Focused on power efficiency, has a "hard" floating-point DSP block, support for 28.3 Gbps and "smart voltage" capability
MAX 10	A non-volatile technology that has programmable logic, DSP blocks and soft DDR3 memory controller and supports dynamic reconfiguration

not widely available at the time of writing, so most of the following discussion is based around the architecture of Stratix® V as many of the architectural features would appear to have remained the same.

The FPGA architecture has evolved from the early Manhattan-style tiling of LUTs/flip-flop cells (see Figure 5.3), into columns of programmable logic, dedicated DSP silicon blocks and scalable memory blocks. Also included are dedicated PLLs, embedded peripheral component interconnect (PCI) express bus standard, transceivers and general-purpose I/Os. The core components include adaptive logic modules (ALMs), DSP blocks and memory, covered below.

5.3.1 ALMs

The ALM is the core programmable unit and extends the basic concept of the four-input LUT and D-type flip-flop which has been the core FPGA programmable part for many years. As shown in Figure 5.4, it contains LUT-based resources that can be divided between two combinational adaptive lookup tables (ALUTs) and four registers, allowing various configurations.

With up to eight inputs for the two combinational ALUTs, one ALM can implement various combinations of two functions, allowing backward compatibility with the older four-input LUT architectures. One ALM can also implement any function with up to six-input and certain seven-input functions. After many years of four-input LUTs, it has now been deemed viable to use large LUT sizes. In addition, the user can configure the ALM as a simple dual-port SRAM in the form of a 64×1 or a 32×2 block. The ALM output can be registered and unregistered versions of the LUT or adder output. The register output can also be fed back into the LUT to reduce fan-out delay.

It also contains four programmable registers, each with the functionality of clocks, synchronous and asynchronous clear, synchronous load and circuitry to drive signals to the clock and clear control signals; this providing a very wide range of functionality. The registers can be bypassed and the output of the LUT can directly drive the ALM outputs.

These ALMs are contained within logic array blocks (LABs), and the LAB contains dedicated logic for driving control signals to its ALMs; it has two unique clock sources and three clock enable signals. LAB-wide signals control the register logic using two clear signals, and the Stratix® V device has a device-wide reset pin that resets all the registers in the device.

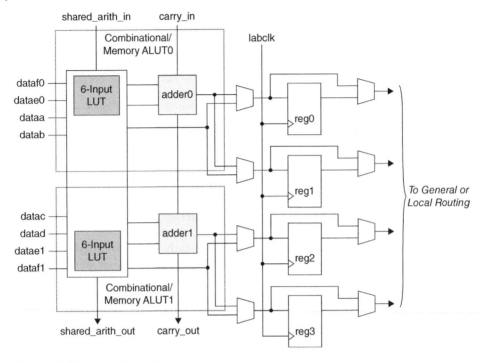

Figure 5.4 Adaptive logic module

5.3.2 Memory Organization

The Stratix® V device contains two types of memory blocks each of which can be clocked at 600 MHz:

- *M20K* The Stratix® V contains up to 11.2 GB of 20 kB blocks of dedicated memory resources. The M20K blocks are ideal for larger memory arrays while still providing a large number of independent ports.
- *640-bit memory LABs (MLABs)* These are memory blocks that are configured from the LABs and can be configured as wide or shallow memory arrays for use as shift registers, wide shallow first-in, first-out (FIFO) buffers, and filter delay lines. For the Stratix® V devices, the ALMs can be configured as ten 32 × 2 blocks, giving a 32 × 20-bit or 64 × 10-bit simple dual-port SRAM block per MLAB.

The mixed-width port configuration is supported in the simple and true dual-port RAM memory modes in various configurations (Altera 2015a). The aim is to provide many modes of operation of the memory hierarchy to support as much functionality as possible, a summary of which is given in Table 5.4. With the streaming nature of DSP and image processing systems, there is a need to delay data, which is typically achieved using shift registers and FIFOs, both of which are supported, and for many DSP operations there is a need to store fixed coefficient values, which the ROM and indeed RAM mode of operation will permit. In more complex modes of operation, there is a need to access and change temporary information, which is supported in the various RAM

Table 5.3 Supported embedded memory block configurations

Memory	Block depth (bits)	Programmable width
MLAB	32	× 16, × 18, or × 20
	64	× 8, × 9, × 10
M20K	512	× 40, × 32
	1K	× 20, × 16
	2K	× 10, × 8
	4K	× 5, × 4
	8K	× 2
	16K	× 1

mode configurations. It is the FPGA vendor's goal to provide underlying support for the many modes of DSP functionality required.

Various clock modes are supported for the memory. This includes *single*, where a single clock controls all registers of the memory block; *read/write*, where a separate clock is available for each read and write port; *input/output*, where a separate clock is available for each input and output port; and *independent*, where a separate clock is available for each port (A and B).

Bit parity checking is supported where the parity bit is the fifth bit associated with every four data bits in data widths of 5, 10, 20, and 40. The error correction code (ECC) support provided allows detection and correction of data errors at the output of the memory, providing single, double-adjacent and triple-adjacent error detection in a 32-bit word.

5.3.3 DSP Processing Blocks

Each variable-precision DSP block spans one LAB row height and offers a range of multiplicative and additive support functionality targeted at support for DSP functionality, specifically 9-bit, 18-bit, 27-bit, and 36-bit wordlength support and even based around an 18×25 multiplier block with built-in addition, subtraction, and 64-bit accumulation unit to combine multiplication results.

Table 5.4 Memory configurations

Memory mode	Brief description
Single-port RAM	Supports one read/one write operation. Read enable port can be used to show the previous held values during the most recent active read enable or to show new data being written
Simple dual-port RAM	One read on port B and one write on port A, i.e. different locations
True dual-port RAM	Any combination of two port operations: two reads, two writes, or one read and one write at two different clock frequencies
Shift register	Can create $w \times m \times n$ shift register for input data width (w), tap size (m), and tap number (n). Can also cascade memory blocks
ROM	Can use memory as ROM. Various configuration of registered and unregistered address lines and outputs
FIFO buffers	Allow single and dual clock asynchronous FIFO buffers

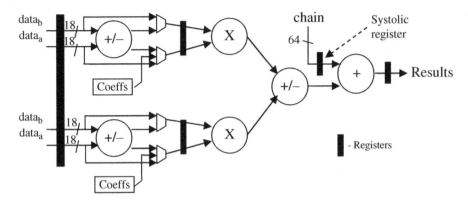

Figure 5.5 Altera DSP block (simplified view)

As illustrated in Figure 5.5, the Stratix® V variable- precision DSP block consists of the following elements:

- input register bank to store the various input data and dynamic control signals;
- pre-adder which supports both addition and subtraction;
- internal coefficient storage which can support up to eight constant coefficients for the multiplicands in 18-bit and 27-bit modes;
- multipliers which can be configured as a single 27 × 27-bit multiplier, two 18 × 18-bit multipliers or three 9 × 9-bit multipliers;
- accumulator and chainout adder which supports a 64-bit accumulator and a 64-bit adder;
- systolic registers, to register the upper multiplier's two 18-bit inputs and to delay the chainout output to the next variable- precision DSP block;
- 64-bit bypassable output register bank.

The processing architecture is highly programmable, allowing various functions to be cascaded together or, in some cases, bypassed. This includes the input and output registers and those referred to as systolic registers which can be used for pipelining, acting to decrease the critical path but increase the latency. This will be achieved during the place process in the synthesis flow using the Altera Quartus® design software. The architecture of the block has been created in such a way as to make it cascadable, thereby allowing core DSP functions to be implemented, such as a FIR filter which comprises a classical multiplier and adder tree function.

Each DSP block can implement one 27 × 27-bit multiplier, two 18 × 18-bit multipliers or three 9 × 9-bit multipliers. It can also be made to perform floating-point arithmetic through the use of additional functionality. The multiplier can support signed and unsigned multiplication and can dynamically switch between the two without any loss of precision. The DSP block can also be configured to operate as a complex multiplier.

The unit can be configured as an adder, a subtracter, or as an accumulator, based on its required mode of operation, and has been designed to automatically switch between adder and subtracter functionality. The DSP blocks have been co-located with the dedicated embedded memory devices to allow coefficients and data to be effectively stored and for memory-intensive DSP applications to be implemented.

5.3.4   Clocks and Interconnect

The Stratix® architecture is organized with three clock networks that are fixed in a hierarchical structure of global clocks which are configured in an H tree structure to balance delay; regional clocks which give low clock skew for logic contained within that quadrant; and periphery clocks which have higher skew than the other clocks. The clock utilizes the Altera MultiTrack interconnect which provides low-skew clock and control signal distribution. It consists of continuous, performance-optimized routing lines of different lengths and speeds used for inter- and intra-design block connectivity.

The FPGA also provides robust clock management and synthesis for device clock management, external system clock management, and high-speed I/O interfaces. It also contains up to 32 fractional phase locked loops (PLLs) that can function as fractional PLLs or integer PLLs, and output counters are dedicated to each fractional PLL that support integer or fractional frequency synthesis. This system of clocking and PLLs should be sufficient to provide low-latency synchronous connections required in many DSP systems.

5.3.5   Stratix® 10 innovations

Stratix® 10 builds on the Stratix® V architecture but with improved DSP performance. It is manufactured on the Intel 14 nm tri-gate process which improves planar transistor technology by creating a "wraparound" gate on the source-to-drain "channel," and gives better performance, reduces active and leakage power, gives better transistor density and a reduction in transistor susceptibility to charged particle single event upsets (SEUs). It offers improved floating-point arithmetic performance and improved bandwidth via the new HyperFlex^{TM} architecture (Altera 2015b). The technology also comprises a 64-bit quad-core ARM^{TM} CortexTM-A53, integrated 28.05- and 14.1-Gbps transceivers, up to 6×72 DDR3 memory interfaces at 933 MHz and 2.5 TMACS of signal processing performance.

HyperFlex^{TM} is a high-speed interconnection architecture that allows users to employ "hyper-registers" which are associated with each individual routing segment in the device and can allow the speed to be improved; they can also be bypassed. The registers are also available at the inputs of all functional blocks such as ALMs, M20K blocks, and DSP blocks. The concept is based around pipelining to avoid long interconnect delays by employing retiming procedures without the need for user effort. It results in an average performance gain of 1.4× for Stratix® 10 devices compared to previous generation high-performance FPGAs.

The Stratix® 10 device offers an improved DSP block which in larger devices can deliver up to 11.5 TMAC or 23 TMAC when using the pre-adder and 9.3 TFLOPS of single-precision, floating-point performance using dedicated hardened circuitry (Altera 2015b). The fixed-point functionality operates at 1 GHz and its floating-point modes operate at 800 MHz. The company argues that this gives a superior power efficiency of 80 GFLOPS/W when compared to GPUs.

5.4   Xilinx Ultrascale^{TM}/Virtex-7 FPGA families

Xilinx's latest FPGA technology is centered on its UltraScale^{TM} family which has been implemented in both a 16 nm and a 20 nm CMOS technology using planar

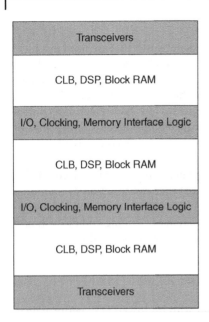

Figure 5.6 Xilinx Ultrascale™ floorplan. Reproduced with permission of Xilinx, Incorp.

and FinFET technologies and 3D scaling. In addition to the classical FPGA offerings, there is a strong focus on multi-processing SoC (MPSoC) technologies. In addition, multiple integrated ASIC-class blocks for 100G Ethernet, 150G Interlaken, and PCIe Gen4 are available to allow fast and efficient data transfer into the FPGA. Static and dynamic power gating is available to address the increasing energy concerns. Advanced encryption standard (AES) bitstream decryption and authentication, key obfuscation, and secure device programming are also included to address security aspects.

The family comprises the Kintex® UltraScale+™ which is focused on low power and the Virtex® UltraScale+™ which is focused on performance. The family also includes the Zynq® UltraScale+™ which incorporates hardware processing technologies in the form of ARM™ processors. UltraRAM has been incorporated to provide larger on-chip SRAM memory. DDR4 can support up to 2666 Mb/s for massive memory interface bandwidth.

The FPGA resources are organized in columns as indicated in Figure 5.6. These include CPUs, DSP blocks called DSP48E2 components and block random access memory (BRAM). High-speed transceivers are used to get the data in and out of the FPGA device quickly, in addition to clocking and memory interfacing circuitry. Xilinx has utilized a new form of routing and an ASIC-like clocking to improve performance. The DSP block has been tweaked to allow better fixed-point and IEEE Standard 754 floating-point arithmetic.

5.4.1 Configurable Logic Block

The Xilinx configurable logic block (CLB) comprises a number of six-input functional generators, flip-flops, fast-carry logic for adder implementation and various programmable hardware to allow various configurations to be created. One quarter of the slice is shown in Figure 5.7 and gives an idea of the core functionality. The six-input

Figure 5.7 Simplified view of 1/4 Xilinx slice functionality

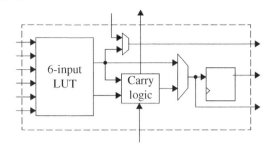

function generator can realize a six-input Boolean function or two arbitrarily defined five-input Boolean functions, as long as these two functions share common inputs. The propagation delay through a LUT is independent of the function implemented, and signals from function generators can connect to slice outputs or to other resources within the slice.

As with the Altera ALM, the six-input functional generator can be configured as a synchronous or distributed RAM. Multiple LUTs in a slice can be combined in various ways to store larger amounts of data up to 256 bits per CLB. Multiple slices can be combined to create larger memories, either single-port RAM ($32 \times 1{-}16$-bit, $64 \times 1{-}8$-bit, $128 \times 1{-}4$-bit, $256 \times 1{-}2$-bit, 512×1-bit), a dual-port RAM ($32 \times 1{-}4$-bit, $64 \times 1{-}4$-bit, 128×2-bit, 256×1-bit), a quad-port RAM ($32 \times 1{-}8$-bit, $64 \times 1{-}2$-bit, 128×1-bit), an octal-port (64×1-bit) and a simple dual-port RAM ($32 \times 1{-}14$-bit, $64 \times 1{-}7$-bit).

The availability of the registers means that sequential circuitry can be implemented by connecting the six-input functional generator to the flip-flops and thus realizing counters or a finite state machine. This allows for a range of controllers to be created that are typically used to organize dataflow in DSP systems.

5.4.2 Memory

As with the other main FPGA vendor, Xilinx offers a range of memory sizes and configurations. The UltraScale™ architecture's memory is organized into 36 kB block RAMs, each with two completely independent ports that share only the stored data (Xilinx 2015a). Like the DSP blocks, the memory is organized into columns as shown in Figure 5.6. Each block can be configured as a single 36 kB RAM or two independent 18 kB RAM blocks. Memory accesses are synchronized to the clock, and all inputs, data, address, clock enables, and write enables are registered with an option to turn off address latching. An optional output data pipeline register allows higher clock rates at the cost of an extra cycle of latency.

BRAMs can be configured vertically to create large, fast memory arrays, and FIFOs with greatly reduced power consumption. BRAM sites that remain unused in the user design are automatically powered down and there is an additional pin on the BRAM to control the dynamic power gating feature. The BRAMs can be configured as $32K \times 1$-bit, $16K \times 2$-bit, $8K \times 4$-bit, $4K \times 9$ (or 8)-bit, $2K \times 18$ (or 16)-bit, $1K \times 36$ (or 32)-bit, or 512×72 (or 64)-bit. The two ports can have different aspect ratios without any constraints. Moreover, as each block RAM can be organized as two 18 kB block RAMs, they be configured to any aspect ratio from $16K \times 1$ to 512×36-bit.

Only in simple dual-port (SDP) mode can data widths greater than 18 bits (18 kB RAM) or 36 bits (36 kB RAM) be accessed. This may have implications for how the memory will be used in the prospective DSP system. In this mode, one port is dedicated to read operation, the other to write operation. In SDP mode, one side (read or write) can be variable, while the other is fixed to 32/36 or 64/72. Both sides of the dual-port 36 kB RAM can be of variable width.

The memory also has an error detection and correction and each 64-bit-wide BRAM can generate, store, and utilize eight additional Hamming code bits and perform single-bit and double-bit error detection during the read process. The ECC logic can also be used when writing to or reading from external 64- to 72-bit-wide memories.

5.4.3 Digital Signal Processing

Each DSP slice fundamentally consists of a dedicated 27 × 18-bit two's complement multiplier and a 48-bit accumulator (Xilinx 2015c) as shown in Figure 5.8. The multiplier can be dynamically bypassed, and two 48-bit inputs can feed a SIMD arithmetic unit (dual 24-bit add/subtract/accumulate or quad 12-bit add/subtract/accumulate), or a logic unit that can generate any one of ten different logic functions of the two operands. The DSP slice includes an additional pre-adder which improves the performance in densely packed designs and reduces the DSP slice count by up to 50%. The 96-bit-wide XOR function, programmable to 12, 24, 48, or 96-bit wide, enables performance improvements when implementing FEC and cyclic redundancy checking algorithms. The DSP also includes a 48-bit-wide pattern detector that can be used for convergent or symmetric rounding. The pattern detector is also capable of implementing 96-bit-wide logic functions when used in conjunction with the logic unit.

The DSP block contains the following components:

- a 27 × 18 two's-complement multiplier with dynamic bypass;
- a power-saving 27-bit pre-adder;
- a 48-bit accumulator that can be cascaded to build 96-bit and larger accumulators, adders, and counters;
- an SIMD arithmetic unit, namely a dual 24-bit or quad 12-bit add/subtract/ accumulate;

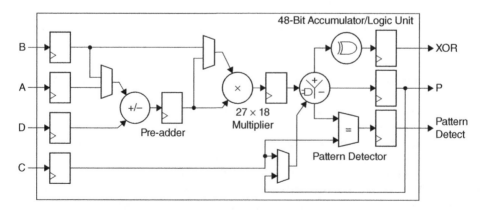

Figure 5.8 Xilinx DSP48E2 DSP block. Reproduced with permission of Xilinx, Incorp.

- a 48-bit logic unit that can implement bitwise AND, OR, NOT, NAND, NOR, XOR, and XNOR functions;
- a pattern detector that allows a number of features including terminal counts, overflow/underflow, convergent/symmetric rounding support, and wide 96-bit wide AND/NOR when combined with logic unit to be performed;
- optional pipeline registers and dedicated buses for cascading multiple DSP slices in a column for larger functions.

Whilst the DSP48E2 slice is backward compatible with older technologies, a number of minor changes have been made to the previous DSP48E1 block, including wider functionality in the slice including increased word size in the multiplier from 25 to 27 bits and accompanying size increase in the pre-adder. The number of operands to the ALU has been increased and means of cascading blocks have been improved.

As has been indicated previously, it is important to have a core understanding of the underlying technology when implementing DSP functionality. For use of the DSP48E2 block, pipelining is encouraged for both performance as it decreases the critical path at the expense of increased latency (see Chapter 9) and power as pipelining reduces the switched capacitance aspect of dynamic power consumption by both reducing the routing length and the switching activity due to less routing (see Chapter 13).

5.5   Xilinx Zynq FPGA Family

The Xilinx Zynq merits a separate section as it represents a new form of FPGA-based system that makes it highly suitable for achieving a new form of computing architecture, historically referred to as an FPGA-based custom computing machine (FCCM) or reconfigurable computing. A simplified version of the Xilinx Zynq architecture is given in Figure 5.9. It comprises a dedicated ARM^{TM} processor environment which has been called a processing system (PS) connected to the standard programmable logic (PL) which is the same as that for the Xilinx UltraScale^{TM}. As the PL has been just described, the description will concentrate on the PS aspects.

The Zynq processing system has an ARM^{TM} processor and a set of resources which form the application processing unit (APU) comprising peripheral interfaces, cache memory, memory interfaces, interconnect, and clock generation circuitry (Crockett *et al.* 2014). The system is shown in Figure 5.10 and is composed of two ARM^{TM} processing cores, each of which has a NEON^{TM} media processing engine (MPE) and

Figure 5.9 Xilinx Zynq architecture

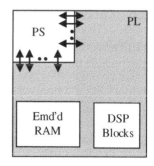

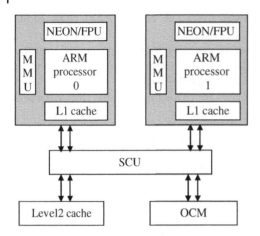

Figure 5.10 Zynq application processing unit

floating-point unit (FPU), a level 1 cache memory and a memory management unit (MMU) unit. The APU also contains a level 2 cache memory, and a further on-chip memory (OCM). Finally, a snoop control unit (SCU) forms a bridge between the ARM™ processors and the level 2 cache and OCM memories and also interfaces to the PL.

The ARM™ can operate at up to 1 GHz, and each has separate level 1 caches for data and instructions, both of which are 32 kB. This allows local storage of frequently required data and instructions for fast processor performance. The two cores additionally share a level 2 cache and there is a further 256 kB of on-chip memory within the APU. The MMU allows translation between virtual and physical addresses.

The Zynq technology has been applied to a number of image processing and evolvable applications. A Zyng-based FPGA-based traffic sign recognition system has been developed for driver assistance applications (Siddiqui *et al.* 2014; Yan and Oruklu 2014). Dobai and Sekanina (2013) argue that the technology has the potential to become "the next revolutionary step in evolvable hardware design."

As FPGAs become more popular in computing systems, there is no doubt that these type of architectures will become popular. Computing vendors develop memory interfaces to directly read from and write to the FPGA directly, such as IBM's coherent accelerator processor interface (CAPI). The use of FPGAs to process data will only become prevalent if data can be quickly loaded and offloaded.

5.6   Lattice iCE40isp FPGA Family

Lattice Semiconductor offers a range of low-power and low-cost products. Its technologies include those listed below:

- MachXO is a non-volatile technology that includes multi-time programmable non-volatile configuration memory and infinitely reconfigurable flash memory. It is contained in small wafer-level chip-scale packaging and is available with low-voltage cores. They comprise mainly LUTs, DRAM and SRAM and various other supporting functionality.

and 7.4, followed by an OpenCL alternative, Altera's SDK for OpenCL, in Section 7.5. Other HLS approaches are briefly reviewed in Section 7.5, including open source C-based tools and those based on a dataflow approach.

7.2 High-Level Synthesis

As the number of transistors on a chip increases, the level of abstraction should increase, as should the design productivity, otherwise design time/effort will increase. The argument is that HLS tools will play an important role in raising productivity. Their main role is to transform an algorithmic description of the behavior of an algorithm into a desired digital hardware solution that implements that behavior. For FPGA designs, the existence of logic synthesis tools from a register transfer level (RTL) description of the design meant that this was a suitable output.

Gajski and Kuhn's Y-chart (Gajski *et al.* 1992; Gajski and Kuhn 1983) describes the HLS design flow (Figure 7.1) A design can be viewed from three perspectives:

- *behavioral*, which describes what the design does, expressed at levels such as transistor functions, Boolean expressions, register transfers, flowcharts/algorithms;
- *structural*, which shows how the design is built using different levels of components such as transistors, gates/flip-flops, registers/ALUs/muxes, processors/memories/buses;
- *physical*, which outlines how the design is physically implemented using different units such as transistor layouts, cells, chips, boards/MCMs

In traditional RTL design flow (Figure 7.1(a)), the behavioral system specifications of the design down to RT level are handled manually by the designer. RTL synthesis and place and route tools is automatically performed, whereas the verification within the automated parts is necessary to match the design against the top-level specifications.

Martin and Smith (2009) provide an outline of the various stages of HLS and suggest that we are now in the third generation after several previous attempts. The first generation was motivated by the observation over two decades ago that RTL-based design

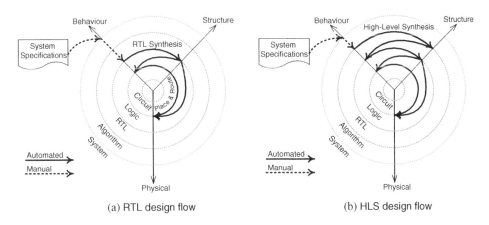

Figure 7.1 High-level synthesis in Gajski and Kuhn's Y-chart

was becoming tedious, complicated, and inefficient as such designs exploded to tens of thousands of lines. To solve this, "behavioral synthesis" was introduced where the detailed description of the architecture was to be replaced with abstract code specifying the design. This required research into sophisticated electronic design automation tools and resulted in the production of multiple behavioral synthesis tools. However, this first attempt failed as the tools were limited and only very small designs were synthesizable; a key example was Synopsys's "Behavioral Compiler." For the second generation, behavioral synthesis was again pursued by academia to fix its problems, and HLS was pursued with improvements such as:

- synthesizing more complex and diverse architectures not just simple data paths;
- synthesizing design I/Os to realize units supporting interface standards and protocols;
- dividing the processing elements into multiple pipelines;
- changing the source language from VHDL or Verilog to the more popular C and C++ languages used by embedded systems designers;
- finding a way to show the trade-offs considered to the designer, namely, speed versus area versus power consumption, latency versus throughput, and memory versus logic.

Since the design space to be explored was vast, thousands of solutions could be created by the tool to find the best one, so it took a long time to develop appropriate HLS-based tools.

While the HLS idea was being developed, another method of raising the abstraction layer was introduced to the market: schematic design to create complicated systems through inserting large reusable IPs, such as processor cores and memory structures, comprising about 80–90% of the design, and 10–20% RTL design including differentiating features. Despite the wide adoption of IP reuse through schematic design, HLS tools reached a level of quality to produce high-performance results for complex algorithms, in particular DSP problems. HLS tools were used to design high-performance IP blocks at high-level languages rather than create the whole system.

With the inclusion of fast processors on chips in SoCs, some designers started to code the 10–20% differentiating part of their design, as software on the processors, hence avoiding any RTL coding in the schematic method. Altera's OpenCL was designed for software engineers who needed software code running on large parallel processor architectures, and also FPGA hardware, as it requires little understanding of hardware.

Given that the previous attempts were viewed to have failed, Martin and Smith (2009) suggest several reasons why this generation of tools would succeed:

- A focus on applications where the tools are being applied in domains where they are expected to have a higher probability of success.
- Algorithm and system designers with the right input languages, allowing them to use languages with which they are comfortable (e.g. C variants, MATLAB®), thus avoiding the learning of special languages.
- Use of compiler-based optimizations which has enabled designers to achieve improved design outputs.
- Performance requirements need significant amounts of signal and multimedia processing and thus need hardware acceleration.
- With FPGAs, the measurement criteria are different than for ASIC as the design has to "fit" into a discrete FPGA size and has to work fast enough, but within the FPGA

speed and size capacity; thus HLS synthesis with FPGA targets is a perfect way of quickly getting an algorithm into hardware.

Overall, the tools appear to be based on C, C++ or C-like tool entry which, as the authors have outlined above, is a major advantage.

7.2.1   HLS from C-Based Languages

Thus there has always been a challenge in using C to model hardware. Given the sequential nature of C, a lot of C-based synthesis tools will translate the C description into some internal data model and then use a series of functionalities to extract the processes, ports and interconnections. This is central to a lot of the design approaches, and a lot of the classical synthesis tools tend to lean on their major investment in classical synthesis tools to achieve the mapping.

Of course, the alternative approach is to adopt C-based languages such as OpenCL, which allows parallelism to be captured. In addition to the ability to capture algorithmic representation, OpenCL defines an application program interface (API) that allows programs running on the host to launch kernels on the compute platform. The language has been driven by the major developments in GPUs which initially developed for processing graphics, but are now applied to a wide range of applications.

The next two sections describe these two varying approaches, one from Xilinx which is based upon C-based synthesis called Vivado and the other from Altera called SDK for OpenCL. While initially Xilinx and Altera created HLS tools which were complementary and created for different groups of users, they have started to add similar functionality to each other's tools.

7.3   Xilinx Vivado

The Xilinx Vivado HLS tool converts a C specification into an RTL implementation synthesizable into a Xilinx FPGA (Feist 2012). C specifications can be written in C, C++, SystemC, or as an OpenCL API C kernel. The company argues that it saves development time, provides quick functional verification, and offers users controlled synthesis and portability.

Figure 7.2 shows that the algorithm can be specified C, C++, or SystemC. These functions are synthesized into RTL blocks and the top-level function arguments are synthesized into RTL I/O ports. Each loop iteration is scheduled to be executed in programmable logic and loops can be unrolled using directives to allow for all iterations to run in parallel. C code arrays are synthesized into block RAM.

Hardware optimized C libraries are available including arbitrary precision data types (to allow optimization to FPGA libraries), HLS stream library, math functions, linear algebra functions, DSP functions, video functions and an IP library. During the synthesis process, a microarchitecture is explored. The tool allows IP system integration and provides RTL generation in VHDL or Verilog.

Vivado creates the optimal implementation based on the default behavior constraints and the user-specified directives. It uses the classical stages of *scheduling* (determining which operations occur when), *binding* (allocating the hardware resource for each

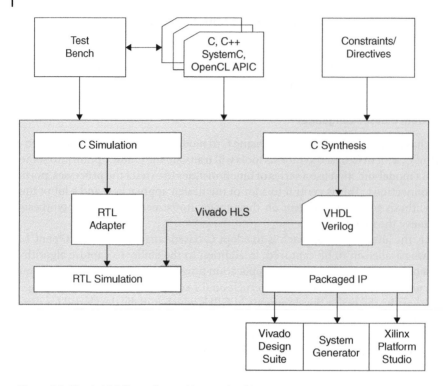

Figure 7.2 Vivado HLS IP creation and integration into a system

scheduled operation) and *control-logic extraction* (which allows the creation of an FSM that sequences the operations).

The user can change the default behavior and allow multiple implementations to be created targeted at reduced area, high-speed, etc. The important performance metrics reported by the HLS are area, latency, loop iteration latency (clock cycles to complete one loop iteration) and loop latency (clock cycles to execute all loop iterations). Generally, the designer can make decisions on functionality, performance including pipeline register allocation, interfaces, storage, design exploration and partitioning into modules.

7.4 Control Logic Extraction Phase Example

The following example shows the control logic extraction and I/O port implementation phase of Vivado. A data computation is placed inside a for loop and two of the function arguments are arrays. The HLS extracts the control logic from the C code and creates an FSM to sequence the operations. The control structure of the C code for loop and the FSM are the same:

$$C0$$
$$C1, C2, C3$$
$$C1, C2, C3$$
$$C1, C2, C3$$
$$C0, \ldots$$

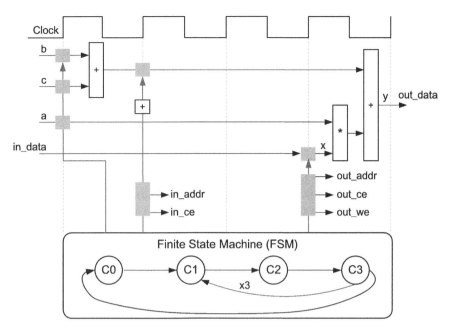

Figure 7.3 Control logic extraction and I/O port implementation example

This is much more easily seen in the timing diagram in Figure 7.3

FSM controls when the registers store data and controls the state of any I/O control signals. Addition of *b* and *c* is moved outside the for loop and into state *C0*. The FSM generates the address for an element in *C1* and an adder increments to count how many times that the design iterates around *C1*, *C2*, and *C3*. In *C2*, the block RAM returns the data for *in_data* and stores as *x*. By default, arrays are synthesized into block RAMs. In *C3*, the calculations are performed and output is generated. Also the address and control signals are generated to store the value outside the block.

```
void F (int in [3], char a, char b, char c, int out[3]){
int x, y;
for (int i = 0; i ¡ 3; i++) {
x = int [i];
y = a * x + b + c;
out [i] = y;
}
}
```

7.5 Altera SDK for OpenCL

Altera makes the case that the OpenCL standard inherently offers the ability to describe parallel algorithms to be implemented on FPGAs at a much higher level of abstraction than HDLs. The company argues that the OpenCL standard more naturally matches the highly parallel nature of FPGAs than do sequential C programs. OpenCL allows the programmer to explicitly specify and control the thread-level parallelism, allowing it to be exploited on FPGAs as the technology offers very high levels of parallelism.

OpenCL is a low-level programming language created by Apple derived from standard ANSI C. It has a lot of the functionality of C but does not have certain headers, function pointers, recursion, variable-length arrays or bit fields. However, it has a number of extensions to extract parallelism and also includes an API which allows the host to communicate with the FPGA-based hardware accelerators, either from one accelerator to another or to the host over PCI Express. In addition, an I/O channel API is needed to stream data into a kernel directly from a streaming I/O interface such as 10Gb Ethernet.

The Altera SDK for OpenCL tools provides the designer with a range of functionality to implement OpenCL on heterogeneous platforms including an emulator to step through the code on an x86, a detailed optimization report to highlight loop dependencies, a profiler and a compiler capable of performing over 300 optimizations on the kernel code and producing the entire FPGA image in one step.

In standard OpenCL, the OpenCL host program is a pure software routine written in standard C/C++. The computationally expensive function which will benefit from acceleration on FPGA is referred to as an OpenCL kernel. Whilst these kernels are written in standard C, they are annotated with constructs to specify parallelism and memory hierarchy. Take, for example, a vector addition of two arrays, *a* and *b*, which produces an output array. Parallel threads will operate on the each element of the vector. If this can be accelerated with a dedicated processing block, then an FPGA offers massive amounts of fine-grained parallelism. The host program has access to standard OpenCL APIs that allow data to be transferred to the FPGA, invoking of the kernel on the FPGA and return of the resulting data.

A pipelined circuit to implement this functionality is given in Figure 7.4. For simplicity, assume the compiler has created three pipeline stages for the kernel. On the first clock cycle, thread 0 is clocked into the two load units and indicates that the first elements of data from arrays *a* and *b* should be fetched. On the second clock cycle, thread 1 is clocked in at the same time that thread 0 has completed its read from memory and stored the results in the registers following the load units. On cycle 3, thread 2 is clocked in, thread 1 captures its returned data, and thread 0 stores the sum of the two values that it loaded. Eventually the pipeline will be filled and numerous computations will be carried out in parallel (Altera 2013).

Figure 7.4 Pipelined processor implementation

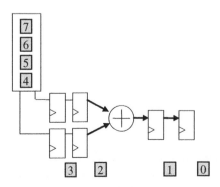

In addition to the kernel pipeline, Altera's OpenCL compiler creates interfaces to external and internal memory. This can include the connections to external memory via a global interconnect structure that arbitrates multiple requests to a group of external DDR memories and also through a specialized interconnect structure to on-chip RAMs. These specialized interconnect structures ensure high performance and best organization of requests to memory.

Altera SDK for OpenCL is in full production release and supports a variety of host CPUs, including the embedded ARM® Cortex®-A9 processor cores. It supports scalable solutions on multiple FPGAs and multiple boards as well as a variety of memory targets, such as DDR SDRAM, QDR SRAM and internal FPGA memory. Half-precision as well as single- and double-precision floating-point is also supported.

7.6 Other HLS Tools

Obviously, FPGA vendors have a key role in producing place and route tools as this is an essential stage in implementing users' designs in their technologies. More recently, this has been extended to HLS tools particularly as the user base of their technology expands to more general applications of computing. Whilst the previous two sections have highlighted the two major FPGA vendors' approach to HLS, there have also been a number of other approaches, both commercial and academic. A sample of such tools is included in this section in no particular order of importance or relevance.

7.6.1 Catapult

Mentor Graphics offers the Catapult® HLS platform which allows users to enter their design in C, SystemC or C++ and produce RTL code to allow targeting to FPGA. The key focus is the tool's ability to produce correct-by-construction, error-free, power-optimized RTL (Mentor Graphics 2014). The tool derives an optimal hardware microarchitecture and uses this to explore multiple options for optimal power, performance and area. This allows design iteration and faster reuse.

Whilst the tool was initially aimed at a wide range of applications, it has now been focused towards power optimization. In addition, the goal of reduction in design and verification has been achieved by having a fixed architecture and optimizing on it. Indeed, the presentation material indicates that the approach of mapping to a fixed architecture has to be undertaken at the design capture stage. This is interesting, given the recent trend of creating soft processors on FPGA as outlined in Chapter 12.

The PowerPro® product allows fast generation of fully verified, power-optimized RTL for ASIC, SOC and FPGA designs. The company argues that this allows analysis of both static and dynamic RTL power usage and allows the user to automatically or manually create power-optimized RTL. From an FPGA perspective, there is clearly a route for trading off parallelism and pipelining as outlined in Chapter 8 as this can be used to alter the power profile as outlined in Chapter 13. Whilst the latter description talks about this from a generated architecture, this approach uses a fixed architecture.

7.6.2 Impulse-C

Impulse-C is available from from Impulse Accelerated Technologies (see Impulse Accelerated Technologies 2011) and provides a C-to-gates workflow. Algorithms are represented using using a communicating sequential process and a library of specific functions, and then described in Impulse-C, which is a standard ANSI C language. The communication between processes is performed mainly by data streams or shared memories, which translates into physical wires or memory storage. Signals can also be transferred to other processes as flags for non-continuous communication. The key is that this allows capture of process parallelization and communication.

The code can be compiled using a standard C compiler and then translated into VHDL or Verilog. As well as the code for the functional blocks, the tools can generate the controller functionality from the communication channels and synchronization mechanisms. Pragma directives are used to control the hardware generation throughout the C code, for example, and to allow loop unrolling, pipelining or primitive instantiation. Also existing IP cores in the form of VHDL code can be incorporated.

Each defined processes is translated to a software thread allowing the algorithm to be debugged and profiled using using standard tools. The co-design environment includes tools for co-simulation and co-execution of the algorithms. The approach is targeted at heterogeneous platforms allowing compilation onto processors and optimization onto a programmable logic platform. Impulse-C code can be implemented in a growing number of hardware platforms and, specifically, Altera and Xilinx FPGA technologies.

The tools have been applied to core DSP functions, many of which were described in Chapter 2, but also more complex systems, specifically, filters and many image processing algorithms including edge enhancement, object recognition, video compression and decompression and hyperspectral imaging. There have also been some financial applications, specifically to high-frequency trading.

7.6.3   GAUT

GAUT is an academic HLS tool (http://www.gaut.fr/) that has been targeted at DSP applications (Coussy *et al.* 2008). It allows the user to start from a C/C++ description of the algorithm and supports fixed-point representation to allow more efficient implementation on FPGAs. The user can set the throughput rate and the clock period as well as other features such as memory mapping (Corre *et al.* 2004) and I/O timing diagrams (Coussy *et al.* 2006). The tool synthesizes an architecture consisting of a processing unit, a memory unit and a communication and interface block. The processing unit is composed of logic and arithmetic operators, storage elements, steering logic and an FSM controller.

The flow for the tool is given in Figure 7.5. It starts with the design of the architecture, which involves selecting the arithmetic operators, then the memory registers and memory banks involving memory optimization, followed by the communication paths such as memory address generators and the communication interfaces. It generates not only VHDL models but also the testbenches and scripts necessary for the Modelsim simulator. It has been applied to a Viterbi decoder and a number of FIR and LMS filters, giving a reduction in code of around two orders of magnitude.

7.6.4   CAL

Sequential software programming depends mostly on HLS tools to automatically extract the parallelism of the code. Other than the automatic detection, to cover the language concurrency-support limitations, some libraries are also introduced or features such as

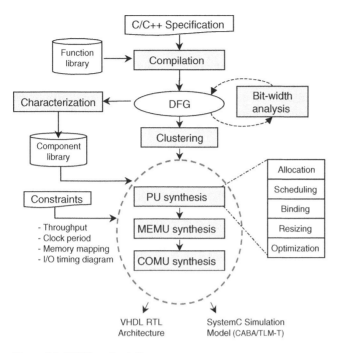

Figure 7.5 GAUT synthesis flow

pragmas are included. These changes in a sequential language to make it executable on hardware have led to different implementations of that language. Development of languages based on the computation model which reflects the hardware specifics and parallel programming seems a better approach than adapting sequential C-like languages to support hardware design.

A language designed to support both parallel and sequential coding constructs and expression of applications as network processes is CAL (Eker and Janneck 2003). The CAL actor language was developed by Eker and Janneck in 2001 as part of the Ptolemy project (Eker *et al.* 2003) at the University of California at Berkeley. CAL is a high-level programming language of the form of a dataflow graph (DFG), for writing actors which transform input streams into output streams. CAL is an ideal language for use as a single behavioral description for software and hardware processing elements.

A subset of the CAL language which has been standardized by the ISO MPEG committee is reconfigurable video coding or RVC-CAL. The main reason for the introduction of RVC is to provide reuse of commonalities among various MPEG standards, and their extension through system-level specifications (Eker *et al.* 2003). This provides a more flexible and faster path to introducing new MPEG standards. The RVC framework is being developed by the MPEG to provide a unified high-level specification of current and future MPEG video coding technologies using dataflow models. In this framework, a decoder is generated by configuring video coding modules which are standard MPEG toolbox libraries or propriety libraries. RVC-CAL is used to write the reference software of library elements. A decoder configuration is defined in XML language by connecting a set of RVC-CAL modules.

The RVC-CAL language limits the advanced features of the CAL language. There are some tools available for development of applications based on RVC-CAL language as summarized in the following.

The Open RVC-CAL Compiler (ORCC) is an open source dataflow development environment and compiler framework which uses RVC-CAL, allows the transcompilation of actors and generates equivalent codes depending on the chosen back-ends (Eker *et al.* 2003). ORCC is developed within the Eclipse-based IDE as a plug-in with graphical interfaces to ease the design of dataflow applications.

CAL2HDL (Janneck *et al.* 2008) was the first implementation of a direct hardware code generation from CAL dataflow programs. A CAL actor language is first converted into an XML language independent model from which Verilog code can be generated using an open source tool. The tool supports a limited subset of CAL actor language such that complex applications cannot be easily expressed or synthesized using this tool. In addition, Xronos (Bezati *et al.* 2013) is an evolution of CAL2HDL and TURNUS (Brunei *et al.* 2013) is a framework used for iterative design space exploration of RVC-CAL programs to find design solutions which satisfy performance constraints or optimize parts of the system.

7.6.5 LegUp

LegUp is an open source HLS tool that allows a standard C source program to be synthesized onto a hybrid FPGA-based hardware/software system (Canis *et al.* 2013). The authors envisage implementing their designs onto an FPGA-based 32-bit MIPS soft processor and synthesized FPGA accelerators. It has been created using modular C++ and uses the state-of-the-art LLVM compiler framework for high-level language parsing and its standard compiler optimizations (LLVM 2010). It also uses a set of benchmark C programs that can be used to a combined hardware/software system and allows specific functionality to be added (Hara *et al.* 2009).

7.7 Conclusions

The chapter has briefly covered some of the tools used to perform HLS. The purpose of the chapter has been to give a brief overview of some relevant tools which may in some cases cover the types of optimizations covered in Chapter 8. In some cases the tools start with a high-level description in C or C++ and can produce HDL output in the form of VHDL or Verilog, allowing vendors' tools to be used to produce the bit files.

C-based tools are particularly useful as many of the FPGA platforms are SoC platforms comprising processors, memory and high-speed on-chip communications. The ability to explore optimization across such platforms is vital and will become increasingly so as future systems requirements evolve as heterogeneous FPGA platforms become ever more complex. Such issues are addressed in the tools outlined in Chapter 10.

Bibliography

Altera Corp. 2013 Implementing FPGA Design with the OpenCL Standard. White paper WP-01173-3.0. Available from www.altera.com (accessed June 11, 2015).

Bezati E, Mattavelli M, Janneck JW 2013 High-level synthesis of dataflow programs for signal processing systems. In *Proc. Int. Symp. on Image and Signal Processing and Analysis*, pp. 750–754.

Brunei SC, Mattavelli M, Janneck JW 2013 TURNUS: A design exploration framework for dataflow system design. In *Proc. IEEE Int. Symp. on Circuits and Systems*. doi 10.1109/ISCAS.2013.6571927

Canis A, Choi J, Aldham M, Zhang V, Kammoona A, Czajkowski T, Brown SD, Anderson JH. 2013. LegUp: An open-source high-level synthesis tool for FPGA-based processor/accelerator systems. *ACM Trans. Embedded Computing Systems*, 13(2), article 24.

Corre G, Senn E, Bomel P, Julien N, Martin E 2004 Memory accesses management during high level synthesis. In *Proc. IEEE Int. Conf. on CODES+ISSS*, pp. 42–47.

Coussy P, Casseau E, Bomel P, Baganne A, Martin E 2006 A formal method for hardware IP design and integration under I/O and timing constraints. *ACM Trans. on Embedded Computing Systems*, 5(1), 29–53.

Coussy P, Chavet C, Bomel P, Heller D, Senn E, Martin E 2008 GAUT: A high-level synthesis tool for DSP applications. In Coussy P, Morawiec A (eds) *High-Level Synthesis: From Algorithm to Digital Circuit*, pp. 147–169. Springer, New York.

Eker J, Janneck J 2003 CAL language report. Technical Report UCB/ERL M 3. University of California at Berkeley.

Eker J, Janneck JW, Lee EA, Liu J, Liu X, Ludvig J, Neuendorffer S, Sachs S, Xiong Y 2003 Taming heterogeneity – the Ptolemy approach. *Proc. IEEE*, 91(1), 127–144.

Feist T 2012 Vivado design suite. White Paper WP416 (v1.1). Available from www.xilinx.com (accessed May 11, 2016).

Gajski DD, Dutt ND, Wu AC, Lin SY-L 1992 *High-Level Synthesis: Introduction to Chip and System Design*. Kluwer Academic, Norwell, MA.

Gajski DD, Kuhn RH 1983 Guest editors' introduction: New VLSI tools. *Computer*, 16(2), 11–14.

Hara Y, Tomiyama H, Honda S, Takada H. 2009 Proposal and quantitative analysis of the CHStone benchmark program suite for practical C-based high-level synthesis. *J. of Information Processing*, 17, 242–254.

Impulse Accelerated Technologies 2011 Impulse codeveloper C-to-FPGA tools. Available from http:// www.impulse accelerated.com/products_universal. htm (accessed May 11, 2016).

Janneck JW, Miller ID, Parlour DB, Roquier G, Wipliez M, Raulet M 2008 Synthesizing hardware from dataflow programs: An MPEG-4 simple profile decoder case study. In *Proc. Workshop on Signal Processing Systems*, pp. 287–292.

LLVM 2010. The LLVM compiler infrastructure project. Available from http://www.llvm.org (accessed May 11, 2016).

Martin G, Smith G 2009 High-level synthesis: Past, present, and future. *IEEE Design & Test of Computers*, 26(4), 18–25.

Mentor Graphics 2014 High-level synthesis report 2014. Available from http://s3.mentor.com/ public_documents/whitepaper/resources/mentorpaper_94095.pdf (accessed May 11, 2016).

8

Architecture Derivation for FPGA-based DSP Systems

8.1 Introduction

The technology review in Chapter 4 and the detailed circuit implementation material in Chapter 5 clearly demonstrated the need to develop a circuit architecture when implementing DSP algorithms in FPGA technology. The circuit architecture allows the performance needs of the application to be captured effectively. One optimization is to implement the high levels of parallelism available in FIR filters directly in hardware, thereby allowing a performance increase to be achieved by replicating the functionality in FPGA hardware. In addition, it is possible to pipeline the SFG or DFG heavily to exploit the plethora of available registers in FPGA; this assumes that the increased latency in terms of clock cycles, incurred as a result of the pipelining (admittedly at a smaller clock period), can be tolerated. It is clear that optimizations made at the hardware level can have direct cost implications for the resulting design. Both of these aspects can be captured in the circuit architecture.

In Chapter 5 it was shown how this trade-off is much easier to explore in "fixed architectural" platforms such as microprocessors, DSP processors or even reconfigurable processors, as appropriate tools can be or have been developed to map the algorithmic requirements efficiently onto the available hardware. As already discussed, the main attraction of using FPGAs is that the available hardware can be developed to meet the specific needs of the algorithm. However, this negates the use of efficient compiler tools as, in effect, the architectural "goalposts" have moved as the architecture is created on demand! This fact was highlighted in Chapter 7 which covered some of the high-level tools that are being developed either commercially or in universities and research labs. Thus, it is typical that a range of architecture solutions are explored with cost factors that are computed at a high level of abstraction.

In this chapter, we will explore the direct mapping of simple DSP systems or, more precisely, DSP components such as FIR or IIR filters, adaptive filters, etc. as these will now form part of more complex systems such as beamformers and echo cancelers. The key aim is to investigate how changes applied to SFG representations can impact the FPGA realizations of such functions, allowing the reader to quickly work in the SFG domain rather than in the circuit architecture domain. This trend will become

FPGA-based Implementation of Signal Processing Systems,
Second Edition. Roger Woods, John McAllister, Gaye Lightbody and Ying Yi.
© 2017 John Wiley & Sons, Ltd. Published 2017 by John Wiley & Sons, Ltd.

increasingly prevalent throughout the book as we attempt to move to a higher-level representation. The later chapters demonstrate how higher levels of abstraction can be employed to allow additional performance improvements by considering system-level implications.

Section 8.2 looks at the DSP characteristics and gives some indication of how these map to FPGA. The various representations of DSP systems are outlined in Section 8.3. Given that a key aspect of FPGA architecture is distributed memory, efficient pipelining is a key optimization and so is explored in detail in Section 8.4. The chapter then goes on to explore how the levels of parallelism can be adjusted in the implementation in order to achieve the necessary speed at both lower or higher area costs; duplicating the hardware is formally known as "unfolding" and sharing the available hardware is called "folding," and both of these techniques are explored in Section 8.5. Throughout the chapter, the techniques are applied to FIR, IIR and lattice filters and explored using the Xilinx Virtex-5 FPGA family. This material relies heavily on the excellent text by Parhi (1999).

8.2 DSP Algorithm Characteristics

By their very nature, DSP algorithms tend to be used in applications where there is a demand to process high volumes of information. As highlighted in Chapter 2, the sampling rates can range from kilohertz, as in speech environments, right through to megahertz, as in the case of image processing applications. It is vital to clearly define a number of parameters with regard to system implementation of DSP systems:

- *Sampling rate* can be defined as the rate at which we need to process the DSP signal samples for the system or algorithm under consideration. For example, in a speech application, the maximum bandwidth of speech is typically judged to be 4 kHz, and the Nyquist rate indicates a sampling rate of 8 kHz.
- *Throughout rate* (TR) defines the rate at which data samples are processed. In some cases, the aim of DSP system implementation is to match the throughput and sampling rates, but in systems with lower sampling rates (speech and audio), this would result in underutilization of the processing hardware. For example, speech sampling rates are 8 kHz, but the speeds of many DSP processors are of the order of hundreds of megahertz. In these cases there is usually a need to perform a large number of computations per second, which means that the throughput rate can be several times (say, p) the sampling rate. In cases where the throughput is high and the computational needs are moderate, there is the possibility of reusing the hardware, say p times. This is a mapping that would need to be applied in FPGA implementation.
- *Clock rate* defines the operating speed of the system implementation. It used to be a performance figure quoted by computing companies, although it is acknowledged that memory size, organization and usage can be more critical in determining performance. In DSP systems, a simple perusal of DSP and FPGA data sheets indicates that clock rates of FPGA families are 550–600 MHz, whereas TI's TMS320C6678 DSP family can run up to 1.25 GHz. It would appear that the DSP processor is faster than the FPGA, but it is the amount of computation that can be performed in a single cycle that it is important (Altera 2014). This is a major factor in determining the throughput rate, which is a much more accurate estimate of performance, but is of course application-dependent.

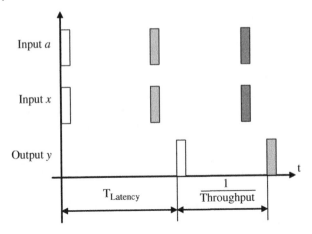

Figure 8.1 Latency and throughput rate relationship for system $y(n) = ax(n)$

Thus, it is clear that we need to design systems ultimately for throughput and therefore sampling rate, as a first measure of performance. This relies heavily on how efficiently we can develop the circuit architecture. As Chapter 5 clearly indicated, this comes from harnessing effectively the underlying hardware resources to meet the performance requirements. In ASIC applications the user can define the processing resources to achieve this, but in FPGAs the processing resources are restrictive in terms of their number and type (e.g. dedicated DSP blocks, scalable adder structures, LUT resources, memory resource (distributed RAM, LUT RAM, registers)) and interconnection (e.g. high-speed Rocket IO, various forms of programmable interconnect). The aim is to match these resources to the computational needs, which we will do here based initially on performance and then trading off area, if throughput is exceeded.

In DSP processors, the fixed nature of the architecture is such that efficient DSP compilers have evolved to allow high-level or C language algorithmic descriptions to be compiled, assembled and implemented onto the platform. Thus, the implementation target is to investigate if the processing resources will allow one iteration of the algorithm to be computed at the required sampling rate. This is done by allocating the processing to the available resources and scheduling the computation in such a way as to achieve the required sampling rate. In effect, this involves reusing the available hardware, but we intend not to think about the process in these terms. In a FPGA implementation, an immediate design consideration is to consider how many times we can reuse the hardware and whether this allows us to achieve the sampling rate. This change of emphasis in creating the hardware resource to match the performance requirements is the reason for a key focus of the chapter.

8.2.1 Further Characterization

Latency is the time required to produce the output, $y(n)$ for the corresponding $x(n)$ input. At first glance, this would appear to equate to the throughput rate, but as the computation of $y(n) = ax(n)$ shown in Figure 8.2 clearly demonstrates, this is not the case, particularly if pipelining is applied. In Figure 8.2, the circuit could have three pipeline stages and thus will produce a first output after three clock cycles, hence known as the

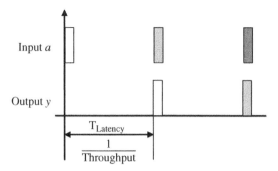

Figure 8.2 Latency and throughput rate relationship for system $y(n) = ay(n-1)$

latency; thereafter, it will produce an output once every cycle which is the throughput rate.

The situation is complicated further in systems with feedback loops. For example, consider the simple recursion $y(n) = ay(n-1)$ shown in Figure 8.2. The present output $y(n)$ is dependent on the previous output $y(n-1)$, and thus the latency determines the throughput rate. This means now that if it takes three clock cycles to produce the first output, then we have to wait three clock cycles for the circuit to produce each output and, for that matter, enter every input. Thus it is clear that any technique such as pipelining that alters both the throughput and the latency must be considered carefully, when deriving the circuit architectures for different algorithms.

There are a number of optimizations that can be carried out in FPGA implementations to perform the required computation, as listed below. Whilst it could be argued that parallelism is naturally available in the algorithmic description and not an optimization, the main definitions here focus on exploitation within FPGA realization; a serial processor implementation does not necessarily exploit this level of parallelism.

Parallelism can either naturally exist in the algorithm description or can be introduced by organizing the computation to allow a parallel implementation. In Figure 8.3, we can

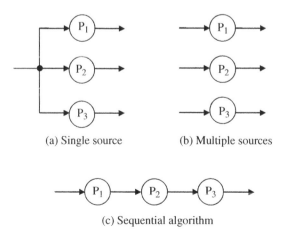

Figure 8.3 Algorithms realizations using three processes P_1, P_2 and P_3

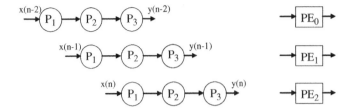

(a) Demonstration of interleaving

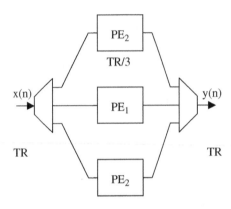

(b) Interleaving realization

Figure 8.4 Interleaving example

realize processes P_1, P_2 and P_3 as three separate processors, PE_1, PE_2 and PE_3, in all three cases. In Figure 8.3(a) the processes are driven from a single source, in Figure 8.3(b) they are from separate sources and in Figure 8.3(c) they are organized sequentially. In the latter case, the processing is inefficient as only one processor will be used at any one time, but it is shown here for completeness.

Interleaving can be employed to speed up computation, by sharing a number of processors to compute iterations of the algorithm in parallel, as illustrated in Figure 8.4 for the sequential algorithm of Figure 8.3(c). In this case, the three processors PE_1, PE_2 and PE_3 perform three iterations of the algorithm in parallel and each row of the outlined computation is mapped to an individual processor, PE_1, PE_2 and PE_3.

Pipelining is effectively another form of concurrency where processes are carried out on separate pieces of data, but at the same time, as illustrated in Figure 8.5. In this case, the three processes PE_1, PE_2 and PE_3 are performed at the same time, but on different

Figure 8.5 Example of pipelining

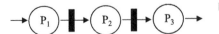

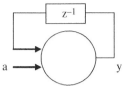

(a) Original recursive computation (clock rate, fc, and throughput rate, f)

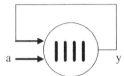

(b) Pipelined version (clock rate, $4\,fc$, and throughput rate, f)

Figure 8.6 Pipelining of recursive computations $y(n) = ay(n-1)$

iterations of the algorithm. Thus the throughput is now given as t_{PE_1} or t_{PE_2} or t_{PE_3} rather than $t_{PE_1} + t_{PE_2} + t_{PE_3}$ as for Figure 8.3(c). However, the application of pipelining is limited for some recursive functions such as the computation $y(n) = ay(n-1)$ given in Figure 8.6. As demonstrated in Figure 8.6(a), the original processor realization would have resulted in an implementation with a clock rate f_c and throughput rate f. Application of four levels of pipelining, as illustrated in Figure 8.6(b), results in an implementation that can be clocked four times faster, but since the next iteration depends on the present output, it will have to wait four clock cycles. This gives a throughput rate of once every four cycles, indicating a nil gain in performance. Indeed, the flip-flop setup and hold times now form a much larger fraction of the critical path and the performance would actually have been degraded in real terms.

It is clear then that these optimizations are not a straightforward application of one technique. For example, it may be possible to employ parallel processing in the final FPGA realization and then employ pipelining within each of the processors. In Figure 8.6(b), pipelining did not give a speed increase, but now four iterations of the algorithm can be interleaved, thereby achieving a fourfold improvement. It is clear that there are a number of choices available to the designer to achieve the required throughput requirements with minimal area requirements such as sequential versus parallel, trade-off between parallelism/pipelining and efficient use of hardware sharing. The focus of this chapter is on demonstrating how the designer can start to explore these trade-offs in an algorithmic representation, by starting with an SFG or DFG description and then carrying out manipulations with the aim of achieving improved performance.

8.3 DSP Algorithm Representations

There are a number of ways of representing DSP algorithms, ranging from mathematical descriptions, to block diagrams, right through to HDL descriptions of implementations.

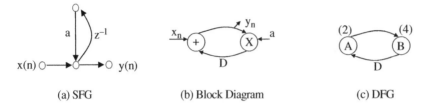

(a) SFG   (b) Block Diagram   (c) DFG

Figure 8.7 Various representations of simple DSP recursion $y(n) = ay(n-1) + x(n)$

In this chapter, we concentrate on an SFG and DFG representation as a starting point for exploring some of the optimizations briefly outlined above. For this reason, it is important to provide more detail on SFG and DFG representations.

8.3.1 SFG Descriptions

The classical description of a DSP system is typically achieved using an SFG representation which is a collection of nodes and directed edges, where a directed edge (j, k) denotes a linear transform from the signal at node j to the signal at node k. Edges are usually restricted to multiplier, adder or delay elements. The classical SFG of the expression $y(n) = ay(n-1) + x(n)$ is given in Figure 8.7(a), while the block diagram is given in Figure 8.7(b). The DFG representation shown in Figure 8.7(c) is often a more useful representation for the retiming optimizations applied later in the chapter.

8.3.2 DFG Descriptions

In DFGs, nodes represent computations or functions and directed edges represent data paths with non-negative numbers associated with them. Dataflow captures the data-driven property of DSP algorithms where the node can fire (perform its computation) when all the input data are available; this creates precedence constraints (Parhi 1999). There is an *intra*-iteration constraint if an edge has no delay; in other words, the order of firing is dictated by DFG arrow direction. The *inter*-iteration constraint applies if the edge has one or more delays and will be translated into a digital delay or register when implemented.

A more practical implementation can be considered for a three-tap FIR filter configuration. The SFG representation is given in Figure 8.8. One of the transformations that can be applied to SFG representation is that of transposition. This is carried out by reversing the directions in all edges, exchanging input and output nodes whilst keeping edge gains or edge delays unchanged as shown in Figure 8.8(b). The reorganized version is shown in Figure 8.8(c). The main difference is that the dataflow of the $x(n)$ input has been reversed without causing any functional change to the resulting SFG. It will be seen later that the SFG of Figure 8.8(c) is a more appropriate structure to which to apply pipelining.

The dataflow representation of the SFG of Figure 8.8(b) is shown in Figure 8.9. In Figure 8.9 the multipliers labeled as a_0, a_1 and a_2 represent pipelined multipliers with two levels of pipeline stages. The adders labeled as A_0 and A_1 represent pipelined adders

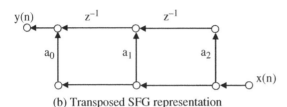

(a) SFG representation of three-tap FIR filter

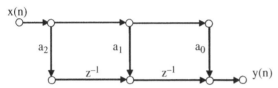

(b) Transposed SFG representation

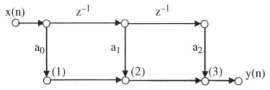

(c) Reorganized transposed SFG representation

Figure 8.8 SFG representation of three-tap FIR filter

with a pipeline stage of 1. The D labels represent single registers with size equal to the wordlength (not indicated on the DFG representation). In this way, the dataflow description gives a good indication of the hardware realization; it is clear that it is largely an issue of developing the appropriate DFG representation for the performance requirements needed. In the case of pipelined architecture, this is largely a case of applying suitable retiming methodologies to develop the correct level of pipelining, to achieve the performance required. The next section is devoted to retiming because, as will be shown, recursive structures, i.e. those involving feedback loops, can present particular problems.

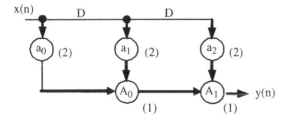

Figure 8.9 Simple DFG

Table 8.1 FIR filter timing

Address Clock	Input	Node 1	Node 2	LUT contents Node 3	Output
0	$x(0)$	$a_0x(0)$	$a_0x(0)$	$a_0x(0)$	$y(0)$
1	$x(1)$	$a_0x(1)$	$a_0x(1) + a_1x(0)$	$a_0x(1) + a_1x(0)$	$y(1)$
2	$x(2)$	$a_0x(2)$	$a_0x(2) + a_1x(1)$	$a_0x(2) + a_1x(1) + a_2x(0)$	$y(2)$
3	$x(3)$	$a_0x(3)$	$a_0x(3) + a_1x(2)$	$a_0x(3) + a_1x(2) + a_2x(1)$	$y(3)$
4	$x(4)$	$a_0x(4)$	$a_0x(4) + a_1x(3)$	$a_0x(4) + a_1x(3) + a_2x(2)$	$y(4)$

8.4 Pipelining DSP Systems

One of the main goals in attaining an FPGA realization is to determine the levels of pipelining needed. The timing of the data through the three-tap FIR filter of Figure 8.8(a) for the nodes labeled (1), (2) and (3) is given in Table 8.1. We can add a delay to each multiplier output as shown in Figure 8.8(a), giving the change in data scheduling shown in Table 8.2. Note that the latency has now increased, as the result is not available for one cycle. However, adding another delay onto the outputs of the adders causes failure, as indicated by Table 8.3. This is because the process by which we are adding these delays has to be carried out in a systematic fashion by the application of a technique known as retiming. Obviously, retiming was applied correctly in the first instance as it did not change the circuit functionality but incorrectly in the second case. Retiming can be applied via the cut theorem as described in Kung (1988).

8.4.1 Retiming

Retiming is a transformation technique used to move delays in a circuit without affecting the input/output characteristics (Leiserson and Saxe 1983). Retiming has been applied in synchronous designs for clock period reduction (Leiserson and Saxe 1983), power consumption reduction (Monteiro et al. 1993), and logical synthesis. The basic process of retiming is given in Figure 8.10 (Parhi 1999). For a circuit with two edges U and V and ω delays between them, as shown in Figure 8.10(a), a retimed circuit can be derived with ω_r delays as shown in Figure 8.10(b), by computing the ω_r value as

$$\omega_r(e) = \omega(e) + r(U) - r(V), \tag{8.1}$$

where $r(U)$ and $r(V)$ are the retimed values for nodes U and V, respectively.

Table 8.2 Revised FIR filter timing

Address Clock	Input	Node 1	Node 2	LUT contents Node 3	Output
0	$x(0)$	$a_0x(0)$			
1	$x(1)$	$a_0x(1)$	$a_1x(0)$	$a_0x(0)$	$y(0)$
2	$x(2)$	$a_0x(2)$	$a_1x(1) + a_1x(0)$	$a_0x(1) + a_1x(0)$	$y(1)$
3	$x(3)$	$a_0x(3)$	$a_1x(2) + a_2x(1)$	$a_0x(2) + a_1x(1) + a_2x(0)$	$y(2)$
4	$x(4)$	$a_0x(4)$	$a_1x(3) + a_2x(2)$	$a_0x(3) + a_1x(2) + a_2x(1)$	$y(3)$

Table 8.3 Faulty application of retiming

Address Clock	Input	Node 1	Node 2	LUT contents Node 3	Output
0	$x(0)$	$a_0x(0)$			
1	$x(1)$	$a_0x(1)$	$a_0x(0)$		
2	$x(2)$	$a_0x(2)$	$a_0x(1) + a_1x(0)$	$a_0x(0)$	$y(0)$
3	$x(3)$	$a_0x(3)$	$a_0x(2) + a_1x(1)$	$a_0x(1) + a_1x(0) + a_2x(0)$	
4	$x(4)$	$a_0x(4)$	$a_0x(3) + a_1x(2)$	$a_0x(2) + a_1x(1) + a_2x(1)$	

Retiming has a number of properties which can be summarized as follows:

1. The weight of any retimed path is given by Equation (8.1).
2. Retiming does not change the number of delays in a cycle.
3. Retiming does not alter the iteration bound (see later) in a DFG as the number of delays in a cycle does not change.
4. Adding the constant value j to the retiming value of each node does not alter the number of delays in the edges of the retimed graph.

Figure 8.11 gives a number of examples of how retiming can be applied to the FIR filter DFG of Figure 8.11(a). For simplicity, we have replaced the labels a_0, a_1, a_2, A_0 and A_1 of Figure 8.9 by 2, 3, 4, 5 and 6, respectively. We have also shown separate connections between the $x(n)$ data source and nodes 2, 3 and 4; the reasons for this will be shown shortly. By applying equation (8.1) to each of the edges, we get the following relationships for each edge:

$$\omega_r(1 \rightarrow 2) = \omega(1 \rightarrow 2) + r(2) - r(1),$$
$$\omega_r(1 \rightarrow 3) = \omega(1 \rightarrow 3) + r(3) - r(1),$$
$$\omega_r(1 \rightarrow 4) = \omega(1 \rightarrow 4) + r(4) - r(1),$$
$$\omega_r(2 \rightarrow 5) = \omega(2 \rightarrow 5) + r(5) - r(2),$$
$$\omega_r(3 \rightarrow 5) = \omega(3 \rightarrow 5) + r(5) - r(3),$$
$$\omega_r(4 \rightarrow 6) = \omega(4 \rightarrow 6) + r(6) - r(4),$$
$$\omega_r(5 \rightarrow 6) = \omega(5 \rightarrow 6) + r(6) - r(5).$$

Figure 8.10 Retiming example

(a) Original SFG

(b) Retimed SFG

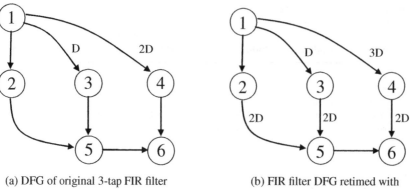

(a) DFG of original 3-tap FIR filter

(b) FIR filter DFG retimed with
$r(1) = -2, r(2) = -2, r(3) = -2, r(4) = -2, r(5) = 0, r(6) = 0$

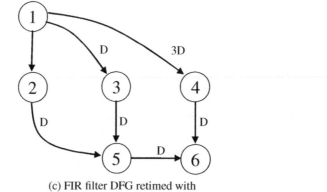

(c) FIR filter DFG retimed with
$r(1) = -2, r(2) = -2, r(3) = -2, r(4) = -1, r(5) = -1, r(6) = 0$

Figure 8.11 Retimed FIR filter

Using a retiming vector $r(1) = -1, r(2) = -1, r(3) = -1, r(4) = -1, r(5) = 0, r(6) = 0$ above, we get the following values:

$$\omega_r(1 \rightarrow 2) = 0 + (-1) - (-1) = 0,$$
$$\omega_r(1 \rightarrow 3) = 1 + (-1) - (-1) = 1,$$
$$\omega_r(1 \rightarrow 4) = 2 + (-1) - (-1) = 2,$$
$$\omega_r(2 \rightarrow 5) = 0 + (0) - (-1) = 1,$$
$$\omega_r(3 \rightarrow 5) = 0 + (0) - (-1) = 1,$$
$$\omega_r(4 \rightarrow 6) = 0 + (0) - (-1) = 1,$$
$$\omega_r(5 \rightarrow 6) = 0 + (0) - (0) = 0.$$

This gives the revised diagram shown in Figure 8.11 which gives a circuit where each multiplier has two pipeline delays at the output edge. A retiming vector could have been applied which provides one delay at the multiplier output, but the reason for this retiming will be seen later. Application of an alternative retiming vector,

$$\omega_r(1 \to 2) = 0 + (-2) - (-2) = 0,$$
$$\omega_r(1 \to 3) = 1 + (-2) - (-2) = 1,$$
$$\omega_r(1 \to 4) = 2 + (-1) - (-2) = 3,$$
$$\omega_r(2 \to 5) = 0 + (-1) - (-2) = 1,$$
$$\omega_r(3 \to 5) = 0 + (-1) - (-2) = 1,$$
$$\omega_r(4 \to 6) = 0 + (0) - (-1) = 1,$$
$$\omega_r(5 \to 6) = 0 + (0) - (-1) = 1,$$

namely $r(1) = -1, r(2) = -1, r(3) = -1, r(4) = -1, r(5) = -1, r(6) = 0$, gives the circuit of Figure 8.11(c) which gives a fully pipelined implementation. It can be seen from this figure that the application of pipelining to the adder stage required an additional delay, D, to be applied to the connection between 1 and 4. It is clear from these two examples that a number of retiming operations can be applied to the FIR filter. A retiming solution is feasible if $w \geq 0$ holds for all edges.

It is clear from the two examples outlined that retiming can be used to introduce inter-iteration constraints to the DFG, manifested as a pipeline delay in the final FPGA implementation (Parhi 1999). However, the major issue would appear to be the determination of the retiming vector which must be such that it moves the delays to the edges needed in the DFG whilst at the same time preserving the viable solution, i.e. $w \geq 0$ holds for all edges. One way of determining the retiming vector is to apply a graphical methodology to the DFG which symbolizes applying retiming. This is known as the cut-set or cut theorem (Kung 1988).

8.4.2 Cut-Set Theorem

A cut-set in an SFG (or DFG) is a minimal set of edges which partitions the SFG into two parts. The procedure is based upon two simple rules.

Rule 1: *Delay scaling.* All delays D presented on the edges of an original SFG may be scaled by D', where $D' \to \alpha D$; the single positive integer α is also known as the pipelining period of the SFG. Correspondingly, the input and output rates also have to be scaled by a factor of α (with respect to the new time unit D'). Time scaling does not alter the overall timing of the SFG.

Rule 2: *Delay transfer* (Leiserson and Saxe 1983). Given any cut-set of the SFG, which partitions the graph into two components, we can group the edges of the cut-set into inbound and outbound, as shown in Figure 8.12, depending upon the direction assigned to the edges. The delay transfer rule states that a number of delay registers, say k, may be transferred from outbound to inbound edges, or vice versa, without affecting the global system timing.

Let us consider the application of Rule 2 to the FIR filter DFG of Figure 8.11(a). The first cut is applied in Figure 8.13(a) where the DFG graph is cut into two distinct regions or sub-graphs: sub-graph #1 comprising nodes 1, 2, 3 and 4; and sub-graph #2 comprising 5 and 6. Since all edges between the regions are outbound from sub-graph #1 to sub-graph #2, a delay can be added to each. This gives Figure 8.13(b). The second cut splits the DFG

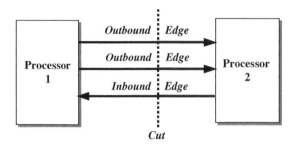

Figure 8.12 Cut-set theorem application

into sub-graph #3, comprising nodes 1, 2, 3 and 5, and sub-graph #4, comprising nodes 4 and 6. The addition of a single delay to this edge leads to the final pipelined design, as shown in Figure 8.11(c).

These rules provide a method of systematically adding, removing and distributing delays in an SFG and therefore adding, removing and distributing registers through-out a circuit, without changing the function. The cut-set retiming procedure is then employed, to cause sufficient delays to appear on the appropriate SFG edges, so that a number of delays can be removed from the graph edges and incorporated into the pro-cessing blocks, in order to model pipelining within the processors; if the delays are left on the edges, then this represents pipelining between the processors.

Of course, the selection of the original algorithmic representation can have a big impact on the resulting performance. Take, for example, the alternative version of the SFG shown initially in Figure 8.8(c) and represented as a DFG in Figure 8.14(a); apply-ing an initial cut-set allows pipelining of the multipliers as before, but now applying the cut-set between nodes 3 and 5, and nodes 4 and 6, allows the delay to be transferred, resulting in a circuit architecture with fewer delay elements as shown in Figure 8.14(c).

8.4.3 Application of Delay Scaling

In order to investigate delay scaling, let us consider a recursive structure such as the second-order IIR filter section given by

$$y(n) = a_0 x(n) + a_1 x(n-1) + a_2 x(n-2) + b_1 y(n-1) + b_2 y(n-2) \tag{8.2}$$

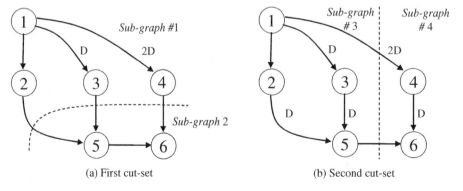

(a) First cut-set   (b) Second cut-set

Figure 8.13 Cut-set timing applied to FIR filter

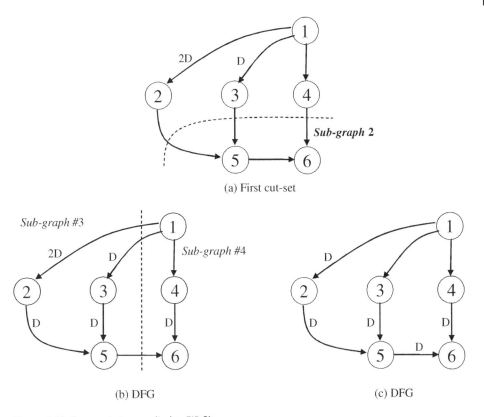

(a) First cut-set

(b) DFG

(c) DFG

Figure 8.14 Cut-set timing applied to FIR filter

The block diagram and the corresponding DFG is given in Figures 8.15(a) and 8.15(b), respectively. The target is to apply pipelining at the processor level, thereby requiring a delay D on each edge. The problem is that there is not sufficient delay in the $2 \rightarrow 3 \rightarrow 2$ loop to apply retiming. For example, if the cut shown in the figure were applied, this would end up moving the delay on edge $3 \rightarrow 2$ to edge $2 \rightarrow 3$. The issue is

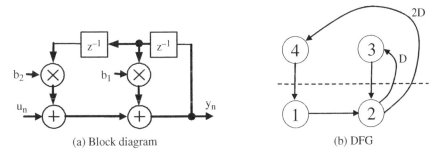

(a) Block diagram

(b) DFG

Figure 8.15 Second-order IIR filter

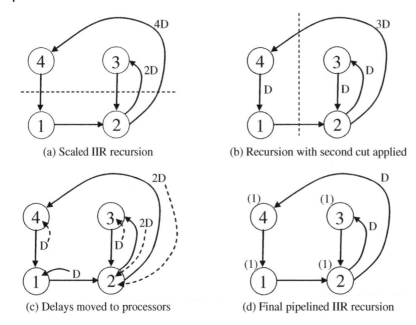

(a) Scaled IIR recursion

(b) Recursion with second cut applied

(c) Delays moved to processors

(d) Final pipelined IIR recursion

Figure 8.16 Pipelining of a second-order IIR filter. Source: Parhi 1999. Reproduced with permission of John Wiley & Sons.

resolved by applying time scaling, by working out the worse-case pipelining period, as defined by

$$\alpha_c = \frac{B_c}{D_c},\tag{8.3}$$

$$\alpha = \max \alpha_c.\tag{8.4}$$

In equation (8.3), the value B_c refers to the delays required for processor pipelining and the value D_c refers to the delays available in the original DFG. The optimal pipelining period is computed using equation (8.4) and is then used as the scaling factor. There are two loops as shown, giving a worst-case loop bound of 2. The loops are given in terms of unit time (u.t.) steps:

$1 \rightarrow 2 \rightarrow 4 \rightarrow 1\ (3u.t.)$

$2 \rightarrow 3 \rightarrow 2\ (2u.t.)$

$Loopbound\#1(3/2 = 1.5u.t.)$

$Loopbound\#2(2/1 = 2u.t.).$

The process of applying the scaling and retiming is given in Figure 8.16. Applying a scaling of 2 gives the retimed DFG of Figure 8.16(a). Applying the cut shown in the figure gives the modified DFG of Figure 8.16(b) which then has another cut applied, giving the DFG of Figure 8.16(c). Mapping of the delays into the processor and adding the numbers to show the pipelining level gives the final pipelined IIR recursion in Figure 8.16(d).

Table 8.4 Retiming performance in the Xilinx Virtex-5

Circuit	Area		Throughput	
	DSP48	Flip-flops	Clock (MHz)	Data rate (MHz)
Figure 8.15(b)	2	20	176	176
Figure 8.16(d)	2	82	377	188

The final implementation has been synthesized using the Xilinx Virtex-5 FPGA and the synthesis results can be viewed for the circuits of Figure 8.15(b) and Figure 8.16(d) in Table 8.4.

8.4.4 Calculation of Pipelining Period

The previous sections have outlined a process for first determining the pipelining period and then allowing scaling of this pipelining period to permit pipelining at the processor level. This is the finest level of pipelining possible within FPGA technology, although, as will be seen in Chapter 13, adding higher levels of pipelining can be beneficial for low-power FPGA implementations. However, the computation of the pipelining period was only carried out on a simple example of an IIR filter second-order section, and therefore much more efficient means of computing the pipelining period are needed. A number of different techniques exist, but the one considered here is the longest path matrix algorithm (Parhi 1999).

8.4.5 Longest Path Matrix Algorithm

A series of matrices is constructed and the iteration bound is found by examining the diagonal elements. If d is the number of delays in DFG, then create $L^{(m)}$ matrices, where $m = 1, 2, \ldots, d$, such that element $l^1_{1,j}$ is the longest path from delay element d which passes through exactly $m - 1$ delays (not including d_i and d_j); if no path exists, then $l^1_{i,j} = -1$. The longest path can be computed using the Bellman–Ford or Floyd–Warshall algorithm (Parhi 1999).

Example 1. Consider the example given in Figure 8.17. Since the aim is to produce a pipelined version of the circuit, we have started with the pipelined version indicated by the (l) expression included in each processor. This can be varied by changing the expression to (0) if the necessary pipelining is not required, or to a higher value, e.g.

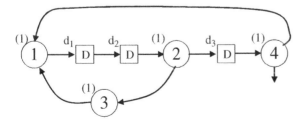

Figure 8.17 Simple DFG example (Parhi 1999)

(2) or (3), if additional pipelined delays are needed in the routing to aid placement and routing or for low-power implementation.

The first stage is to compute the $L^{(m)}$ matrices, beginning with $L^{(1)}$. This is done by generating each term, namely $l^1_{i,j}$, which is given as the path from delay d_i through to d_j. For example, d_1 to d_1 passes through either 1 ($d_1 \rightarrow d_2 \rightarrow 2 \rightarrow 3 \rightarrow 1 \rightarrow d_1$) or 2 delays ($d_1 \rightarrow d_2 \rightarrow 2 \rightarrow d_4 \rightarrow 1 \rightarrow d_1$), therefore $l^1_(1, 1) = -1$. For $l^1_{3,1}$, the path d_3 to d_1 passes through nodes (4) and (1), giving a delay of 2; therefore, $l^1_{3,1} = 2$. For $l^1_{2,1}$, the path d_2 to d_1 passes through nodes (2), (3) and (1), therefore $l^1_{2,1} = 3$. This gives the matrix

$$
\begin{pmatrix}
-1 & 0 & -1 \\
7 & -1 & 3 \\
3 & -1 & -1
\end{pmatrix}.
$$

The higher-order matrices do not need to be derived from the DFG. They can be recursively computed as

$$
l^{m+1}_{i,j} = \max_{k \in K}(-1, l^1_{i,j} + l^m_{k,j}),
$$

where K is the set of integers k in the interval $[1, d]$ such that neither $l^1_{i,k} = -1$ nor $l^m_{i,k} = -1$ holds. Thus for $l^2_{1,1}$ we can consider $K = 1, 2, 3$ but $K = 1, 3$ include -1, so only $K = 2$ is valid. Thus

$$
l^2_{1,1} = \max_{k \in 3}(-1, 0 + 7).
$$

The whole of $L^{(2)}$ is generated is this way as shown below:

$$
\underbrace{\begin{pmatrix}
-1 & 0 & -1 \\
7 & -1 & 3 \\
3 & -1 & -1
\end{pmatrix}}_{L^{(1)}}
\underbrace{\begin{pmatrix}
-1 & 0 & -1 \\
7 & -1 & 3 \\
3 & -1 & -1
\end{pmatrix}}_{L^{(1)}}
=
\underbrace{\begin{pmatrix}
7 & -1 & 3 \\
6 & 7 & -1 \\
-1 & 3 & -1
\end{pmatrix}}_{L^{(2)}}.
$$

While $L^{(2)}$ was computed using only $L^{(1)}$, the matrix $L^{(3)}$, is computed using both $L^{(1)}$ and $L^{(2)}$ as shown below, with the computation for each element given as

$$
l^3_{i,j} = \max_{k \in K}(-1, l^1_{i,j} + l^2_{k,j})
$$

as before. This gives the computation of $L^{(3)}$ as

$$
\underbrace{\begin{pmatrix}
-1 & 0 & -1 \\
7 & -1 & 3 \\
3 & -1 & -1
\end{pmatrix}}_{L^{(1)}}
\underbrace{\begin{pmatrix}
7 & -1 & 3 \\
6 & 7 & -1 \\
-1 & 3 & -1
\end{pmatrix}}_{L^{(2)}}
=
\underbrace{\begin{pmatrix}
6 & 7 & -1 \\
14 & 6 & 10 \\
10 & -1 & 6
\end{pmatrix}}_{L^{(3)}}.
$$

Once the matrix $L^{(m)}$ is created, then the iteration bound can be determined from the equation

$$T_\infty = \max_{i,m \in 1,2,\ldots,D} \left\{ \frac{l_{1,l}^m}{m} \right\}. \tag{8.5}$$

In this case, $m = 3$ as there are three delays, therefore $L^{(3)}$ represents the final iteration. For this example, this gives

$$T_\infty = \left\{ \frac{7}{2}, \frac{7}{2}, \frac{6}{3}, \frac{6}{3}, \frac{6}{3} \right\}.$$

Example 2. Consider the lattice filter DFG structure given in Figure 8.18(a). Once again, a pipelined version has been chosen by selecting a single delay (1) for each processor.

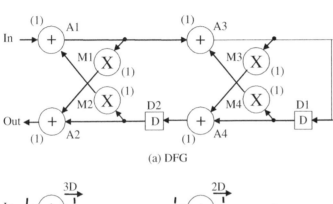

(a) DFG

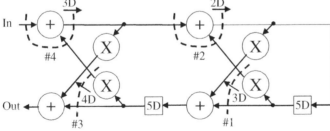

(b) DFG with cuts indicated

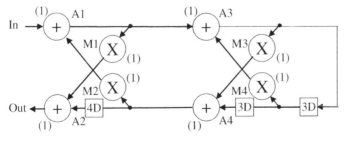

(c) Delays moved to processors

Figure 8.18 Lattice filter

The four possible matrix values are determined as follows:

$$D1 \rightarrow M3 \rightarrow A3 \rightarrow D1$$
$$D1 \rightarrow A4 \rightarrow D2 \text{ and } D1 \rightarrow M4 \rightarrow A3 \rightarrow M3 \rightarrow A4 \rightarrow D2$$
$$D2 \rightarrow M2 \rightarrow A1 \rightarrow A3 \rightarrow D1$$
$$D2 \rightarrow M2 \rightarrow A1 \rightarrow A3 \rightarrow M2 \rightarrow A4 \rightarrow D2,$$

thereby giving

$$\begin{pmatrix} 2 & 4 \\ 3 & 5 \end{pmatrix}.$$

The higher-order matrix L^2 is then calculated as shown below:

$$\underbrace{\begin{pmatrix} 2 & 4 \\ 3 & 5 \end{pmatrix}}_{L^{(1)}} \underbrace{\begin{pmatrix} 2 & 4 \\ 3 & 5 \end{pmatrix}}_{L^{(1)}} = \underbrace{\begin{pmatrix} 7 & 9 \\ 8 & 10 \end{pmatrix}}_{L^{(2)}}$$

This gives the iteration bound

$$T_\infty = \max_{i,m \in 1,2} \left\{ \frac{l^m_{1,l}}{m} \right\}. \tag{8.6}$$

For this example, this gives

$$T_\infty = \left\{ \frac{2}{1}, \frac{5}{1}, \frac{7}{2}, \frac{10}{2} \right\} = 5.$$

Applying this scaling factor to the lattice filter DFG structure of Figure 8.18(b) gives the final structure of Figure 8.18(c), which has pipelined processors as indicated by the (1) expression added to each processor. This final circuit was created by applying delays across the various cuts and applying retiming at the processor level to transfer delays from input to output.

8.5 Parallel Operation

The previous section has highlighted methods to allow levels of pipelining to be applied to an existing DFG representation, mostly based on applying processor-level pipelining as this represents the greatest level applicable in FPGA realizations. This works on the principle that increased speed is required, as demonstrated by the results in Table 8.4, and more clearly speed improvements with FIR filter implementations. Another way to improve performance is to parallelize up the hardware (Figure 8.19). This is done by converting the SISO system such as that in Figure 8.19(a) into a MIMO system such as that illustrated in Figure 8.19(b).

This is considered for the simple FIR filter given earlier. Consider the four-tap delay line filter given by

$$y(n) = a_0 x(n) + a_1 x(n-1) + a_2 x(n-2) + a_3 x(n-3). \tag{8.7}$$

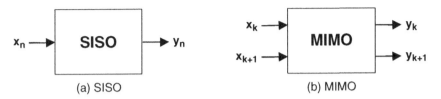

(a) SISO   (b) MIMO

Figure 8.19 Manipulation of parallelism

Assuming blocks of two samples per clock cycle, we get the following iterations performed on one cycle:

$$y(k) = a_0x(k) + a_1x(k-1) + a_2x(k-2) + a_3x(k-3),$$
$$y(k+1) = a_0x(k+1) + a_1x(k) + a_2x(k-1) + a_3x(k-2).$$

In these expressions, two inputs, $x(k)$ and $x(k+1)$, are processed and corresponding outputs, $y(k)$ and $y(k+1)$, produced at the same rate. The data are effectively being processed in blocks and so the process is known as block processing, where k is given as the block size. Block diagrams for the two cycles are given in Figure 8.20. Note that in these structures any delay is interpreted as being k delays as the data are fed at twice the clock rate. As the same data are required at different parts of the filter at the same time, this can be exploited to reduce some of the delay elements, resulting in the circuit of Figure 8.20(b).

The FIR filter has a critical path of $T_M + (N-1)T_A$ where N is the number of filter taps which determines the clock cycle. In the revised implementation, however, two samples are being produced per cycle, thus the throughput rate is $2/T_M + (N-1)T_A$. In this way, block size can be varied as required, but this results in increased hardware cost.

Parhi (1999) introduced a technique where the computation could be reduced by reordering the computation as

$$y(k) = a_0x(k) + a_2x(k-2) + z^{-1}(a_1x(k+1) + a_3x(k-1)).$$

By creating two tap filters, given as $y(1k) = a_0x(k) + a_2x(k-2)$ and $y(2k) = a_1x(k+1) + a_3x(k-1)$, we recast the expressions for $y(k)$ and $y(k+1)$ as

$$y(k) = y(1k) + z^{-1}(y(2(k+1))),$$
$$y(k+1) = (a_0 + a_1)(x(k+1) + x(k)) + (a_2 + a_3)(x(k-1) + x(k-2))$$
$$- a_0x(k) - a_1x(k+1) - a_2x(k-2) - a_3x(k-1).$$

This results in a single two-tap filter given in Figure 8.21, comprising a structure with coefficients $a_0 + a_1$ and $a_2 + a_3$, thereby reducing the complexity of the original four-tap filter. It does involve the subtraction of two terms, namely $y(k)$ and $y(2k+1)$, but these were created earlier for the computation of $y(k)$. The impact is to reduce the overall multiplications by two at the expense of one addition/subtraction. This is probably not as important for an FPGA implementation where multiplication cost is comparable to addition for typical wordlengths. More importantly, though, the top and bottom filters are reduced in length by $2(N/2)$ taps and an extra $2 - (N/2)$-tap filter is created to realize the first line in each expression. In general terms, filters have been halved, thus

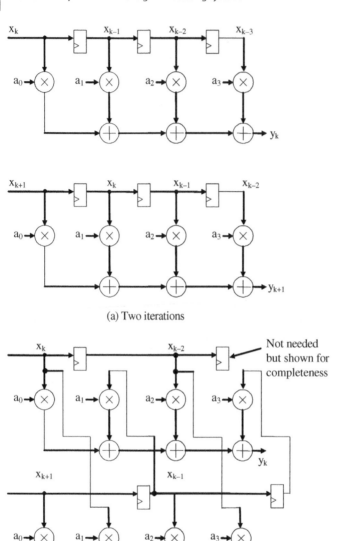

(a) Two iterations

(b) Combined operation

Figure 8.20 Block FIR filter

the critical path is given as $T_M + (N/2)T_A + 3T_A$ with three adders, one to compute $x(k) + x(k+1)$, one to subtract $y(1k)$ and one to subtract $y(2(k+1))$:

$$y(k+1) = (a_0 + a_1)(x(k+1) + x(k)) + (a_2 + a_3)(x(k-1) + x(k-2))$$
$$-y(1k) - y(2(k+1)).$$

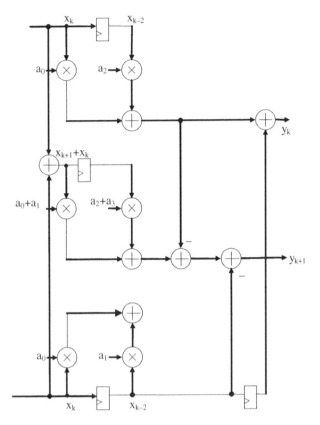

Figure 8.21 Reduced block-based FIR filter

8.5.1 Unfolding

The previous section indicated how we could perform parallel computations in blocks. Strictly speaking, this is known as a transformation technique called unfolding, which is applied to a DSP program to create a new program that performs more than one iteration of the original program. It is typically described using an unfolding factor J which describes the number of iterations by which it is unfolded. For example, consider unfolding the first-order IIR filter section, $y(n) = x(n) + by(n-1)$ by three, giving the expressions below:

$$y(k) = x(k) + by(k-1),$$
$$y(k+1) = x(k+1) + by(k),$$
$$y(k+2) = x(k+2) + by(k+1).$$

The SFG and DFG representation is given in Figure 8.22(a), where the adder is replaced by processor A and the multiplier by B. The unfolded version is given in Figure 8.22(b), where A_0, A_1 and A_2 represent the hardware for computing the three additions and B_0, B_1 and B_2 that for computing the three multiplications. With unlooped expressions, each delay is now equivalent to three clock cycles. For example, the previous value needed at

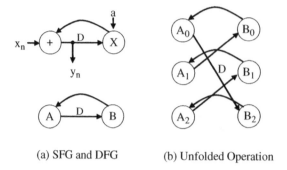

(a) SFG and DFG   (b) Unfolded Operation

Figure 8.22 Unfolded first-order recursion

processor B_0 is $y(n-1)$ which is generated by delaying the output of A_0, namely $y(n+2)$, by an effective delay of 3. When compared with the original SFG, the delays would appear to have been redistributed between the various arcs for $A_0 - B_0$, $A_1 - B_1$ and $A_2 - B_2$.

An algorithm for automatically performing unfolding is based on the fact that the kth iteration of the node $U(i)$ in the J-unfolded DFG executes the $J(k+i)$th iteration of the node U in the original DFG (Parhi 1999):

1. For each node U in the original DFG, draw the J nodes $U(0), U(1), \ldots, U(J-1)$.
2. For each edge $U \rightarrow V$ with ω delays in the original DFG, draw the J edges $U(i) \rightarrow V(i+\omega)/J$ with $(i+w\%J)$ delays for $i = 0, 1, \ldots, J-1$, where % is the remainder.

Consider the FIR filter DFG, a DFG representation of the FIR filter block diagram of Figure 8.23(a). Computations of the new edges in the transformed graphs, along with the computation of the various delays, are given below for each edge:

$$X0 \rightarrow A(0+0)\%2 = A(0), \quad \text{Delay} = \lfloor 0/2 \rfloor = 0$$
$$X1 \rightarrow A(1+0)\%2 = A(1), \quad \text{Delay} = \lfloor 1/2 \rfloor = 0$$
$$X0 \rightarrow B(0+1)\%2 = B(1), \quad \text{Delay} = \lfloor 1/2 \rfloor = 0$$
$$X1 \rightarrow B(1+1)\%2 = A(2), \quad \text{Delay} = \lfloor 2/2 \rfloor = 1$$
$$X0 \rightarrow C(0+2)\%2 = C(0), \quad \text{Delay} = \lfloor 2/2 \rfloor = 1$$
$$X1 \rightarrow C(1+2)\%2 = C(1), \quad \text{Delay} = \lfloor 3/2 \rfloor = 1$$
$$X0 \rightarrow D(0+3)\%2 = D(1), \quad \text{Delay} = \lfloor 3/2 \rfloor = 1$$
$$X1 \rightarrow D(1+3)\%2 = D(0), \quad \text{Delay} = \lfloor 4/2. \rfloor = 2$$

This gives the unfolded DFG of Figure 8.23(b) which equates to the folded circuit given in Figure 8.23(a).

8.5.2 Folding

The previous section outlined a technique for a parallel implementation of the FIR filter structure. However, in some cases, there is a desire to perform hardware sharing, i.e. folding, to reduce the amount of hardware by a factor, say k, and thus also reduce the sampling rate. Consider the FIR filter block diagram of Figure 8.24(a). By collapsing the

Figure 8.23 Unfolded FIR filter-block

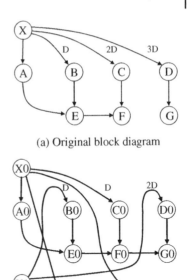

(a) Original block diagram

(b) Unfolded operation

filter structure onto itself four times, i.e. folding by four, the circuit of Figure 8.24(b) is derived. In the revised circuit, the hardware requirements have been reduced by four with the operation scheduled onto the single hardware units, as illustrated in Table 8.5.

The timing of the data in terms of the cycle number number is given by 0, 1, 2 and 3, respectively, which repeats every four cycles (strictly, this should by k, $k + 1$, $k + 2$ and

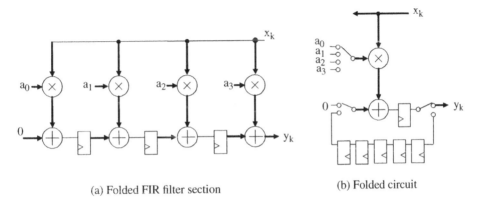

(a) Folded FIR filter section

(b) Folded circuit

Figure 8.24 Folded FIR filter section

Table 8.5 Scheduling for Figure 8.24(b)

Cycle clock	Adder input	Adder input	Adder output	System output
0	a_3	0	$a_3x(0)$	$y(3)'''$
1	a_2	0	$a_2x(0)$	$y(2)''$
2	a_1	0	$a_1x(0)$	$y(1)'$
3	a_0	0	$a_0x(0)$	$y(0)$
4	a_3	0	$a_3x(1)$	$y(4)'''$
5	a_2	$a_2x(1)$	$a_2x(1) + a_3x(0)$	$y(3)''$
6	a_1	$a_1x(1)$	$a_1x(1) + a_2x(0)$	$y(2)'$
7	a_0	$a_0x(1)$	$a_1x(1) + a_2x(0)$	$y(1)$
8	a_3	0	$a_3x(2)$	$y(5)'''$
9	a_2	$a_2x(1) + a_3x(0)$	$a_2x(1) + a_2x(1) + a_3x(0)$	$y(4)''$

$k + 3$). It is clear from the table that a result is only generated once every four cycles, in this case on the 4th, 8th, ..., cycle. The partial results are shown in brackets as they are not generated as an output. The expression $y(3)'''$ signifies the generation of the third part of $y(3)$, $y(3)''$ means the second part of $y(3)$, etc.

This folding equation is given by

$$D_F(U \overset{e}{\rightarrow} V) = Nw(e) - P_u + v - u, \tag{8.8}$$

where all inputs of a simplex component arrive at the same time and the pipelining levels from each input to an output are the same (Parhi 1999). In equation (8.8), $w(e)$ is the number of delays in the edge $U \overset{e}{\rightarrow} V$, N is the pipelining period, P_u is the pipelining stages of the H_u output pin, and u and v are folding orders of the nodes U and V that satisfy $0 \leq u, v \leq N - 1$. Consider the edge e connecting the nodes U and V with $w(e)$ delays shown in Figure 8.25(a), where the nodes U and V may be hierarchical blocks. Let the executions of the ith iteration of the nodes U and V be scheduled at time units $NL + u$ and $NL + v$ respectively, where u and v are folding orders of the nodes U and V that satisfy $0 \leq u, v \leq N - 1$.

The folding order of a node is the time partition to which the node is scheduled to execute in hardware (Parhi 1999). H_u and H_v are the functional units that execute the nodes U and V, respectively. N is the folding factor and is defined as the number of operations folded onto a single functional unit. Consider the lth iteration of the node U. If the H_u output pin is pipelined by P_u stages, then the result of the node U is available at the time unit $Nl + u + P_u$, and is used by the $(l + w(e))$th iteration of the node V. If the minimum value of the data time format of H_u input pin is A_v, this input pin of the node V is executed at $N(l + w(e)) + v + A_v$. Therefore, the result must be stored for $D_F(U \overset{e}{\rightarrow} V) = [N(l + w(e)) + v + A_v] - [Nl + P_u + A_v + u]$ time units. The path from H_u to H_v needs

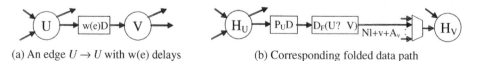

(a) An edge $U \rightarrow U$ with w(e) delays   (b) Corresponding folded data path

Figure 8.25 Folding transformation

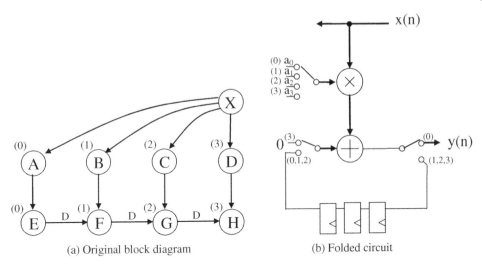

(a) Original block diagram   (b) Folded circuit

Figure 8.26 Folding process

$D'_F(U \xrightarrow{e} V)$ delays, and data on this path are inputs H_v at $Nl + v + A_v$, as illustrated in Figure 8.25(b). Therefore, the folding equation for hierarchical complexity component is given by

$$D_F(U \xrightarrow{e} V) = Nw(e) - P_u + A_v + v - u. \tag{8.9}$$

This expression can be systematically applied to the block diagram of Figure 8.25(a) to derive the circuit of Figure 8.25(b). For ease of demonstration, the DFG of Figure 8.26(a) is used. In the figure, an additional adder, H has been added for simplicity of folding. In Figure 8.26(a), we have used a number of brackets to indicate the desired ordering of the processing elements. Thus, the goal indicated is that we want to use one adder to implement the computations $a_3x(n)$, $a_2x(n)$, $a_1x(n)$ and $a_0x(n)$ in the order listed. Thus, these timings indicate the schedule order values u and v. The following computations are created as below, giving the delays and timings required as shown in Figure 8.26(a):

$$D_{F(A \to H)} = 4(0) - 0 + 0 - 0 = 0$$
$$D_{F(B \to E)} = 4(0) - 0 + 1 - 1 = 0$$
$$D_{F(C \to F)} = 4(0) - 0 + 3 - 3 = 0$$
$$D_{F(D \to G)} = 4(0) - 0 + 4 - 4 = 0$$
$$D_{F(H \to E)} = 4(1) - 0 + 1 - 2 = 3$$
$$D_{F(E \to F)} = 4(1) - 0 + 2 - 3 = 3$$
$$D_{F(F \to G)} = 4(1) - 0 + 3 - 4 = 3.$$

Figure 8.27(a) shows how a reverse in the timing ordering leads to a slightly different folded circuit in Figure 8.27(b) where the delays on the feedback loop have been changed and the timings on the multiplexers have also been altered accordingly. This example

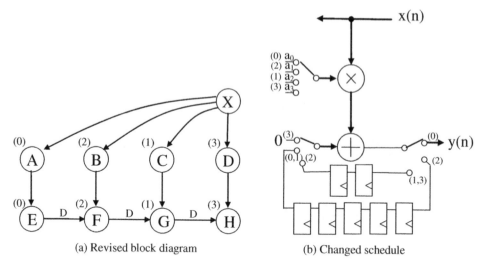

(a) Revised block diagram   (b) Changed schedule

Figure 8.27 Alternative folding

demonstrates the impact of changing the time ordering on the computation. The various timing calculations are shown below:

$$D_{F(A \to H)} = 4(0) - 0 + 0 - 0 = 0$$
$$D_{F(B \to E)} = 4(0) - 0 + 2 - 2 = 0$$
$$D_{F(C \to F)} = 4(0) - 0 + 1 - 1 = 0$$
$$D_{F(D \to G)} = 4(0) - 0 + 3 - 3 = 0$$
$$D_{F(H \to E)} = 4(0) - 0 + 0 - 2 = 2$$
$$D_{F(E \to F)} = 4(0) - 0 + 2 - 1 = 5$$
$$D_{F(F \to G)} = 4(0) - 0 + 1 - 3 = 2.$$

The example works on a set of order operations given as (1), (3), (2) and (4), respectively, and requires two different connections between adder output and input with different delays, namely 3 and 6.

The application of the technique becomes more complex in recursive computations, as demonstrated using the second-order IIR filter example given in Parhi (1999). In this example, the author demonstrates how the natural redundancy involved when a recursive computation is pipelined, can be exploited to allow hardware sharing to improve efficiency.

8.6 Conclusions

The chapter has briefly covered some techniques for mapping algorithmic descriptions, in the form of DFGs, into circuit architectures. The initial material demonstrates how we could apply delay scaling to first introduce enough delays into the DFGs to allow retiming to be applied. This translates to FPGA implementations where the number of registers can be varied as required.

In the design examples presented, a pipelining of 1 was chosen as this represents the level of pipelining possible in FPGAs at the processor level. However, if you consider the Xilinx DSP48E2 or the Altera DSP block as a single processing unit, these will allow a number of layers of pipelining as outlined in Chapter 5. Mapping to these types of processor can then be achieved by altering the levels of pipelining accordingly, i.e. by ensuring inter-iteration constraints on the edges which can then be mapped into the nodes to represent pipelining. The delays remaining on the edges then represent the registers needed to ensure correct retiming of the DFGs.

The chapter also reviews how to incorporate parallelism into the DFG representation, which again is a realistic optimization to apply to FPGAs, given the hardware resources available. In reality, a mixture of parallelism and pipelining is usually employed in order to allow the best implementation in terms of area and power that meets the throughput requirement.

These techniques are particularly suitable in generating IP core functionality for specific DSP functionality. As Chapter 11 illustrates, these techniques are now becoming mature, and the focus is moving to creating efficient system implementations from high-level descriptions where the node functionality may already have been captured in the form of IP cores. Thus, the rest of the book concentrates on this higher-level problem.

Bibliography

Altera Corp. 2014 Understanding peak floating-point performance claims. Techical White Paper WP-01222-1.0. Available from www.altera.com (accessed May 11, 2016)].

Leiserson CE, Saxe JB 1983 Optimizing synchronous circuitry by retiming. In *Proc. 3rd Caltech Conf. on VLSI*, pp. 87–116.

Kung SY 1988 *VLSI Array Processors*. Prentice Hall, Englewood Cliffs, NJ.

Monteiro J, Devadas S, Ghosh A 1993 Retiming sequential circuits for low power. *Proc. IEEE Int. Conf. on CAD*, pp. 398–402.

Parhi KK 1999 *VLSI Digital Signal Processing Systems: Design and Implementation*. John Wiley & Sons, New York.

9

Complex DSP Core Design for FPGA

9.1 Introduction

It is feasible to incorporate many billions of gates on a single chip, permitting extremely complex functions to be built as a complete SoC. This offers advantages of lower power, greater reliability and reduced cost of manufacture and has enabled an expansion of FPGA capabilities with devices such as Altera's Stratix® 10 and Xilinx's UltraScale™ FPGA families. With their vast expanse of usable logic comes the problem of implementing increasingly complex systems on these devices.

This problem has been coined the "design productivity gap" (ITRS 1999) and has increasingly become of major concern within the electronics industry. Whilst Moore's law predicts that the number of available transistors will grow at a 58% annual growth rate, there will only be a 21% annual growth rate in design productivity. This highlights a divergence that will not be closed by incremental improvements in design productivity. Instead a complete shift in the methodology of designing and implementing multi-million-gate chips is needed that will allow designers to concentrate on higher levels of abstraction within the designs.

As the silicon density increases, the design complexity increases at a far greater rate since silicon systems are now composed of more facets of the full system design and may combine components from a range of technological disciplines. Working more at the system level, designers become more heavily involved with integrating the key components without the freedom to delve deeply into the design functionality. Existing design and verification methodologies have not progressed at the same pace, consequently adding to the widening gap between design productivity and silicon fabrication capacity.

Testing and verification have become a major aspect of electronic design. Verification of such complex systems has now become the bottleneck in system-level design as the difficulties scale exponentially with the chip complexity. Design teams may often spend as much as 90% of their development effort on block or system-level verification (Rowen 2002). Verification engineers now often outnumber design engineers. There are many design and test strategies being investigated to develop systems to accelerate chip testing and verification. With the increasing level of components on a single piece of silicon

FPGA-based Implementation of Signal Processing Systems,
Second Edition. Roger Woods, John McAllister, Gaye Lightbody and Ying Yi.
© 2017 John Wiley & Sons, Ltd. Published 2017 by John Wiley & Sons, Ltd.

there is an increasing risk involved in the verification of the device. Added to this is the increased difficulty in testing design components integrated from a third party. So much more is at stake, with both time and monetary implications. The industry consensus on the subject is well encapsulated by Rowen (2002): "Analysts widely view earlier and faster hardware and software validation as a critical risk-reducer for new product development projects."

This chapter will cover the evolution of reusable design processes, concentrating on FPGA-based IP core generation. Section 9.2 discusses design for reuse, and Section 9.3 goes on to to talk about reusable IP cores. Section 9.4 discusses the evolution of IP cores, and Section 9.5 goes on to talk about parameterizable IP cores. Section 9.6 describes IP core integration and Section 9.7 covers current FPGA-based IP cores. Section 9.8 presents watermarking. Concluding comments are made in Section 9.9.

9.2 Motivation for Design for Reuse

There is a need to develop design and verification methodologies that will accelerate the current design process so that the design productivity gap will be narrowed (Bricaud 2002). To enable such an achievement, a great effort is needed to research the mechanics of the design, testing and verification processes, an area that to date has so often has been neglected. *Design for reuse* is heralded to be one of the key drivers in enhancing productivity, particularly aiding system-level design.

In addition to exponentially increased transistor counts, the systems themselves have become increasingly complex due to the combination of complete systems on a single device, with component heterogeneity bringing with it a host of issues regarding logic design and, in particular, testing and verification. Involving full system design means that designers need to know how to combine all the different components building up to a full system-level design. The sheer complexity of the full system design impacts the design productivity and creates ever more demanding time-to-market deadlines. It is a multidimensional problem trying to balance productivity with design issues such as power management and manufacturability.

Design productivity can be enhanced by employing design-for-reuse strategies throughout the entire span of the project development from initial design through to functional testing and final verification. By increasing the level of abstraction, the design team can focus on pulling together the key components of the system-level design, using a hierarchical design approach.

The 2005 International Technology Roadmap for Semiconductors report covers the need for design for reuse in great depth (ITRS 2005). To increase overall productivity and keep pace with each technology generation, the amount of reuse within a system design must increase at the same rate, and the level of abstraction must rise. A summary of one of the tables is given in Table 9.1. Productivity gains by employing reuse strategies for high-level functional blocks are estimated to exceed 200% (ITRS 2005). These reusable components need to be pre-verified with their own independent test harness that can be incorporated into the higher-level test environment. This can be achieved by incorporating IP cores from legacy designs or third-party vendors. The need for such cores has driven the growth of an IP core market, with ever greater percentages of chip components coming from IP cores.

Table 9.1 SoC design productivity trends (normalized to 2005)

	2012	2014	2016	2018	2020
Design needed to be reused (%)	58	66	74	82	90
Trend SoC total logic size	5.5	8.5	13.8	20.6	34.2
Required productivity for new designs	4.6	6.7	10.2	14.3	22.1
Required productivity for reused designs	9.2	13.5	20.4	28.6	44.2

In 2006, the ITRS reported that the percentage of logic from reused blocks was at 33%, and this figure is expected to reach 90% by 2020. It is hard to determine if this ambitious target will be achieved, but in 2016 the Design & Reuse website (http://www.design-reuse.com/) boasted 16,000 IP cores from 450 vendors. Thus there seems to be an active community involved in producing IP cores but, of course, this does not translate into reuse activity.

The discussion to date regarding design for reuse has focused on ASIC design. However, some of the key concerns are becoming increasingly relevant with FPGA design. With the onset of microprocessors and other additional auxiliary components on an FPGA, the drive is now for system-level design on a single device. With this advance comes the need to drive design reuse methodologies for FPGA technologies and to close the design productivity gap.

9.3 Intellectual Property Cores

One of the most favorable solutions for enhancing productivity is the strategy of using pre-designed functional blocks known as silicon IP cores. The term "IP core" applies to a range of implementations ranging from hard cores, which are given in the form of circuit layout, through to soft cores, which can be in the form of efficient code targeted at programmable DSP or RISC processors, or dedicated cores captured in an HDL.

The flexibility inherent in DSP processor solutions has often been cited as a key reason for their widespread use within industry, despite the obvious reduction in overall performance criteria such as speed, area and power. At the other end of the spectrum application-specific hardware designs provide unrivaled performance capabilities at the cost of design flexibility.

Design-for-reuse methodologies provide the flexibility allowing designs targeted to one project to be regenerated for another; the key is how to develop the initial design so that high performance can be obtained meeting the changing needs of the project specifications. Within the realms of ASIC and FPGA implementations, IP cores are often partitioned into three categories: hard, firm and soft. Hard IP refers to designs represented as mask layouts, whereas firm IP refers to synthesized netlists for a particular technology. Soft IP refers to the HDL version of the core that will have scalability and parameterization built in. For the latter, the term that has evolved is parameterizable IP. They can be designed so that they may be synthesized in hardware for a range of specifications and processes. For DSP applications parameters such as filter tap size, DCT point size, and wordlength may be made flexible (Ding *et al.* 1999). Parameters controlling these features would be fed into the code during the synthesis, resulting in the

Figure 9.1 Benefits of IP types

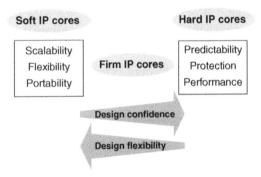

desired hardware for the application. There are advantages and disadvantages with each type of IP core, as shown in Figure 9.1.

Some flexibility can still be included from the outset in firm and hard IP devices. In these cases, the IP parameters that define the core are termed static IP (Junchao *et al.* 2001), whereby registers internal to the final design can be set to allow a multiplexing of internal circuits so as to reconfigure the functionality of the design. Reconfiguration has been a subject of great interest for FPGAs, particularly with their increasing capabilities (see Alaraje and DeGroat 2005).

In contrast, the IP parameters within soft IP cores are termed dynamic IP parameters. They are often local or global parameters such as data widths, memory sizes and timing delays. Control circuitry may also be parameterized, allowing scalability of the design. Parameters may be set to allow the same primary code to optimize for different target technologies from ASIC libraries to different FPGA implementations.

Many companies offer IP products based around DSP solutions, that is, where the IP code is embedded onto DSP processors. This offers full flexibility, but with the obvious reduction in performance in terms of area, power and speed. Texas Instruments and particularly ARM^{TM} are two examples of successful companies supplying chipsets with supporting libraries of embedded components.

In a similar manner, there are now many companies delivering firm and soft IP cores. Several FPGA companies not only sell the chips on which the user's designs can be implemented, but can also provide many of the fundamental building blocks needed to create these designs. The availability of such varied libraries of functions and the blank canvas of the FPGA brings great power to even the smallest design team. They no longer have to rely on internal experts in certain areas, allowing them to concentrate on the overall design, with the confidence that the cores provided by the FPGA vendors have been tested through use by previous companies. The following list of current IP vendors (Davis 2006) shows the diversity of IP products:

CEVA: The CEVA families of silicon IP cores are fully programmable low-power architectures for signal processing and communications (http://www.ceva-dsp.com/).
Barco-Silex: IP cores in RTL HDL form or netlist for cryptography functions including AES, Data Encryption Standard (DES) and hashing, public key and video products including JPEG 2000, JPEG, MPEG-2 and VC-2 LD (http://www.barco-silex.com).
OpenCores: OpenCores is the world's largest site for development of hardware IP cores as open source (www.opencores.org).

Digital Blocks: VHDL and Verilog core for various network functionality including UDP, IPv4, IPv6 and Transmission Control Protocol (TCP) (http://www.digitalblocks.com/).

Within hard IP cores, components can be further defined (Chiang 2001), although arguably the definitions could be applied across all variations of IP cores, with the presilicon stage relating more to the soft IP core and the production stage relating to a pre-implemented fixed design hard IP core:

Pre-silicon: Given a one-star rating if design verified through simulation.
Foundry verified: Given a three-star rating if verified on a particular process.
Production: Given a five-star rating if the core is production proven.

When developing IP, vendors often offer low-cost deals so as to attract system designers to use their new product and prove its success. Once silicon proven, the product offers a market edge over competing products.

9.4 Evolution of IP cores

As technology advances, the complexity of the granularity of the cores blocks increases. This section gives a summary of the evolution of IP cores.

Within the realms of ASICs, families of libraries evolved bringing a high level of granularity to synthesis. At the lowest level the libraries define gated functions and registers. With increased granularity, qualified functional blocks were available within the libraries for functions such as UARTs, Ethernet and USBs. Meanwhile, within the domain of DSP processors, companies such as TI were successfully producing software solutions for implementation on their own devices.

The development of families of arithmetic functions is where the role of IP cores in design for reuse for ASIC and FPGA designs came to play. It was a natural progression from the basic building blocks that supported ASIC synthesis. The wealth of dedicated research into complex and efficient ways of performing some of the most fundamental arithmetic operations lent itself to the design of highly sophisticated IP cores operating with appealing performance criteria.

Figure 9.2 illustrates the evolution of IP cores and how they have increased in complexity, with lower-level blocks forming key components for the higher levels of abstraction. The arithmetic components block shows a number of key mathematical operations, such as addition, multiplication and division, solved and implemented using the techniques described in Chapter 3. The list is far from conclusive.

With greater chip complexity on the horizon, arithmetic components became the building blocks for the next level in the complexity hierarchy, for designs such as filter banks consisting of a large array of multiply and accumulate blocks. This led to the development of fundamental DSP functions such as FFT and DCT. These examples are matrix-based operations consisting of a large number of repetitive calculations that are performed poorly in software. They may be built up from a number of key building blocks based on multiply and accumulate operations.

The structured nature of the algorithms lends itself to scalability, allowing a number of parameters to control the resulting architecture for the design. Obvious examples

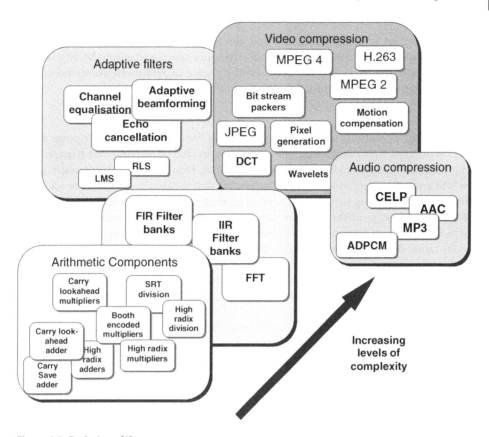

Figure 9.2 Evolution of IP cores

of parameters are wordlength and truncation. Other examples would be based on the dimensions of the matrix operations, relating, for example, to the number of taps on a filter. The work devoted to a single application could be expanded to meet the needs of a range of applications.

Other more complicated foundation blocks were developed from the basic arithmetic functions. More complicated filter-based examples followed such as adaptive filters implemented by the rudimentary LMS algorithm or the more extravagant QR-RLS algorithm (see Chapter 11). Highly mathematical operations lend themselves well to IP core design. Other examples, such as FEC chains and encryption, whereby there is a highly convoluted manipulation of values have also been immensely successful.

IP cores have now matured to the level of full functions that might previously have been implemented on independent devices. Again there is an increased level of complexity. Within image processing, the DCT is a key algorithm for JPEG and MPEG functions. Each of these will be covered in more detail below.

9.4.1 Arithmetic Libraries

Figure 9.2 lists a number of basic mathematical operations, namely addition, multiplication, division and square root. The efficient hardware implementation of even the most basic of these, addition, has driven an area of research, breaking down the operations to their lowest bit level of abstraction and cleverly manipulating these operations to enhance the overall performance in terms of area, clock speed and output latency (Koren 1993). This subsection gives some detail on the choice of arithmetic components and how parameters could be included within the code.

Fixed-Point and Floating-Point Arithmetic

The arithmetic operations may be performed using fixed-point or floating-point arithmetic. With fixed-point arithmetic, the bit width is divided into a fixed-width magnitude component and a fixed-width fractional component. Due to the fixed bit widths, overflow and underflow detection are vital to ensuring that the resulting values are accurate. Truncation or rounding would be needed to protect against such problems.

With floating-point arithmetic, the numbers are stored in a sign–magnitude format. The most significant bit represents the sign. The next component represents an exponential value. Biasing is used to enable the exponent to represent very small and very large number. The remaining data width is the mantissa, which represents the fractional component of the number and is given the boundaries of greater than or equal to 1 but less than 2. The greater flexibility of floating-point enables a wider range of achievable values.

Although number representation within the data width differs for fixed-point and floating-point design, there is overlap in how the main functionality of the operation is performed, as illustrated for multiplication in Figure 9.3; there has been research into automating the conversion from fixed-point to floating-point.

Addition, Multiplication, Division and Square Root

There has been an extensive body of work devoted to high-performance implementations of arithmetic components as indicated in Chapter 3. At was clear from the description given in Chapter 5, dedicated hardware functionality has been included in many

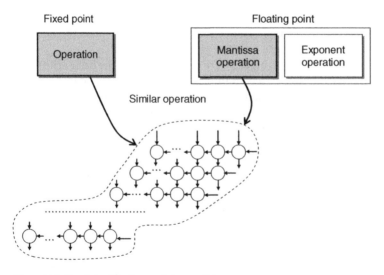

Figure 9.3 Fixed- and floating-point operations

FPGA families to support fixed-point addition/subtraction and multiplication and is supported in high- level synthesis tools. Division and square root are more complex techniques, and procedures for their implementation were described in Chapter 3. Most FPGA vendors supply IP libraries to support such functions.

9.4.2   Complex DSP Functions

More complex DSP cores can be created from lower-level arithmetic modules as illustrated in Figure 9.2, leading to the creation of systems for audio and video. For example, FFT cores can be used to create OFDMA systems and a power modulation/demodulation scheme for communications applications such as wireless (802.11a/g) or broadcasting (DVB-T/H). DCT and wavelet cores are used for a wide range of image processing cores, and LMS and RLS filtering cores applied to a range of adaptive beamformers and echo cancelation systems.

9.4.3   Future of IP Cores

As the level of abstraction within the core building blocks in designs increases, the role of the designer moves toward that of a system integrator, particularly with development using current FPGA devices enabling full system functionality on a single device. For the growth in IP core usage to continue, other aspects of the design flow will need to be addressed. This has driven developments in higher-level languages along with associated synthesis tools.

9.5   Parameterizable (Soft) IP Cores

This section covers the development of parameterizable IP cores for DSP functions. The starting point for the hardware design of a mathematical component may be the SFG representation of the algorithm. Here, a graphical depiction of the algorithm shows the components required within the design and their interdependence. The representation could be at different levels, from the bit-level arithmetic operations through to the cell-level functions.

Figure 9.4 shows the conventional design flow for a DSP-based circuit design, starting from the SFG representation of the algorithm. If a certain aspect of the specification were to be changed, such as wordlength, then the traditional full design flow would need to be repeated. The development of the IP core where the HDL is parameterized allows this flow to be dramatically altered, as shown in Figure 9.5.

The IP core design process needs to encompass the initial studies on data performance on the effects of wordlength and truncation, etc. Effort is needed to ensure that operation scheduling would still be accurate if additional, pipeline stages are included. The aim is for the parameterization of the core to lead seamlessly to a library of accurate cores targeted to a range of specifications, without the need to alter the internal workings of the code.

The system should effectively allow a number of parameters to be fed into the top level of the code. These would then be passed down through the different levels of abstraction of the code to the lowest levels. Obviously, considerable effort is needed at the architecture level to develop this parameterizable circuit architecture. This initial expense in

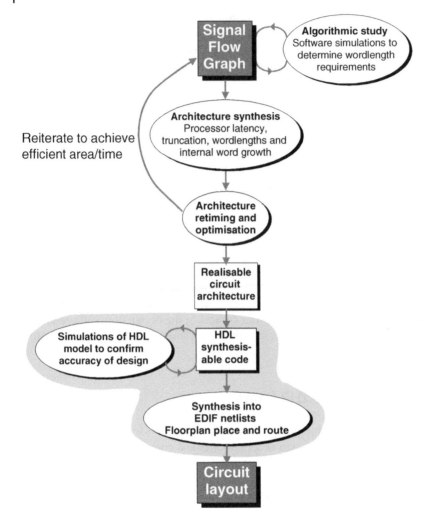

Figure 9.4 Circuit design flow

terms of time and effort undoubtedly hinders the expanded use of design-for-reuse principles. However, with this initial outlay great savings in company resources of time and money may be obtained. The choice of design components on which to base further designs and develop as IP is vitally important for this success. The initial expenditure must, in the long run, result in a saving of resources.

Future design engineers need to be taught how to encompass a full design-for-reuse methodology from the project outset to its close. The design process needs to consider issues such as wordlength effects, hardware mapping, latency and other timing issues before the HDL model of the circuit can be generated. The aspects that need to be considered create a whole new dimension to the design process, and designers need to keep in mind reusability of whatever they produce whether for development or test purposes.

If a design is developed in a parameterized fashion then initial analysis stages can be eliminated from the design flow, as illustrated in Figure 9.5, allowing additional circuits

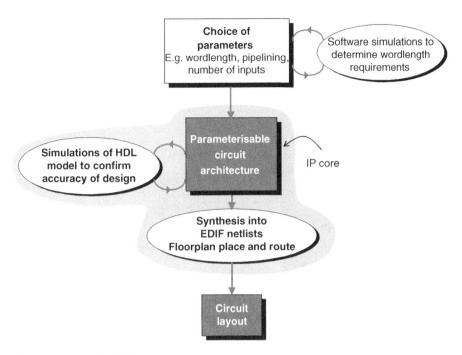

Figure 9.5 Rapid design flow

to be developed and floorplanned extremely quickly, typically in days as opposed to months. This activity represents a clear market for IP core developers (Howes 1998) as it can considerably accelerate the design flow for their customers. However, it requires a different design approach on behalf of the IP core companies to develop designs that are parameterizable and will deliver a quality solution across a range of applications.

9.5.1 Identifying Design Components Suitable for Development as IP

Within a company structure, it is vital that the roadmap is considered within the development of IP libraries as there is a greater initial overhead when introducing design-for-reuse concepts. Greater success can be achieved by taking an objective look at possible future applications so that a pipeline of developments can evolve from the initial ground work. If design for reuse is incorporated from the outset then there can be immense benefits in the development of a library of functions from the initial design.

It is often possible to develop a family of products from the same seed design by including parameterization in terms of wordlength and level of pipelining, and by allowing scalability of memory resources and inputs.

Larger designs may need to be broken down into manageable sections that will form the reusable components. This is particularly true for large design such as MPEG video compression whereby a range of different applications would require slightly different implementations and capabilities. By picking out the key components that remain unchanged throughout the different MPEG profiles and using these as the key hardware accelerators for all of the designs, vast improvements in time to market can be made. Furthermore, existing blocks from previous implementations have the advantage

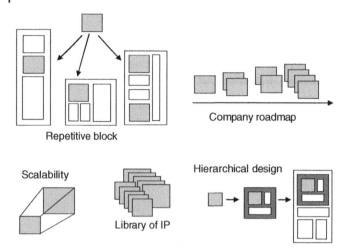

Repetitive block

Company roadmap

Scalability

Library of IP

Hierarchical design

Figure 9.6 Components suitable for IP

of having been fully tested, particularly if they have gone to fabrication or deployment on FPGA. Reusing such blocks adds confidence to the overall design. Existing IP cores may also form the key building blocks for higher-level IP design, creating a hierarchical design. These key points are illustrated in Figure 9.6.

9.5.2 Identifying Parameters for IP Cores

Identifying the key parameters when developing an IP core requires a detailed understanding of the range of implementations in which the core may be used. It is important to isolate what variables exist within possible specifications. The aim is to create as much flexibility in the design as possible, but only to the extent that the additional work will be of benefit in the long run. Overparameterization of a design affects not only the development but also the verification and testing time needed to ensure that all permutations of the core have been considered. In other words, consider the impact on the design time and design performance by adding an additional variable and weigh this up with thoughts on how the added flexibility will broaden the scope of the IP core.

Figure 9.7 lists some of example parameters: modules/architecture, wordlength, memory, pipelining, control circuitry, and test environment. Aspects such as wordlength or truncation can be parameterized. Other features can be used to allow full scalability, such as scaling the number of taps in an FIR filter. The diagram highlights the flexibility of allowing different modules depending on the application, or enabling the level of pipelining to be varied. Scalable parameters such as wordlength and level of pipelining affect the timing and the operations of the IP core and therefore need to be accounted for within initial development, so that the code can rapidly be re-synthesized for a new architecture. This a key factor for the success of an IP core.

It is crucial that the resulting core has performance figures (in terms of area, power and speed) comparable to a handcrafted design. As usual, the process comes down to a balance between time and money resources and the performance criteria of the core.

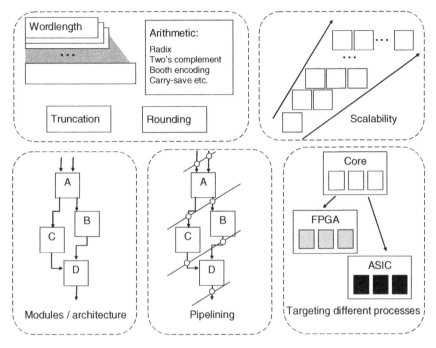

Figure 9.7 IP parameters

With time to market being a critical aspect to a product's success, the continuing use of IP components within the electronics industry has major benefits.

With further parameterization, cores can be developed to support a range of different technologies enabling the same design to be re-targeted from ASIC to FPGA, as highlighted in Figure 9.7. Allowing the same code to be altered between the two technologies has the obvious advantage of code reuse; however, it also allows for a verification framework whereby cores are prototyped on FPGA and then the same code is re-targeted to ASIC. There is obviously no guarantee that the code conversion from FPGA to ASIC implementations will not in itself incur errors. However, the ability to verify the code on a real-time FPGA platform brings great confidence to the design process and enables even the functional design to be enhanced to better meet the needs of the specification.

Consideration must be given to the level of parameterization as it makes the design more flexible and widens the market potential for the IP core. Gajski *et al.* (2000) highlight the issue of overparameterization, as increasing the number of variables complicates the task of verifying the full functionality of each permutation of the design. There is also the aspect that designs have been made so generic that they may not match the performance requirements for a specific application. Gajski *et al.* argue that increasing the number of parameters decreases the quality and characterizability of the design, that is to say, how well the design meets the needs of the user. There are also the added complications with verification and testing. These points are highlighted in Figure 9.8.

An obvious parameter is wordlength, which ultimately represents the trade-off between SNR and performance criteria such as area and critical path. Figure 9.9 gives an illustration of such analysis by plotting SNR against a range of wordlengths. It can

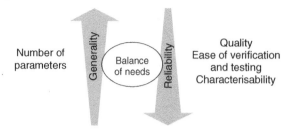

Number of parameters

Generality

Balance of needs

Reliability

Quality
Ease of verification
and testing
Characterisability

Figure 9.8 Effect of generalization on design reliability

be seen that increasing the wordlength does not significantly improve the overall performance. For addition an increase of one bit will linearly scale the area of the resulting design, whereas it has an exponential effect for multiplication and division. As with area, additional bits will affect critical path, possibly resulting in the need to introduce further pipeline stages.

Existing designs may have relied on carefully crafted libraries of arithmetic functions that were scalable in terms of bit width and level of pipelining, providing optimum performance in terms of area and speed. However, the impact of introducing processing blocks has a granular impact on area and performance when adjusting wordlengths. Obviously, there will be a need to add parameters to allow the wordlengths to be varied from the module boundary without having to manually edit the code.

Memory will also need to be scalable to account for the different wordlengths, but also for variations in the number of inputs or stored values. In addition, flexibility will need to be included within the code to allow different types of memory blocks to be employed in accordance with the choice of target technology.

In Verilog, one of two solutions can be used. Either instantiations of BRAMs for the target device can be scripted with DEFINEs at the top level of the code pointing to the memory of choice. Alternatively, the code can be written in such a way as to "imply" the application of a memory, which will be picked up during synthesis and will instantiate the memories accordingly. However, slight improvements may be still be obtained if the memory instantiations are hand-crafted but this will result in more complex code.

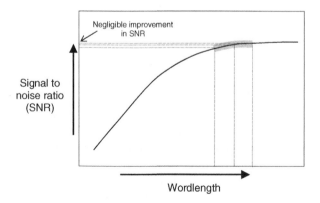

Negligible improvement
in SNR

Signal to
noise ratio
(SNR)

Wordlength

Figure 9.9 Wordlength analysis

Allowing flexibility in the data rate and associated clock rate performance for an application requires the ability to vary the level of pipelining to meet critical path requirements. This may tie in to changes in the wordlength and often a variation in the number if pipeline stages are to be part of the reusable arithmetic cores. Obviously, an increase or decrease in the number of pipeline cuts in one module will have a knock-on effect on timing for the associated modules and the higher level of hierarchy as discussed in Chapters 8 and 10.

Thus, control over the pipeline cuts within the lower-level components must be accessible from the higher level of the module design, so that lower-level code will not need to be edited manually. Control circuitry is coupled strongly with the level of pipelining and therefore must include some level of scalability to ensure that the design is totally parameterizable.

An alternative method of developing parameterizable cores can be used to develop a software code to automate the scripting of the HDL version of the module. This is particularly useful with Verilog as it does not have the same flexibility in producing scalable designs as VHDL does.

Parameterized Design and Test Environment

All associated code accompanying the IP core should be designed with scalability in mind. Bit-accurate C-models used for functional verification should have the capability to vary bit widths to match the IP core. For cycle accurate testing, the timing must also be considered. Testbenches and test data derivation are also required to be parameterizable, allowing for a fully automation generation of an IP core and it associated test hardness. The use of software such as a C-model to generate the test hardness and test data files may be advantageous in the development of the IP core. This is illustrated in Figure 9.10.

9.5.3 Development of Parameterizable Features

Many of the IP designs applied for ASIC design can be expanded for FPGA implementations. Each family of devices has its own memory components and methods for instantiating the built-in modules. The principle would be to design the code so as to allow the core to be re-targeted at the top level to the family FPGA devices of choice. This is particularly important as FPGAs are rapidly progressing, thus legacy code needs to accommodate additions for future devices and packages.

Arithmetic Block Instantiation

One example of the variations between FPGA devices is memory blocks. Each family has its own architecture for these blocks as outlined in Chapter 5. They can either be instantiated directly, or the memories can be inferred by synthesis tools. The latter allows the synthesis tool to pick up the memory blocks directly from the library and map the register values to this memory or even to ROM blocks. This has obvious benefits in that the code does not become FPGA family-specific.

There may still be a benefit in manually instantiating the memory blocks as a slight improvement in usage of the blocks can sometimes be achieved. However, the code is specified for the target device.

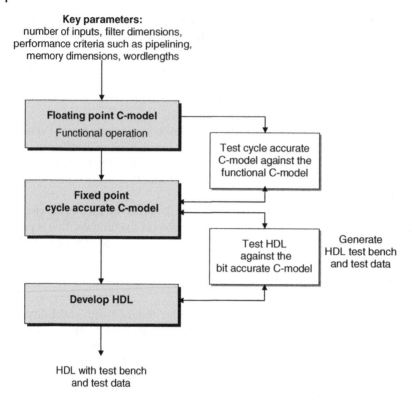

Figure 9.10 Design flow

Arithmetic Block Instantiation

Different target FPGAs may have variants of arithmetic operations available for the user. Typically the FPGA will contain a number of high-performance DSP blocks for instantiation. If the code is to be employed over a range of FPGA families and even between FPGA and ASIC, then there needs to be a facility to define the operator choice at the top level. Within Verilog, this would be done through the use of DEFINEs held in a top-level file, allowing the user to tailor the design to their current requirements.

9.5.4 Parameterizable Control Circuitry

For complex modules, there may be a need to allow for scalable control circuitry, i.e. a framework that will allow for the changes in parameters, such as the knock-on effect from additional pipelining delays. Any increase in the number of inputs or wordlength may have an effect on the scheduling and timing of the module. It may be possible to develop the control circuitry to cope with these variations.

9.5.5 Application to Simple FIR Filter

This section concludes with an example of parametric design applied to a simple FIR filter. The key parameters for the design will be highlighted and suggestions made

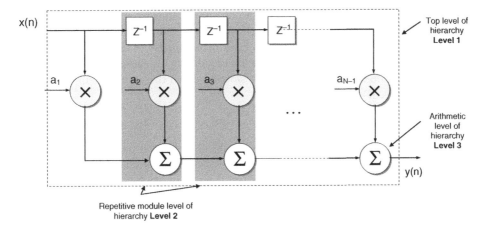

Figure 9.11 FIR filter hierarchy example

concerning how they can be incorporated into the code. Recall the difference equation for an FIR filter (equation (4.2)).

In this example, the FIR filter has three levels of hierarchy as depicted in Figure 9.11:

- Level 1: This is the top level of the filter structure with input $x()$ and output $y(n)$.
- Level 2: The top level of the FIR filter can be composed of a single DSP processing block to compute $a_0x(n)$ followed by an addition.
- Level 3: This is the arithmetic operation level, consisting of a single multiply, add and delay modules.

Another dimension of the design may be the folding of the FIR operations onto a reduced architecture as described in Chapter 8 where the hardware modules (shown as level 2 in Figure 9.11) are reused for different operations within the filter. Of course, multiplexers and programmable register blocks need to be added, and in this example all the MAC operations are performed on one set of multiplier and adder modules. The multiplexers are used to control the flow of data from the output of the MAC operations and back into the arithmetic blocks. The choice of level of hardware reduction will depend on the performance requirements for the application.

9.6 IP Core Integration

One of the key challenges of successful design reuse is with the integration of the IP cores within a user's system design. This can often be a stumbling block within a development. Investigations have been carried out to highlight these issues (Gajski *et al.* 2000), and guidelines have been set out to try to standardize this process (Birnbaum 2001; Coussy *et al.* 2001).

For the successful integration of an IP core into a current design project, certain design strategies must be employed to make the process as smooth as possible. This section highlights some of the pitfalls that might be met and provides some guidance when dealing with IP core sources externally to the design team, whether within or outside the company.

One of the considerations that may need to be addressed is whether to outsource IP components externally, or even source the IP from within a company or organization but from different departments. Successful intra-company use of IP requires adequate libraries and code management structures. Incorporating IP components from other design teams can often be the main barrier slowing down the employment of design-for-reuse strategies in system-level design.

9.6.1 Design Issues

Greater success can be obtained by taking an objective look at possible future applications so that a pipeline of developments can evolve from the initial ground work. If design for reuse is incorporated from the outset then there can be immense benefits in the development of a library of functions from the initial design.

- Need to determine the parts of the design that will be useful in future developments.
- What are possible future applications?
- Study the roadmap for the product.
- Is there a possibility of development of a family of products from the same seed design?
- How can a larger design be partitioned into manageable re-usable sections?
- Find existing level of granularity, i.e. is there any previous IP available that could provide a starting level for development?

Outsourcing IP

One of the limiting factors of using outsourced IP is the lack of confidence in the IP. The IP can be thought of as having different grades, with one star relating to a core verified by simulation, and three stars relating to a core that has been verified through simulation on the technology (i.e. a gate-level simulation). A five-star IP core provides the most confidence as it has been verified through implementation (Chiang 2001).

FPGA vendors have collaborated with the IP design houses to provide a library of functions for implementation on their devices. The backing of the FPGA vendors brings a level of confidence to the user. The aspect of core reliability is not as crucial for FPGA as it is for ASIC. However, it is still important. Time wasted on issues of IP integration into the user's product may be critical to the success of the project.

Certain questions could be answered to help determine the reliability of an IP vendor:

- Has the core been in previous implementations for other users?
- Do the company supply user guide and data book documentation?
- Does the core come supplied with its own testbench and some test data?
- Will the company supply support with the integration, and will this incur an added cost?

In-house IP

For a small company within the same location, it would be a much easier task to share and distribute internal IP. However, this task is logistically difficult for larger companies spanning a number of locations, some of which may be affected by time zones as well as physical distance.

It would be wise for the company to introduce a structure for IP core design and give guidelines on the top-level design format. Stipulating a standard format for the IP cores could be worthwhile and create greater ease of integration. Forming a central repository for the cores once they have been verified to an acceptable level would be necessary to enable the company's full access to the IP. Most companies already employ some method of code management to protect their products.

Interface Standardization

IP integration is a key issue as it ensures IP reuse and core configurability. IP cores are becoming more complex and configurable and can have numerous ports and hundreds of different configurations. Thus the provision of standardized interfaces is vital and the IP-XACT standard provides a mechanism for standardizing IP interfaces. It is now an IEEE standard (IEEE 2014) and is a mechanism to express and exchange information about design IP and its required configuration.

IP-XACT was developed by the Spirit consortium to enable sharing of standard component descriptions from multiple component vendors (Murray and Rance 2015). It defines an XML schema that is very easy to process and has the ability to make IP more "integration-ready" through interface standardization. It is argued that it can result in a 30% improvement in the time and cost of SoC integration.

It allows the creation of an interface on the component that contains a well- known set of ports called a bus interface; it can generally have high- level transactional or dataflow characteristics and behave as master or slave and also have different variants like direction and size. Once defined, these bus definitions can be used in conjunction with IP-XACT component descriptions to describe hardware interfaces which define the physical ports and allow mapping of these ports to the standardized definition.

Once this mapping has been defined, it is a case of checking that all of the required ports in a bus definition have been mapped, all directions are correct, all port widths are consistent with the bus definition, and there is no illegal mapping of ports.

9.7 Current FPGA-based IP cores

There are a number of cores available, both open source and commercial offerings. The open source cores tend to be available from the OpenCores website (opencores.org) and commercial offerings are available from a range of sites (see www.design-reuse.com/). The FPGA vendors also have their own IP repositories.

- Xilinx through LogiCore and their partner offer IP cores for DSP and math, embedded communications, memory interfaces and controllers and video and imaging.
- Altera's MegaCore® outlines cores for some FIR filters, FFTs, DRAM and SRAM controllers and their third-party providers offer a wide range of cores for communications, interfaces, DSP and video processing.
- Microsemi CompanionCore's portfolio offers a comprehensive collection of data security, image and vision processing, communications and processors, bus interfaces and memory controller cores.
- Lattice Semiconductor's IP portfolio comprises cores for DSP, Ethernet, PCI Express and video processing and display.

The FPGA companies aim to develop IP core technology to ensure a better relationship with their customer base and may look to provide this IP to ensure FPGA sales.

Thus in many cases, the IP requirements tend to be driven by the customer base. Also, FPGA companies will sometimes provide IP through their university programmes to stimulate technical interest in this areas.

9.8 Watermarking IP

Protecting FPGA IP has become a key research area both in terms of protecting the investment that the IP vendors have made but also as a reassurance to their customers that their investment was indeed worthwhile and has not been tampered with. For these reasons, an increasing amount of attention has been paid to developing techniques for ensuring this protection (Teich and Ziener 2011).

One solution is to hide a unique signature in the core, essentially termed *watermarking*, although there are also techniques for validating the core with no additional signature (Teich and Ziener 2011). Identification methods are based on the extraction of unique characteristics of the IP core, e.g. LUT contents for FPGA IP cores allowing the core author to be identified.

The concept of digital watermarking FPGA was first proposed by Lach *et al.* (1998). The owner's digital signature is embedded into an unused LUT located in a constrained area of unused slices in the FPGA at the place and route level of the implementation. This area is then obfuscated in the design using unused interconnect and "don't care" inputs of neighboring LUTs. The approach uses additional area and may impact timing and be vulnerable to attacks that look to remove the signature.

An alternative approach in Jain *et al.* (2003) embeds the watermark at the place and route stage by modifying the non-critical path delay between non-synchronous registers. It does not need additional hardware resources but can impact the path delay, and thus the performance of the design. The DesignTag is a novel, patented, security tag by Kean *et al.* (2008) which is used to verify the authenticity of a semiconductor device. It comprises a small, digital circuit which communicates through the package with an external sensor.

9.9 Summary

This chapter began by highlighting the need for design for reuse to address the challenges of building increasingly complex SoC devices. The increasing levels of silicon technology have stressed the need to reuse good designs from previous projects.

Design for reuse has been achieved by the creation of IP cores either in the form of pre-designed functional layout such as the ARM cores which present the user with a hardware platform on which they can develop software to implement the required functionality, or parameterized HDL code which can produce highly efficient code for programmable logic implementation. The aim of the HDL code is to capture the good design practice and procedures in such a way that HDL code is provided with a series of parameters which can be set and produce efficient implementation across a range of performance needs.

The process requires the creation of a base design from which a range of implementations can be derived where the area and speed will scale with change in parameters, otherwise it is frustrating for the designer to optimize the parameters for the best design. This process is demonstrated in detail in Chapter 11 for a QR-based RLS filter.

Bibliography

Alaraje N, DeGroat JE 2005 Evolution of reconfigurable architectures to SoFPGA. In *Proc. 48th Midwest Symp. on Circuits and Systems*, 1, pp. 818–821.

Birnbaum M 2001 VSIA quality metrics for IP and SoC. In *Proc. Int. Symp. on Quality Electronic Design*, pp. 279–283.

Bricaud P 2002 *Reuse Methodology Manual for System-On-A-Chip Designs*. Springer, New York.

Chiang S 2001 Foundries and the dawn of an open IP era. *Computer*, 34(4), 43–46.

Coussy P, Casseau E, Bomel P, Baganne A, Martin E. 2006. A formal method for hardware IP design and integration under I/O and timing constraints. *ACM Trans. Embedded Computing Systems*, 5(1), 29–53.

Davis L 2006 Hardware components, semiconductor-digital-programmable logic IP cores. Available from www.interfacebus.com/IPCoreVendors.html (accessed May 11, 2016).

Ding TJ, McCanny JV, Hu Y 1999 Rapid design of application specific FFT cores. *IEEE Trans. Signal Processing*, 47(5), 1371–1381.

Gajski DD, Wu ACH, Chaiyakul V, Mori S, Nukiyama T, Bricaud P 2000 Essential issues for IP reuse. In *Proc. Design Automation Conf.*, pp. 37–42.

Howes J 1998 IP new year. *New Electronics*, 31(1), 41–42.

IEEE 2014 IEEE Standard for IP-XACT, Standard Structure for Packaging, Integrating, and Reusing IP within Tool Flows. Available from http://standards.ieee.org/ (accessed May 11, 2016).

ITRS 1999 International Technology Roadmap for Semiconductors, Semiconductor Industry Association. http://public.itrs.net (accessed May 11, 2016).

ITRS 2005 International Technology Roadmap for Semiconductors: Design. available from http://www.itrs.net/Links/2005ITRS/Design2005.pdf (accessed May 11, 2016).

Jain AK, Yuan L, Pari PR, Qu G 2003 Zero overhead watermarking technique for FPGA designs. In *Proc. 13th ACM Great Lakes Symp. on Very Large Scale Integration*, pp. 147–152.

Junchao Z, Weiliang C, Shaojun W 2001 Parameterized IP core design. In *Proc. 4th Int. Conf. on Application Specific Integrated Circuits*, pp. 744–747.

Kean T, McLaren D, Marsh C 2008 Verifying the authenticity of chip designs with the DesignTag system. In *Proc. IEEE Int. Workshop on Hardware-Oriented Security and Trust*, pp. 59–64.

Koren I 1993 *Computer Arithmetic Algorithms*. Prentice Hall, Englewood Cliffs, NJ.

Lach J, Mangione-Smith WH, Potkonjak M 1998 Signature hiding techniques for FPGA intellectual property protection. In *Proc. IEEE/ACM Int. Conf. on Computer-Aided Design*, pp. 186–189.

Murray D and Rance S 2015 Leveraging IP-XACT standardized IP interfaces for rapid IP integration. ARM White paper. Available from https://www.arm.com/ (accessed May 11, 2016).

Rowen C 2002 Reducing SoC simulation and development time. *Computer*. 35(12), 29–34.

Teich J, Ziener D 2011 Verifying the authorship of embedded IP cores: Watermarking and core identification techniques. Plenary talk. *Int. Conf. on Engineering of Reconfigurable Systems and Algorithms*. Available at http://ersaconf.org/ersa11/program/teich.php

10

Advanced Model-Based FPGA Accelerator Design

10.1 Introduction

As described in Chapter 8, architectural synthesis of SFG models is a powerful approach to the design of high-throughput custom circuit accelerators for FPGA. This approach is one particular case of a wider trend toward design of high-performance embedded systems via the use of a model of computation (MoC), where a domain-specific modeling language is used to express the behavior or a system such that it is semantically precise, well suited to the application at hand and which emphasizes characteristics of its behavior such as timeliness (how the system deals with the concept of time), concurrency, liveness, heterogeneity, interfacing and reactivity in a manner that may be readily exploited for efficient implementation.

A plethora of MoCs have been proposed for modeling of different types of system (Lee and Sangiovanni-Vincentelli 1998), and determining the appropriate MoC for certain types of system should be based on the specific characteristics of that system. For instance, a general characterization of DSP systems could describe systems of repetitive intensive computation on streams of input data. Given this characterization, the dataflow MoC (Najjar *et al.* 1999) has been widely adopted and is a key enabling feature of a range of industry-leading design environments, such as National Instruments' LabVIEW and Keysight Technologies' SystemVUE.

This chapter addresses dataflow modeling and synthesis approaches for advanced accelerator architectures which fall into either of two classes. The first is that of *multidimensional* accelerators: those which operate on complex multidimensional data objects, or multiple channels of data. The second focuses on accelerators with an issue largely ignored by the SFG synthesis techniques of Chapter 8, where it is a heavy demand for high- capacity memory resource which must be accessed at a high rate.

The dataflow modeling of DSP systems is the subject of Section 10.2. The synthesis of custom accelerators is covered in Section 10.3, and this is extended to multidimensional versions in Section 10.4. Memory-intensive accelerators are covered in Section 10.5. A summary is given in Section 10.6.

FPGA-based Implementation of Signal Processing Systems,
Second Edition. Roger Woods, John McAllister, Gaye Lightbody and Ying Yi.
© 2017 John Wiley & Sons, Ltd. Published 2017 by John Wiley & Sons, Ltd.

10.2   Dataflow Modeling of DSP Systems

10.2.1   Process Networks

The roots of the most popular current dataflow languages lie in the Kahn process network (KPN) model (Kahn 1974). The KPN model describes a set of parallel processes (or "computing stations") communicating via unidirectional FIFO queues – the general structure of a KPN is shown in Figure 10.1. A computing station maps streams of data tokens impinging along its input lines, using localized memory, onto streams on its output lines.

In DSP systems, the tokens are usually digitized input data values. Continuous input to the system generates streams of input data, prompting the computing stations to produce streams of data on the system outputs. The semantics of mapping between streams of data in KPN makes this modeling approach a good match with the behavior of DSP systems. A KPN structure can be described as a graph $G = (V, E)$, where V is a set of vertices (the computing stations) and E a set of directed edges connecting the vertices. An edge connecting source and sink computing stations a and b respectively is uniquely identified using the tuple (a, b).

Lee and Parks (1995) developed this modeling framework further into the dataflow process network (DPN) domain. DPN models augment KPN computing stations with semantics which define how and under what conditions mapping between streams occurs. Specifically, a stream is said to be composed of a series of data tokens by invocation or *firing* of a dataflow actor; tokens input to an actor are translated to tokens output. Firing only occurs when one of a series of rules is satisfied. Each rule defines a pattern, such as the available number of tokens at the head of an edge FIFO, and when the pre-specified pattern for each input edge is satisfied, the actor may fire. When it does so, tokens are *consumed* from incoming edge FIFOs and resulting tokens *produced* on outgoing edges. Via repeated firing, each actor maps a succession, or a stream, of tokens on its input edges to streams on its output edges. Combined, the KPN and DPN models provide a functional modeling foundation with important properties, such as determinism (Lee and Parks 1995), a foundation upon which a series of more refined dataflow dialects have been devised. Three refinements of specific importance in this section are synchronous dataflow (SDF), cyclo-static dataflow (CSDF) and multidimensional synchronous dataflow (MSDF).

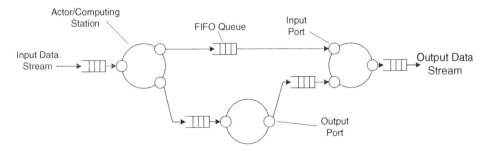

Figure 10.1 Simple KPN structure

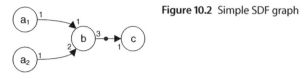

Figure 10.2 Simple SDF graph

10.2.2 Synchronous Dataflow

An SDF model is a DPN with highly restricted semantics. It specifies, for each actor, a single firing rule which states as a condition for actor firing a fixed, integer number of tokens required on its incoming edges (Lee and Parks 1995). Hence SDF is a domain where "we can specify a priori the number of input samples consumed on each input and the number of output samples produced on each output each time the block is invoked" (Lee and Messerschmitt 1987a). This restriction permits compile-time graph analysis with three powerful capabilities:

1. **Consistency**: It can be determined whether a graph is consistent, i.e. whether a program realizing the graph can be constructed which operates on infinite input streams of data within bounded memory.
2. **Deadlock Detection**: It may be determined whether a program realizing the graph operates without deadlock .
3. **Compile-Time Optimization**: Not only can a program implementing the graph be constructed in compile time, the schedule can be analyzed and optimized as regards, for example, buffer and code memory costs or communications costs (Bhattacharyya *et al.* 1999; Sriram and Bhattacharyya 2000).

These capabilities allow compile-time derivation of very low-overhead, efficient programs realizing the SDF model whose buffer memory cost may be highly tuned to the target platform. This capability has pioneered a large body of research into dataflow system modeling, analysis and implementation techniques (Bhattacharyya *et al.* 1999; Sriram and Bhattacharyya 2000). However, this advantage is gained at the expense of expressive power since the SDF forbids data-dependent dataflow behavior.

Each SDF actor exhibits a set of ports, via which it connects to and exchanges tokens with an edge. The number of tokens consumed or produced at a port for each firing of the actor is known as that port's *rate, r*. This value is quoted adjacent to the port, as illustrated in Figure10.2.[1] When all ports in the graph are equi-rate, the graph is known as a single-rate or *homogeneous* or single-rate dataflow graph (SR-DFG). Otherwise, the DFG is known as a multi-rate dataflow graph (MR-DFG). A simple SDF model is shown in Figure 10.2. Note the black dot on the edge (b, c); this denotes a delay, which in dataflow terms represents an initial token, i.e. a token resident in the inferred FIFO before any has been produced by the source actor.

If, for actor j connected to edge i, x_j^i (y_j^i) is the rate of the connecting port, an SDF graph can be characterized by a topology matrix Γ, given by

$$\Gamma_{ij} = \begin{cases} x_j^i & \text{if task } j \text{ produces on edge } i \\ -y_j^i & \text{if task } j \text{ consumes from edge } i \\ 0 & \text{otherwise.} \end{cases} \tag{10.1}$$

---

1 By convention, this annotation is omitted in cases where the rate is 1.

This topology matrix permits compile-time verification of consistency, specifically by determining a number of firings of each actor so that a program schedule may be derived which is balanced, i.e. it may repeat an infinite number of times within bounded memory. It does so by ensuring that the net gain in the number of tokens on each edge, as a result of executing an iteration of the schedule, is zero (Lee 1991). This is achieved by balancing the relative number of firings of each actor according to the rates of the ports via which they are connected. Specifically, for every actor a, which fires proportionally q_a times in an iteration of the schedule and produces r_a tokens per firing, connected to actor b, which fires proportionally q_b times and consumes r_b tokens per firing, since for operation in bounded memory an iteration of the schedule must see all FIFO queues return to their initial state (Lee and Messerschmitt 1987b), the equation

$$q_a r_a = q_b r_b \tag{10.2}$$

holds. Collecting such an equation for each edge in the graph, a system of balance equations is constructed, which is written compactly as

$$\Gamma\mathbf{q} = \mathbf{0}, \tag{10.3}$$

where the *repetitions vector*, $\mathbf{q}$, describes the number of firings of each actor in an iteration of the execution schedule of the graph and where q_i is the number of firings of actor i in the schedule.

10.2.3 Cyclo-static Dataflow

The CSDF model (Bilsen *et al.* 1996) notes the limitation of SDF actors to a single firing rule pre-specifying the availability of an integer number of tokens on each input edge. Due to this restriction, SDF actors can only perform one fixed behavior on each firing. CSDF attempts to broaden this capability to allow an actor to perform a multitude of predefined behaviors whilst maintaining the powerful compile-time analysis features of SDF. In CSDF, actors have cyclically changing actor behavior, whereby an actor j defines a firing *sequence* $\gamma = \{f_j(1), f_j(2), \ldots, f_j(P_j)\}$. Given this sequence, it is then said that the actor operates in one of P_j *phases* with the behavior of γ_i invoked during firing $i(\bmod j)$.

In addition, the restriction imposed by SDF that the rate of each port be a scalar integer is similarly extended in CSDF to permit rates to be *sequences* of integer scalars. A simple example in shown in Figure 10.3. In this case, whilst b and c are SDF actors (or, more generally, CSDF actors with single-phase firing and rate sequences), a in this case is cyclic, operating a three-phase schedule, with the rate of its output port iterating over the sequence $\{1, 0, 0\}$.

Figure 10.3 Simple CSDF graph

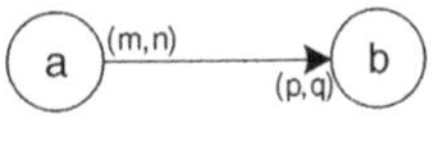

Figure 10.4 Simple MSDF graph

In general, for CSDF actor j connected to edge i, if $X_j^i(n)$ is the total number of tokens produced and $Y_j^i(n)$ the total number consumed during the first n firings of the actor, a CSDF topology matrix Γ is defined by

$$\Gamma_{ij} = \begin{cases} X_j^i(P_j) & \text{if task } j \text{ produces on edge } i \\ -Y_j^i(P_j) & \text{if task } j \text{ consumes from edge } i \\ 0 & \text{otherwise.} \end{cases} \tag{10.4}$$

10.2.4 Multidimensional Synchronous Dataflow

Both SDF and CSDF operate on the assumption that tokens are atomic: a firing of an actor cannot consume anything other than an integer number of tokens traversing along an edge. This restriction is alleviated in MSDF, a domain at its most beneficial for complex multidimensional tokens, first via work which elaborates a single MSDF graph into equivalent SDF structures based on rectangular lattice-shaped problems, such as matrices (Lee 1993a,b), but later further to arbitrary shaped lattices (Murthy and Lee 2002). In MSDF, rates are specified as M-tuples of integers and the number of balance equations per edge increased from 1 to M. An example MSDF graph is shown in Figure 10.4. Note that the form of a multidimensional token is expressed using braces. The balance equations for this graph are given by

$$\begin{aligned} q_{a,1}m &= q_{b,1}p, \\ q_{a,2}n &= q_{b,2}q. \end{aligned} \tag{10.5}$$

The generalization to multiple dimensions inherent in the MSDF model has a similar effect on the mathematical representations of rates and repetitions structures, both of which are generalized to matrices. MSDF provides an elegant solution to multidimensional scheduling problems in SDF graphs, and exposes additional intra-token parallelism for higher order dimension tokens (Lee 1993a).

10.3 Architectural Synthesis of Custom Circuit Accelerators from DFGs

Previous chapters have described how the register-rich programmable logic present in FPGAs makes them ideal for hosting pipelined custom circuit accelerator architectures for high-throughput DSP functions. Furthermore, the substantial body of research into automatically deriving and optimizing these structures from SFGs (SR-DFGs where all ports have a rate of 1) presents a perfect opportunity to enable automatic accelerator synthesis for DFG-based design approaches. A typical architectural synthesis approach deriving such accelerators from generalized MR-DFG models is outlined in Figure10.5.

As this shows, the MR-DFG is first converted to a single-rate equivalent, before undergoing architectural synthesis. This initial conversion is important. SFGs are more

Figure 10.5 MR-DFG accelerator architectural synthesis

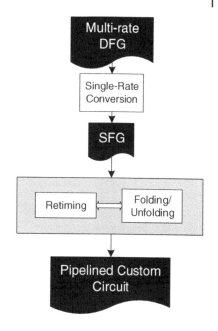

restricted than general MR-DFG models, including SDF, CSDF and MSDF. There are three key restrictions of note:

1. The port rates of all ports in the DFG are fixed at unity.
2. The each actor fires only once in an iteration of the schedule.
3. Port tokens are atomic.

Since MR-DFG models are semantically more expressive than SFGs, on conversion to variations in port rates, actor repetitions or token dimensions are manifest explicitly in the structure of the SFG and hence any accelerator derived from it. This means that an SFG accelerator can only realize one configuration of MR-DFG actor and that the designer does not have explicit control over the structure of their accelerator from the MR-DFG structure. In complex FPGA system designs, it is often desired to reuse components in multiple designs; but if an accelerator derived via the SFG route can only realize one MR-DFG actor configuration, how can such reuse be enabled and controlled? Given traditional SFG architectural synthesis techniques, it cannot.

10.4 Model-Based Development of Multi-Channel Dataflow Accelerators

Changing the MR-DFG operating context for an SFG accelerator, e.g. altering token dimensions or port rates, requires re-generation of the accelerator. In many cases, this is unavoidable, but in many others there may be an opportunity, given an augmented synthesis approach, to derive components which are reusable in numerous contexts.

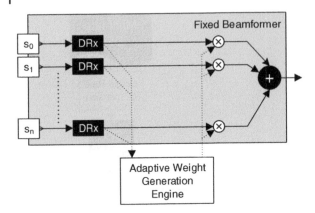

Figure 10.6 Beamformer architecture

Consider one such example. Beamformers are versatile array-processing compo-nents for spatial filtering for radar, sonar, biomedical and communications applica-tions (Haykin 2013). A beamformer is typically used with an array of sensors which are positioned at different locations so that they are able to "listen" for a received signal by taking spatial samples of the received propagating wave fields. A block diagram repre-sentation of a beamformer structure is shown in Figure 10.6.

As shown, the signals emanating from each antenna element are filtered by a digital receiver (DRx) and scaled before being summed to produce the output signal. The upper portion, consisting of the DRx, scaling and sum components is known as a fixed beam-former (FBF), with the addition of an adaptive weight generation engine producing an adaptive beamformer system.

Consider the case where a custom circuit accelerator is created to realize the DRx component in an FBF. FBF systems can be of different scales and have differing through-put and latency constraints; for example, the number of antenna elements, n, may vary. In order to make sure these requirements are met with minimum cost, it is desirable to use m DRx accelerators, with the same DRx required to process multiple channels of data in the case where $m < n$. But since the original DRx is created to service only one channel, there is no guarantee that it can be reused for multiple channels.

A similar situation may arise in, for example, matrix multiplication. To demon-strate, consider multiplication of two matrices, M_1 and M_2 (of dimensions (m,n) and (n,p), respectively). In this case, assuming that an accelerator has been created to form the product of 3×3 matrices, how may that accelerator be used to mul-tiply M_1 and M_2 when $(m, n, p) = (3, 3, 12)$? One possible approach is illustrated in Figure 10.7.

As this shows, by interpreting M_2 as a sequence of parallel column vectors, groups of columns of arbitrary size can be formed and individually multiplied by M_1 to derive M_3 by concurrent multiplication of M_1 by an array of y matrices $\{M_2^0, M_2^1 \cdots M_2^{y-1}\}$ where M_2^i is composed of the p column vectors $\{i \times \frac{p}{y}, \dots, ((i + 1) \times \frac{p}{y}) - 1\}$. The subdivision of M_2 into parallel submatrices for $p = 4$ is given in Figure 10.7. Note the regular relation-ship between the number of multipliers and the size of submatrix consumed by each. This kind of relationship could be exploited to regularly change the structure of the DFG,

Figure 10.7 Parallel matrix multiplication

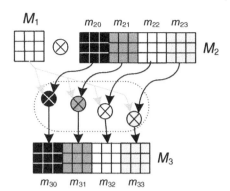

trading off the number of actors and the token dimensions processed at the ports of each; given an appropriate one-to-one correspondence between actors and FPGA accelerators, this would then permit explicit control of the number of accelerators and the token dimensions processed by each. It demands, however, accelerators which are sufficiently flexible to process multiple streams of data, trading resource usage with performance without accelerator redesign.

This capability is dependent on two enabling features:

1. Expressing a dataflow application in such a way that the number of actors and the channels processed by each are under designer control without variation affecting the behavior of the application.
2. Synthesizing accelerators which can support varying multi-channel configurations.

10.4.1 Multidimensional Arrayed Dataflow

The key issue with lack of explicit designer control on the structure of the implementation is the lack of structural flexibility in the MR-DFG itself. A single actor in standard dataflow languages like SDF or MSDF can represent any number of tasks in the implementation, rather than employing a close relationship between the number of DFG actors and number of accelerators in the solution. To overcome this structural inflexibility, the multidimensional arrayed dataflow (MADF) domain may be used (McAllister *et al.* 2006).

To demonstrate the semantics of the domain, consider the same matrix multiplication problem described at the beginning of this section. The MADF graph of this problem is given in Figure 10.8. In this formulation, M_1 and M_2 are sources for the operand matrices, whilst *mm* is the matrix multiply actor and M_3 is a sink for the product. In MADF, the notions of DFG actors and edges are extended to arrays. Hence an MADF graph

Figure 10.8 Matrix multiplication MADF

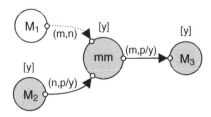

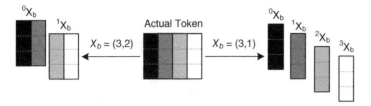

Figure 10.9 Matrix decomposition for fixed token size processing

$G = \{V_a, E_a\}$ describes arrays of actors connected by arrays of edges. Actor arrays are gray, as opposed to single actors (or actor arrays of size 1) which are white. Edge arrays are solid, as opposed to single edges (or edge arrays of size 1) which are dashed. The size of an actor array is quoted in brackets above the actor array.

In such a graph, the system designer controls parameters such as y in Figure 10.8. This is used to define the size of the M_1, M_2, mm and M_3 actor arrays, as well as the dimensions of the tokens produced/consumed by M_2, mm and M_3. Under a one-to-one translation between the number of, for example, mm actors and the number of accelerators, this enables direct graph-level control of the number of accelerators and token dimensions for each. However, as outlined, accelerators derived from SFGs have fixed port token dimensions and a mechanism must be established to allow processing of higher-order tokens.

Consider the case of the array of submatrices of M_2 input to mm in the matrix multiplication example of Figure 10.8. How may a single accelerator be made flexible enough to implement any size of input matrix on this input, given that the pipelined accelerator produced from an SFG description has fixed token dimensions?

As outlined at the beginning of this section, each of the y submatrices can be interpreted as a series of p column vectors, with the ith submatrix composed of the column vectors $\{i \times \frac{p}{y}, \ldots, ((i+1) \times \frac{p}{y}) - 1\}$ of M_2. As such, for the case where $y = 4$, the submatrix can be interpreted in two ways, as illustrated in Figure 10.9. As this shows, the matrix can be interpreted as an aggregation of *base* tokens. If the actor to process the submatrix can only process the base tokens, then the aggregate may be processed by using multiple firings of the actor, each of which processes a different component base token. In a sense, then, the actor is treating the aggregate as an array of base tokens over which it iterates.

To support this concept, MADF support variable-sized arrays of actor ports, each of which consumes identical base tokens, with the resulting accelerator derived to process the base token. To enable multiple iterations of the actor to process the multiple base tokens in the actual token, MADF actors may be cyclic (Section 10.2.3), with individual firings consuming one or more base tokens through each port in the array in turn.

Using this formulation, Figure 10.10 illustrates the full, fixed token processing version of the MADF matrix multiplication problem. Note the presence of differentiated arrays

Figure 10.10 Full MADF matrix multiplication

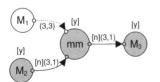

Figure 10.11 Block processing matrix multiplication

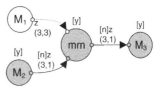

of ports (gray) and individual ports (white). In the case of an array of ports, note that the size of the array is annotated on the port using brackets; for instance, the array of ports on M_3 is of size dimension $[n]$.

10.4.2 Block and Interleaved Processing in MADF

Having exposed intra-token parallelism by separating the actual token processed across multiple streams transporting base tokens, further implementation exploration may be enabled. In the case where the port array is used to process a single token, *interleaved* processing of each port in the array is required, i.e. a single base token is consumed through each port in turn to form the full token. In this case, the rate of each port array element is 1. However, having opened up the token processing into a multi-stream processing problem, the generalized multi-rate nature of dataflow languages can be exploited to enable *block* processing via rates greater than 1 at each element of the port array.

At a port array, the ith element has a production/consumption vector of length p_{size} (the size of the port array) with all entries zero except the ith. These vectors exhibit a diagonal relationship (i.e. for the port array a, all entries in the consumption vector of a_0 are zero except the zeroth, all entries in the consumption vector for a_1 are zero except the first, and so forth. A generalized version of this pattern, for a port array with n elements with thresholds z is denoted by $[n]z$, as illustrated in Figure 10.11 for mm when $y = 3$. The value of z, the rate of each port array element, indicates whether interleaved or block processing is used ($z = 1$ for interleaved, $z > 1$ for block processing).

Given a one-to-one correspondence between the number of actors in an MADF graph, the designer then has the capability to control the number of accelerators in the realization. However, the number of accelerators and the characteristics of each are interlinked. For instance, in the case of the matrix multiplication arrangement in Figure 10.10, if the MADF model is to form the product of M_1 and M_2 when $(m, n, p) = (3, 3, 12)$ and mm has a $(3, 3)$ base token, then depending on the number of mm accelerators, y, the characteristics of each will change; in particular, n will vary as $\frac{12}{y}$. Similarly, the behavior will change with the rate of each port. As such, the traditional SFG architectural synthesis approach in Figure 10.5 needs to be augmented to produce accelerators which can process a variable number of streams at each port, and operate on the variable number of streams in either an interleaved or block-processed manner.

10.4.3 MADF Accelerators

When realized on FPGA, an array of MADF actors translates to an equi-sized array of dataflow accelerators (DFAs). The general structure of a DFA is illustrated in Figure 10.12.

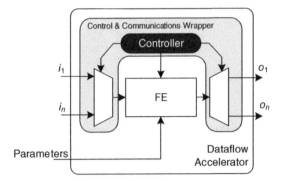

Figure 10.12 Dataflow accelerator architecture

The dataflow accelerator object is composed of three main elements:

1. **Functional engine (FE)**: The FE implements the functionality of the actor and is the accelerator portion of the unit. It can take any architecture, but here is a high-throughput pipelined accelerator. It is created to realize a specific MR-DFG actor, but is capable of performing that functionality on multiple streams of data in either an interleaved or block-processed manner.
2. **Control and communications wrapper (CCW)**: This implements a cyclic schedule to realize multi-stream operation and handles the arbitration of multiple data streams through the FE, in either an interleaved or block processed manner depending on the MADF actor port configuration. The read unit here also implements the necessary edge FIFO buffering for the MADF network.
3. **Parameter bank (PB)**: The PB provides local data storage for run-time constants for the accelerator, e.g. FIR filter tap weights. It is not an active part of the streaming application (i.e. the data stored here can be created and inserted off-line) and so it is not discussed further.

The pipelined FE accelerator part is flexible for reuse across multiple applications and MADF actor configurations, and as such may only require creation and reuse in an accelerator-based design strategy. Efficient generation of FIFO buffers and controllers for automatic generation of dedicated dataflow hardware is a well-researched area (Dalcolmo *et al.* 1998; Harriss *et al.* 2002; Jung and Ha 2004; Williamson and Lee, 1996). The remainder of this section addresses realization of the FE.

10.4.4 Pipelined FE Derivation for MADF Accelerators

The FE part of a dataflow accelerator is a pipelined accelerator, designed as a white box component (WBC) (Yi and Woods 2006). It is one whose structure is parameterized such that it may be reused in various forms in various systems. In the case of MADF accelerators, these parameterized factors are summarized in Table 10.1.

The WBC is derived via architectural synthesis of an SFG representing a specific multi-rate actor instance. To understand how this structure may be reused for multi-stream operation, consider an example two-stage FIR filter WBC, as illustrated in Figure 10.13. The WBC is composed of a computational portion, composed of all of the arithmetic

Table 10.1 WBC parameterization

Parameter	Significance
Streams	Number of streams realized by accelerator is to be shared
Blocking Factor	Blocking (factor > 1) or interleaved (factor = 1) modes.

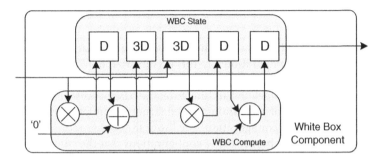

Figure 10.13 Two-stage FIR WBC

operators, and a state space including any delay elements, shift registers or memories.[2] A standard SFG synthesis process such as that in Yi and Woods (2006) will create both, but it is only the state space which restricts the result to a specific MADF actor configuration. The key to designing reusable, configurable accelerators lies in the proper arbitration of the state space and appropriate design of the circuitry such that the data relevant to multiple streams may be stored and properly arbitrated onto the computation portion to enable block or interleaved processing.

To create WBC structures, SFG architectural synthesis is undertaken to create compute and state-space portions for the base configuration, which is then augmented to give the WBC a flexible internal structure which may be regularly changed without redesign to achieve regular changes in MADF actor configuration.

The pipelined WBC architecture resulting from SFG architectural synthesis is merely a retimed version of the original SFG algorithm. The computational resource of the resource must effectively be time-multiplexed between each of the elements of the input stream array, with the entire computation resource of the SFG dedicated to a single stream for a single cycle in the case of interleaved processing, and for multiple cycles in the case of block processing.

To enable interleaved processing, the first stage in the WBC state space augmentation process requires k-slowing (Parhi 1999), where the delay length on every edge resulting from SFG architectural synthesis is scaled by a factor k, and in the case of interleaved processing of n input streams, $k = n$. This type of manipulation is known as *vertical*.

In the case where block processing is required, base tokens are consumed/produced from a single-port array element for a sustained number of cycles. Accordingly, the dataflow accelerator state space should have enough state capacity for all s streams,

---

2 Henceforth only delay elements are considered.

activating the state space associated with a single stream in turn, processing for an arbitrary number of tokens, before loading the state space for the next stream. This kind of load–compute–store behavior is most suited to implementation as a distributed memory component, with the active memory locations determined by controller schedule. This is known here as *lateral* delay scaling, where each SFG delay is scaled into an s-element disRAM.

Given the two general themes of lateral and vertical delay scalability, an architectural synthesis process for reusable WBC accelerators to allow multi-stream actor and accelerator reuse involves four steps:

1. **Perform MADF actor SFG architectural synthesis.** For a chosen MADF actor, C is fixed and defined as the base configuration C_b. This is converted to SFG for architectural synthesis. The MADF actor C_b is the minimum possible set of configuration values for which the resulting pipelined architecture, the base processor P_b, may be used, but by regular alteration of the parameterized structure the processor can implement integer supersets of the configuration. The lower the configuration values in the base, the greater the range of higher-order configurations that the component can implement. To more efficiently implement higher-order configurations, C_b can be raised to a higher value. For a two-stage FIR, $C_b = \{1, 1, 1\}$, the WBC of the P_b is shown in Figure 10.13.

2. **Vertical delay scalability for interleaved processing.** To implement k-slowing for variable interleaved operation, the length of all delays must be scaled by a constant factor m. All the lowest-level components (adder/multipliers) are built from pre-designed accelerators which have fixed pipelined depths (in the case of Figure 10.13 these are all 1) which cannot be altered by the designer. To enable the scaling of these delays, these are augmented with delays on their outputs to complete the scaling of the single pipeline stages to that of length m. The resulting FIR circuit architecture for the pipelined FIR of Figure 10.13 is shown in Figure 10.14(a). The notation (m) D refers to an array of delays with dimensions $(1, m)$. Note that all delay lengths are now a factor of m, the vertical scaling factor, and note the presence of the added delay chains on the outputs of the lowest-level components. This type of manipulation is ideally suited to FPGA where long delays are efficiently implemented as shift registers (Xilinx 2005).

3. **Lateral delay scalability for block processing.** For block processing the circuit delays are scaled by a vertical scaling factor n post lateral scaling to allow combined interleaved/block processing if required. This results in arrays of delays with dimensions (m, n). The resulting FIR circuit architecture when this is applied to the circuit of Figure 10.14(a) is shown in Figure 10.14(b). Note the presence of the vertical scaling factor on all delay arrays. This kind of miniature embedded-RAM-based behavior is ideally suited to FPGA implementation, since these can implement small disRAM in programmable logic. These disRAMs have the same timing profile as a simple delay (Xilinx 2005), and as such do not upset edge weights in the circuit architecture.

4. **Retime structure to minimize lateral delay scalability.** When P_b is configured to implement a much higher-order MADF actor configuration than C_b, very large delay lengths can result. To minimize these, retiming is applied to the augmented processor architecture.

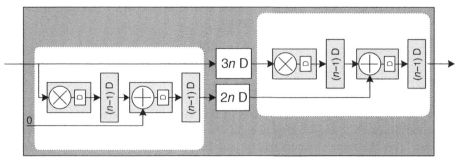

(a) Vertically scaled two-stage FIR WBC

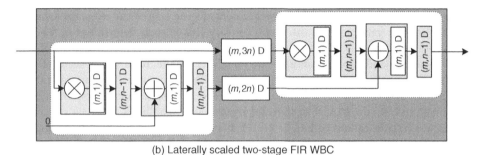

(b) Laterally scaled two-stage FIR WBC

Figure 10.14 Scaled variants of two-stage FIR WBC

10.4.5 WBC Configuration

After creation of the WBC architecture P_b, it must be configured for use for specific MADF actor configuration. Consider a base processor created via SFG architectural synthesis P_b realizing a MADF actor with configuration $C_b = (\mathbf{r}_b, \mathbf{x}_b, \mathbf{s}_b)$, where $\mathbf{r}_b, \mathbf{x}_b, \mathbf{s}_b$ respectively represent the rates, token dimensions and number of streams processed by the actor in question, with pipeline period α_c created using SFG architectural synthesis (Parhi 1999). To realize an MADF actor P, where X is an n-dimensional token of size $x(i)$ in the ith dimension, the following procedure is used:

1. Determine the vertical scaling factor m, given by

$$m = \left\lceil \frac{1}{\alpha_c} \prod_{i=0}^{n-1} \frac{x(i)}{x_b(i)} \right\rceil. \tag{10.6}$$

2. k-slow P_b by the factor m.
3. Scale primitive output delays to length $(m-1) \times l$, where l is the number of pipeline stages in the primitive.
4. Scale all delays laterally by the scaling factor n, given by

$$n = \frac{s}{s_b}. \tag{10.7}$$

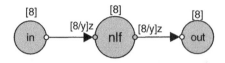

Figure 10.15 Eight-channel NLF filter bank MADF graph

10.4.6   Design Example: Normalized Lattice Filter

In order to demonstrate the effectiveness of this kind of architectural synthesis and exploration approach, it is applied to an eight-channel filter bank design problem, where each filter takes the form of a normalized lattice filter (NLF) (Parhi 1999). The MADF graph is shown in Figure 10.15, with accelerators realizing the *nlf* actors to be created.

As Figure 10.15 shows, *in* and *out* arrays generate an array of eight scalar tokens which are processed by the *nlf* array. The designer controls the size of the *nlf* actor array by manipulating the variable *y* on the graph canvas. This in turn determines *n*, the size of the port array of each element of the *nlf* actor array. To test the efficiency of this MADF synthesis and exploration approach the SFG architectural synthesis capability for P_b synthesis is limited to retiming (i.e. advanced architectural explorations such as folding/unfolding are not performed), placing the emphasis for implementation optimization entirely on the MADF design and exploration capabilities. The base processor P_b operates on scalar tokens with $C_b = (1, 1, 1)$ to maximize flexibility by maximizing the number of achievable configurations. The target device is the smallest possible member of the Virtex-II Pro^{TM} family which can support the implementation. This enables two target-device-specific design rules for efficient synthesis:

$$D_{\text{type}} = \begin{cases} \text{FDE} & \text{if } (P, Q) = (1, 1) \\ \text{LUT RAM} & \text{if } P > 1 \\ \text{SRL16+FDE} & \text{otherwise.} \end{cases} \quad (10.8)$$

The SFG of the base NLF actor is shown in Figure 10.16(a), with the SFG of the NLF stage shown in Figure 10.17(a). If the lowest-level components (adders and multipliers) from which the structure is to be constructed are implemented using single-stage pipelined black box components (a common occurrence in modern FPGA), then a particular feature of the NLF structure is the presence of 36 recursive loops in the structure, with the critical loop (Parhi 1999) occurring when two pipelined stages are connected. For single-stage pipelined adders and multipliers, this has a pipeline period, α, of 4 clock cycles. Hence, by equation (10.7), $n = \frac{x_i}{4x_b}$.

The base processor P_b is created via hierarchical SFG architectural synthesis (Yi and Woods 2006), and produces the pipelined architecture of Figure 10.16(b), with the architecture of each stage as in Figure 10.17(b). After lateral and vertical delay scaling and retiming, the NLF and stage WBC architectures are as shown in Figure 10.16(c) and Figure 10.17(c), respectively.

Synthesis of the given architecture for three different values of *y* has been performed. $^8_1BS\text{-}NLF$, $^2_4BS\text{-}NLF$ and $^1_8BS\text{-}NLF$ are the structures generated when *y* is 1, 2 and 8 respectively, and each dataflow accelerator performs interleaved sharing over the impinging data streams, whilst results for a single dataflow accelerator processing a 68-element vector ($^{68}_1BS\text{-}NLF$) are also quoted to illustrate the flexibility of the WBC architectures. A block-processing illustration of 16 streams of four-element vector tokens ($^1_4BS\text{-}NLF_{16}$)

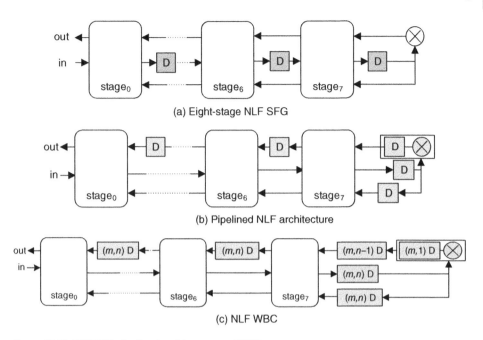

(a) Eight-stage NLF SFG

(b) Pipelined NLF architecture

(c) NLF WBC

Figure 10.16 NLF SFG, pipelined architecture and WBC

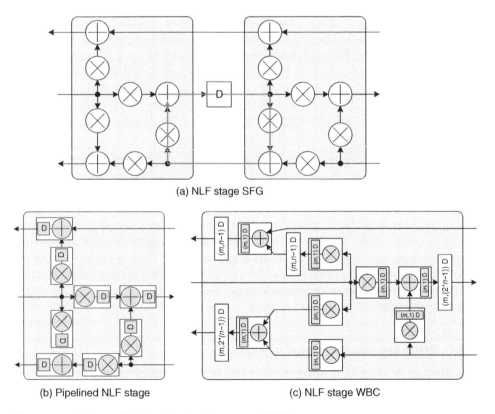

(a) NLF stage SFG

(b) Pipelined NLF stage

(c) NLF stage WBC

Figure 10.17 NLF stage SFG, pipelined architecture and WBC

Table 10.2 NLF post place and route synthesis results on Virtex-II Pro FPGA

	Logic				Throughput
	LUTs	SRL	DisRAM	mult18	(MSamples/s)
$^{8}_{1}BS\text{-}NLF$	1472	–	–	312	397.4
$^{2}_{4}BS\text{-}NLF$	368	–	–	78	377.9
$^{1}_{8}BS\text{-}NLF$	186	207	–	39	208.6
$^{68}_{1}BS\text{-}NLF$	186	207	–	39	208.6
$^{1}_{4}BS\text{-}NLF_{16}$	188	7	576	39	135.8

is also quoted in Table 10.2. These illustrate the effectiveness of this approach for accelerator generation and high-level architecture exploration. Transforming the MADF specification by trading off number of actors in the family, token size per actor, and number of functions in the MADF actor cyclic schedule has enabled an effective optimization approach without redesign.

The initial implementation ($y = 8$, $^{1}_{8}BS\text{-}NLF$) created an right-element dataflow accelerator array. Given the large number of multipliers (mult18 in Xilinx Virtex-II) required for implementation, the smallest device on which this architecture can be implemented is an XCV2DA70. However, given the pipeline period inefficiency in the original WBC architecture, reducing y to 2 produces two four-element vector processors ($^{2}_{4}BS\text{-}NLF$) with almost identical throughput, and enables a significant reduction in required hardware resource with little effect on throughput rate. This amounts to a throughput increase by a factor of 3.9 for each dataflow accelerator with no extra hardware required in the WBC. The large reduction in required number of embedded multipliers also allows implementation on a much smaller XC2DA20 device. Decreasing y still further to 1 produces a single eight-element vector processor ($^{1}_{8}BS\text{-}NLF$). Whilst the throughput has decreased, a significant hardware saving has been made. The NLF array can now be implemented on a smaller XC2DA7 device.

This example shows that the MADF synthesis approach can achieve impressive implementation results via simple system-level design space exploration. Using a single pipelined accelerator, this approach has enabled highly efficient architectures (3.9 times more efficient than one-to-one mappings) to be easily generated, in a much simpler and more coherent manner than in SFG architectural synthesis. Furthermore, by manipulating a single DFG-level parameter, this design example can automatically generate implementations with wildly varying implementation requirements, offering an order-of-magnitude reduction in device complexity required to implement the design is desired. This illustrates the power of this approach as a system-level, accelerator-based design approach with highly efficient implementation results and rapid design space exploration capabilities.

10.4.7 Design Example: Fixed Beamformer System

The structure of the FBF is highly regular and can be represented in a very compact fashion using MADF, as shown in Figure 10.18. The structure of a beamformer is also included in Figure 10.19 for comparison.

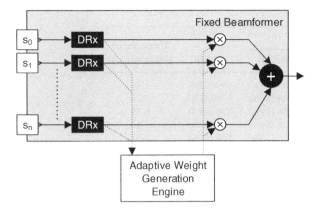

Figure 10.18 Fixed beamformer MADF graph

The MADF graph consists of an array of *n* inputs, one for each sensor in the array. This is tightly correlated with the number of members in the *DRx* and *gain* actor families, as well as the size of the port array *i* on the *sum* actor (again a port array is denoted in grey). Hence by altering the value of *n*, parameterized control of the algorithm structure is harnessed for a variable number of sensors. By coupling the implementation structure tightly to the algorithm structure, this gives close control of the number of *DRx* and *gain* accelerators in the implementation.

For the purposes of this design example, $n = 128$ and the design process targets a Xilinx Virtex-II Pro^{TM} 100 FPGA (Xilinx 2005). The accelerator library consists only of complex multiplication, addition and sum accelerators, and hence the entire system is to be composed from these. The length of the *DRx* filters is taken as 32 taps. Given that this structure then requires 16,896 multipliers, and it is desirable to utilize the provided 18-bit multipliers on the target device (of which only 444 are available) this presents a highly resource-constrained design problem.

To enable the exploration of the number of channels processed by each accelerator in the implementation, each actor must be able to process multiple channels in the MADF algorithm. This is enabled using the MADF structure of Figure 10.20. Here, a second parameter, *m*, has been introduced to denote the number of actors used to process the *n* channels of data. Note that the ports on the *DRx* and *multK* actors are now both families

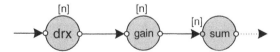

Figure 10.19 Fixed beamformer overview

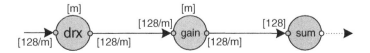

Figure 10.20 Fixed beamformer MADF graph

Table 10.3 FBF post place and route synthesis results on Virtex-II Pro FPGA

	Logic				Throughput
m(i)	LUTs	SRL	DisRAM	mult18	(MSamples/s)
1(i)	3493 (8%)	16128 (37%)	8448 (8%)	99 (22%)	1.45
2(i)	4813 (11%)	16128 (37%)	8448 (19%)	198 (45%)	3.18
4(i)	8544 (19%)	16128 (37%)	8448 (19%)	396 (89%)	6.19
1(b)	3490 (8%)	0 (0%)	24576 (56%)	99 (22%)	1.45
2(b)	4812 (11%)	0 (0%)	24576 (56%)	198 (45%)	3.51
4(b)	8554 (19%)	0 (0%)	24576 (56%)	396 (89%)	1.45

of size m to denote the sharing of the actor amongst $\frac{n}{m}$ data streams processed in a cyclic fashion (McAllister *et al.* 2006). On synthesis, a wide range of synthesis options are available for the FBF custom circuit system on a chosen device, with an accompanying wide range of real-time performance capabilities and resource requirements, and these are summarized in Table 10.3. The breakdown of the proportion of the programmable logic (LUT/FDE) by dataflow accelerator function (WBC, PB or CCW) is given in Table 10.4.

From an initial implementation consisting of a single accelerator process a 128-element vector (i.e. interleave shared across the 128 input streams), increasing the value of m by 2 and 4 has produced corresponding increases in throughput by factors of 2.2 and 4.3 respectively, and it should be noted that the architectures used for accelerator sharing amongst multiple streams exhibit minimal resource differences. This is a direct result of the abstraction of the accelerator architectures for target portability. Whilst the overheads in terms of LUTs (which may be configured as 16-bit SRL or disRAMs) for the WBC wrapping in the dataflow accelerator are high (up to 35%), the major part of this is required entirely for storage of on-chip filter tap and multiplier weights in the SFO parameter banks. This storage penalty is unavoidable without exploiting on-chip embedded BRAMs. In addition, the overhead levels decrease with increasing values of m since the number of tap weights remains constant independent of m. The CCW incurs little LUT overhead, instead exploiting the embedded *muxF5/muxF6/muxF7/muxF8* fabric of the FPGA (Xilinx 2005) to implement the switching. These are not used at all anywhere else in the design and hence are plentiful. Finally, it should be noted that all the accelerators in the system are 100% utilized depending on input data.

Table 10.4 FBF implementation resource breakdown

	LUT			FDE		
m(i)	%WBC	%CCW	%PB	%WBC	%CCW	%PB
1(i)	3.7	31.3	6.5	1.3	0	98.7
2(i)	3.6	28.7	69.5	0.9	0	99.1
4(i)	3.2	25.5	71.3	0.3	0	99.7
1(b)	3.7	31.3	6.5	1.3	0	98.7
2(b)	3.6	28.7	69.5	1.0	0	99.0
4(b)	3.2	25.5	71.3		0.3	99.7

Table 10.5 Memory resources available on Virtex-6

Number		Off-chip 1	Pixel BRAM 215
Capacity		2 GB	8 K
Access rate (pixels/s)	600 MHz	230 MB/s	600 Mpixel/s
	200 MHz	230 MB/s	200 Mpixel/s

10.5 Model-Based Development for Memory-Intensive Accelerators

The previous section addressed techniques for representing, deriving and optimizing FPGA accelerators for pipelined streaming operators. It is notable, though, that these operations are all performed on scalar data elements, which imposes a small overhead for storage of memory; for the most part memory can be realized using registers, SRLs or disRAMs. However, in a great many cases the demands for buffer memory are high, particularly in applications such as image and video processing where large frames of data are to be handled. In these situations, how are large quantities of buffer memory realized, handled and optimized?

Consider the memory resources on a Xilinx Virtex-6 FPGA in the context of a typical memory-intensive operation for video processing: full search motion estimation (FSME) on CIF 352×288 video frames at 30 frames per second. The resources at the designer's disposal with which this operation may be realized are summarized in Table 10.5.[3] When gauging the anticipated access rate of an on-chip BRAM, it is necessary to take into account the anticipated clock rate of the final synthesize accelerator and the entire system in which it resides; hence, whilst peak clock rates may reach 600 MHz, operation at around 200 MHz is much more reasonable. The designer's challenge is to collect these resources in such a fashion that: data are hosted off-chip where possible to minimize the on-chip buffering cost, and the final accelerator architecture meets real-time performance requirements.

10.5.1 Synchronous Dataflow Representation of FSME

A block diagram of the FSME operation is shown in Figure 10.21(a), and an SDF representation of the FSME operation is shown in, is shown in Figure 10.21(b).

Consider the behavior of the SDF model in Figure 10.21(b). The actors C and R are source actors to represent the stream of current and reference frames incident on the FSME operator. The dimensions of the output tokens indicate the size of the respective video frames. The current frame C is decomposed into 396 $(16, 16)$ non-overlapping CBs via cb, with R decomposed into corresponding $(48, 48)$ SWs, each of which is centered around the same center pixel as the corresponding CB. From each SW are extracted 1089 $(16, 16)$ sub-blocks, with each compared with the cb in turn via a minimum absolute difference (MAD) operation, with the lowest of these 1089 metrics selected for

---

3 Assuming BRAM configuration as 8K $\times$ 4-bit BRAM, six of which are used to construct a three-byte "pixel BRAM".

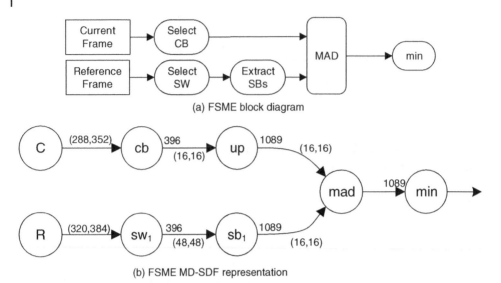

(a) FSME block diagram

(b) FSME MD-SDF representation

Figure 10.21 Full search motion estimation representations

computation of motion vectors. Accordingly, in order to ensure balanced dataflow, each CB is replicated 1089 times by *up*.

Consider now the memory costs of naive implementation of a model such as this on FPGA, where the FIFO buffering associated with each edge is realized using the available memory resource. The capacity and access rate requirements for the buffer relating to each edge are described in Table 10.6.

As Table 10.6 shows, the capacity and access requirements of each of $e_1, \ldots, e_4$ are such that each could be hosted in off-chip RAM, or indeed on-chip BRAM. However, in the case of e_5 and e_6 there is a major issue: whilst the access rate of these buffers can be realized using 17-pixel BRAM (102 BRAMs in total), many more BRAMs would be required to satisfy its capacity requirements than are available on the device. Similarly, the access rate requirements are such that these buffers cannot be realized using off-chip DRAM. Hence, an architecture such as this could not be realized using the Virtex-6 FPGA, necessitating refinement to overcome the capacity issues on (*up, mad*) and (*sb, mad*).

Table 10.6 SDF FSME memory requirements

Edge	Capacity		Rate	
	pixels	BRAM	pixels/s	BRAM
$e_1 : (C, cb)$	101.4 K	13	3.04 M	1
$e_2 : (R, sw)$	122.9 K	16	3.96 M	1
$e_3 : (cb, up)$	101.4 K	13	3.04 M	1
$e_4 : (sw, sb)$	912.4 K	115	27.37 M	1
$e_5 : (up, mad)$	110.4 M	13,800	3.3 G	17
$e_6 : (sb, mad)$	110.4 M	13,800	3.3 G	17

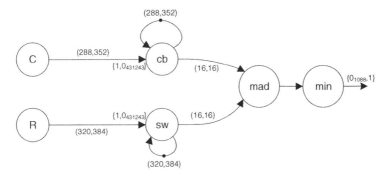

Figure 10.22 FSME modeled using CSDF

Table 10.7 CSDF FSME memory requirements

Edge	Capacity		Rate	
	pixels	BRAM	pixels/s	BRAM
e_1 : (C, cb)	101.4 K	13	3.04 M	1
e_2 : (R, sw)	122.9 K	16	3.96 M	1
e_3 : (cb, cb)	101.4 K	13	1.3 T	6500
e_4 : (sw, sw)	122.9 K	115	1.59 T	8000
e_5 : (cb, mad)	256	1	3.3 G	17
e_6 : (sw, mad)	256	1	3.3 G	17

10.5.2 Cyclo-static Representation of FSME

Despite being concise, a major issue in the SDF model in Figure 10.21 is duplication of information; in particular, the excessive capacity requirements of the edges (up, mad) and (sb, mad) mask the fact that the former houses 1089 copies of the same $(16, 16)$ CB, whilst the latter contains successive SBs containing substantial amounts of duplicated pixels. In order to address this issue of duplication of pixels, consider the use of CSDF for modeling. A CSDF representation of FSME is shown in Figure 10.22.[4]

Note that cb and sw are now cyclic actors, each of which operates over 431,244 phases. During the first firing of each, the respective frames are consumed, with the appropriate sequences of CBs for mad producing over 431,244 firings. In order to satisfy the need to have access to the entire current and reference frames during each phase for extraction of the relevant CB and SW, these are "recycled" using the self-loops on each of cb and sw. Consider the capacities and access rates of the buffers associated with each edge in Figure 10.22, as detailed in Table 10.7.

As Table 10.7 shows, the capacity and access rates of e_1 and e_2 are such that these can be realized either off-chip or on-chip, whilst the access rate demands of e_5 and e_6 demand BRAM storage. In addition, the issue of very large-capacity buffers encountered in the SDF model has been avoided by exploiting the multi-phase production capabilities of CSDF – the edges impinging on mad now require sufficient capacity for only a single

---

4 Note that, henceforth, the notation m_n, as seen in Figure 10.22, represents a length-n sequence of elements, each of which takes the value m.

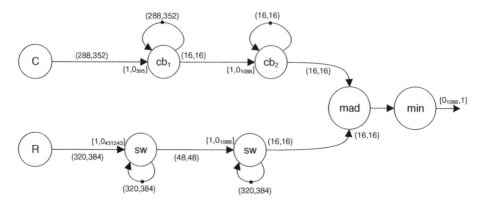

Figure 10.23 Modified FSME CSDF model

CB/SB: very large-capacity savings. Whilst this has come at the cost of an extra buffer for each of the self-loops on *cb* and *sw*, it still represents a very substantial capacity reduction.

However, it has also come at the cost of very high access rate requirements for the buffers representing e_3 and e_4. Indeed, these access rates cannot be realized either off-chip or on-chip and as such a capacity constraint encountered for the SDF network has been replaced by an access rate constraint in this CSDF formulation. Consider the alternative CSDF formulation, shown in Figure 10.23.

In this graph, *cb* and *sw* have both been subdivided into a sequence of two actors. cb_1 is a 396-phase CSDF actor which consumes *C* during the first phase and recycles it via a self-loop during the remaining 395 phases allowing it to produce a single CB per phase. Subsequently, cb_2 operates over 1089 phases, with a CB consumed during the first phase and recycled via a self-loop during the remaining 1088, allowing a single copy to be produced during each phase. The combination of sw_1 and sw_2 performs a single function on *R*, except that sw_2 extracts the distinct SBs from each SW rather than simply duplicating the input, as is the case in cb_2. Consider the memory costs for this CSDF formulation, detailed in Table 10.8.

As described, this approach has had a profound impact on the buffer structure for FSME. In this case, all buffers can be hosted off-chip, up to and including e_6 can be

Table 10.8 CSDF (2) full search motion estimation memory requirements

Edge	Capacity		Rate	
	pixels	BRAM	pixels/s	BRAM
$e_1 : (C, cb_1)$	101.4 K	13	3.04 M	1
$e_2 : (R, sw_1)$	122.9 K	16	3.96 M	1
$e_3 : (cb_1, cb_1)$	101.4 K	13	1.2 G	6
$e_4 : (sw_1, sw_1)$	122.9 K	115	1.5 G	8
$e_5 : (cb_1, cb_2)$	256	1	3.04 M	1
$e_6 : (sw_1, sw_2)$	2304	1	27.37 M	1
$e_7 : (cb_2, cb_2)$	256	1	3.3 G	17
$e_8 : (sw_2, cw_2)$	2304	1	29.8 G	149
$e_9 : (cb_2, mad)$	256	1	3.3 G	17
$e_{10} : (sw_2, mad)$	256	1	3.3 G	17

hosted off-chip, with the access rates of $e_7, \ldots, e_{10}$ dictating on-chip realization. Whilst the access rate of e_8 in particular is quite high and imposes a high BRAM cost, the situation is now that a realization of this form is at least feasible; this was not previously the case. In addition, the use of more advanced dataflow modeling approaches exploiting non-destructive read capabilities can be used to reduce the on-chip cost even further (Fischaber *et al.* 2010; Denolf *et al.* 2007).

10.6 Summary

This section has described techniques to enable system-level design and optimization of custom circuit accelerators for FPGA using dataflow application models.

The use of MADF as a modeling approach for DSP systems helps encapsulate the required aspects of system flexibility for DSP systems, in particular the ability to exploit data-level parallelism, and control how this influences the implementation. This has been shown to be an effective approach; for an NLF filter design example, impressive gains in the productivity of the design approach were achieved. In this example, this included an almost fourfold increase in the efficiency of the implementation via simple transformations at the DFG level, negating the need for complex SFG architectural manipulations.

Otherwise, this approach has proven effective at rapid design space exploration, producing NLF implementations of varying throughput and drastically different physical resource requirements (on order-of-magnitude variation in device complexity) simply by manipulating a single parameter at the graph level. Further, in an FBF design example the effectiveness of this approach was demonstrated by enabling rapid design space exploration, producing a variety of implementations for a specific device via manipulation of a single DFG parameter.

Similarly, the use of CSDF modeling has been shown to make feasible realization of FPGA accelerators which otherwise could not have been achieved. In particular, for memory-intensive accelerators which access large amounts of memory at high access rates, careful design is required to ensure that both the capacity and access rate requirements are met, whilst reducing cost if possible. Frequently, this means devising combinations of on-chip BRAM and off-chip DRAM in multi-level memory structures customized to the application and performance requirements. The design of an FSME accelerator has highlighted both capacity and access rate issues which have been overcome by intuitive employment of CSDF modeling, allowing an otherwise infeasible accelerator to be realized on the Virtex-6 FPGA.

Bibliography

Bilsen G, Engels M, Lauwereins R, Peperstraete J 1996 Cyclo-static dataflow. *IEEE Trans. on Signal Processing*, 44(2), 397–408.

Bhattacharyya SS, Murthy PK, Lee EA 1999 Synthesis of embedded software from synchronous dataflow specifications. *J. of VLSI Signal Processing*, 21(2), 151–166.

Dalcolmo J, Lauwereins R, Ade M 1998 Code generation of data dominated DSP applications for FPGA targets. In *Proc. Int. Workshop on Rapid System Prototyping*, pp. 162–167.

Denolf K, Bekooij M, Gerrit J, Cockx J, Verkest D, Corporaal H 2007 Exploiting the expressiveness of cyclo-static dataflow to model multimedia implementations. *EURASIP J. on Advances in Signal Processing*, 2007, 084078.

Fischaber S, Woods R, McAllister J 2010 SoC memory hierarchy derivation from dataflow graphs. *J. of VLSI Signal Processing*, 60(3), 345–361.

Harriss T, Walke R, Kienhuis B, Deprettere EF 2002 Compilation from Matlab to process networks realised in FPGA. *Design Automation for Embedded Systems*, 7(4), 385–403.

Haykin S 2013 *Adaptive Filter Theory*, 5th edn. Pearson, Upper Saddle River, NJ.

Jung H, Ha S 2004 Hardware synthesis from coarse-grained dataflow specification for fast HW/SW cosynthesis. In *Proc. Int. Conf. on Hardware/Software Codesign and System Synthesis*, pp. 24–29.

Kahn G 1974 The semantics of a simple language for parallel programming. *Proc. IFIP Congress*, pp. 471–475.

Lee EA 1991 Consistency in dataflow graphs. *IEEE Trans. on Parallel and Distributed Systems*, 2(2), 223–235.

Lee EA 1993a Multidimensional streams rooted in dataflow. *IFIP Transactions: Proceedings of the IFIP WG10.3. Working Conf. on Architectures and Compilation Techniques for Fine and Medium Grain Parallelism*, A-23, pp. 295–306.

Lee EA 1993b Representing and exploiting data parallelism using multidimensional dataflow diagrams. In *Proc. IEEE Int. Conf. on Acoustics, Speech, and Signal Processing*, pp. 453–456.

Lee EA, Messerschmitt DG 1987a Synchronous data flow. *Proc. of the IEEE*, 75(9), 1235–1245.

Lee EA, Messerschmitt DG 1987b Static scheduling of synchronous data flow programs for digital signal processing. *IEEE Trans. on Computers*, 36(1), 24–35.

Lee EA, Parks TM 1995 Dataflow process networks. *Proc. of the IEEE*, 83(5), 773–801.

Lee EA, Sangiovanni-Vincentelli A 1998 A framework for comparing models of computation. *IEEE Trans. on Computer-Aided Design of Integrated Circuits and Systems*, 17(12), 1217–1229.

McAllister J, Woods R, Walke R,d Reilly D 2006 Multidimensional DSP core synthesis for FPGA. *J. of VLSI Signal Processing Systems for Signal, Image and Video Technology*, 43(2), 207–221.

Murthy PK, Lee EA 2002 Multidimensional synchronous dataflow. *IEEE Trans. on Signal Processing*, 50(8), 2064–2079.

Najjar WA, Lee EA, Gao GR 1999 Advances in the dataflow computational model. *Parallel Computing*, 25(4), 1907–1929.

Parhi, KK 1999 *VLSI Digital Signal Processing Systems: Design and Implementation*. John Wiley & Sons, New York.

Sriram S, Bhattacharyya SS 2000 *Embedded Multiprocessors: Scheduling and Synchronization*. Marcel-Dekker, New York.

Williamson MC, Lee EA 1996 Synthesis of parallel hardware implementations from synchronous dataflow graph specifications. In *Proc. 30th Asilomar Conf. on Signals, Systems and Computers*, 2, pp. 1340–1343.

Xilinx Inc. 2005 Virtex-II Pro and Virtex-II Pro X Platform FPGAs: Complete Data Sheet. Available from http://www.xilinx.com (accessed June 11, 2015).

Yi Y, Woods R 2006 Hierarchical synthesis of complex DSP functions using IRIS. *IEEE Trans. on Computer-Aided Design of Integrated Circuits and Systems*, 25(5), 806–820.

11

Adaptive Beamformer Example

The material in Chapter 9 highlighted the importance of IP in the design of complex FPGA-based systems. In particular, the use of soft IP has had a major impact in creating such systems. This has created a new market, as witnessed by the Design & Reuse website (http://www.design-reuse.com/) which has 16,000 IP cores from 450 vendors, and the open source cores available from the OpenCores website (opencores.org). This, along with the major FPGA vendors' cores (LogiCore from Xilinx and MegaCore® from Altera) as well as their partners' programs, represents a core body of work.

A lot of companies and FPGA developers will have invested a lot of effort into creating designs which well match their specific application. It may then seem relatively straightforward to extend this effort to create soft IP for a range of application domains for this function. However, the designer may have undertaken a number of optimizations specific to a FPGA family which will not transfer well to other vendors. Moreover, the design may not necessarily scale well to the functional parameters.

The ability to create an IP core requires a number of key stages. Firstly, the designer needs to generate the list of parameters to which the core design should scale. The architecture should then be designed such that it scales effectively across these parameters; to be done effectively, this requires a detailed design process. The description is considered in this chapter for a QR-based IP core for adaptive beamforming. It is shown how an original architecture developed for the design can then be mapped and folded to achieve an efficient scalable implementation based on the system requirements.

Section 11.1 provides an introduction to the topic of adaptive beamforming. Section 11.2 outlines the generic process and then how it is applied to adaptive beamforming. Section 11.3 discusses how the algorithm is mapped to the architecture and shows how it is applied to the squared Givens rotations for RLS filtering. The efficient architecture design is then outlined in Section 11.4 and applied to the QR design example. Section 11.5 outlines the design of a generic QR architecture. A key aspect of the operation is the retiming the generic architecture which is covered in Section 11.6. Section 11.7 covers the parameterizable QR architecture, and the generic control is then covered in Section 11.8. The application to the beamformer design is then addressed in Section 11.9, with final comments given in Section 11.10.

FPGA-based Implementation of Signal Processing Systems,
Second Edition. Roger Woods, John McAllister, Gaye Lightbody and Ying Yi.
© 2017 John Wiley & Sons, Ltd. Published 2017 by John Wiley & Sons, Ltd.

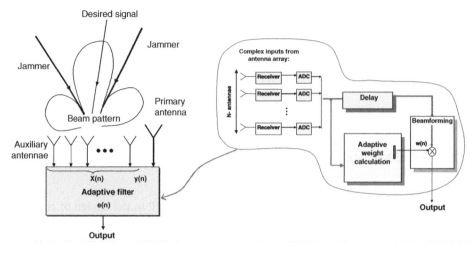

Figure 11.1 Diagram of an adaptive beamformer for interference canceling

11.1 Introduction to Adaptive Beamforming

Adaptive beamforming is a form of filtering whereby input signals are received from a number of spatially separated antennae, referred to as an antenna array. Typically, its function is to suppress signals from every direction other than the desired "look direction" by introducing deep nulls in the beam pattern in the direction of the interference. The beamformer output is a weighted linear combination of input signals from the antenna array represented by complex numbers, therefore allowing an optimization both in amplitude and phase due to the spatial element of the incoming data.

Figure 11.1 shows an example with one primary antenna and a number of auxiliary antennae. The primary signal constitutes the input from the main antennae, which has high directivity. The auxiliary signals contain samples of interference threatening to swamp the desired signal. The filter eliminates this interference by removing any signals in common with the primary input signal. The input data from the auxiliary and primary antennae are fed into the adaptive filter, from which the weights are calculated. These weights are then applied on the delayed input data to produce the output beam.

There are a range of applications for adaptive beamforming, from military radar applications to communications and medical applications (Athanasiadis *et al.* 2005; Baxter and McWhirter 2003; Choi and Shim 2000; de Lathauwer *et al.* 2000; Hudson 1981; Shan and Kailath 1985; Wiltgen 2007). Due to the possible applications for such a core, this chapter investigates the development of an IP core to perform the key computation found in a number of such adaptive beamforming applications.

11.2 Generic Design Process

Figure 11.2 gives a summary of a typical design process followed in the development of a single use implementation. It also gives the additional considerations required in generic IP core design. In both cases, the process begins with a detailed specification

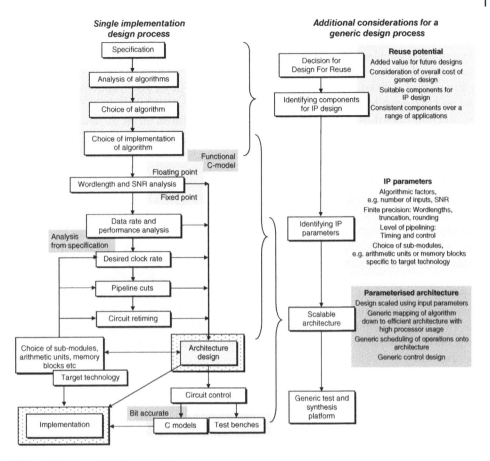

Figure 11.2 Generic design process

of the problem. At this point, consideration may be given to employing a design-for-reuse strategy to develop a generic product. This is based on a number of factors, the most prevalent being whether such a generic core would be worthwhile to have. Would it be applicable over a range of applications, or is it a one-off requirement? There is an initial extra cost in terms of money and time in the development of a generic core, so it is essential that this cost will be more than recouped if the IP core is used in future designs.

Once defined, an analysis is performed to determine the most suitable algorithm. Once chosen, the method for implementation of this algorithm is key as, even with simple addition, there are a number of different arithmetic techniques; these impact on the overall performance of a circuit, in terms, for example, of area compactness, critical path, and power dissipation.

Within a design, there may only be a number of key components that will be suitable to implement as IP cores. These are the parts of the design that will have a level of consistency from application to application. It is imperative to determine expected variations for future specifications. Are all these variables definable with parameters within the generic design, or would other techniques be required to create the flexibility of the design? An example of this could be hardware and software co-design. Here, the fixed components could be implemented as IP cores driven by a software harness adding the needed flexibility for further developments.

The choice of fixed-point or floating-point arithmetic is also vital. From the data rate and performance analysis, a decision can be made regarding certain issues for the architecture design. A desired clock rate may be required to meet certain data rate requirements. Again this will relate to the target technology or specific FPGA device. Clock rate and area criteria will also influence the choice of submodules within the design and the level at which they may need to be pipelined so as to meet circuit speeds.

What we have is an interlinked loop, as depicted in Figure 11.2, with each factor influencing a number of others. With additional pipeline cuts there will be effects on circuit timing and area as well as the desired improvement in clock rate. All these factors influence the final architecture design. It is a multidimensional optimization with no one parameter operating in isolation.

Within a generic design, different allowable ranges may be set on the parameters defining the generated architectures. Different wordlength parameters will then have a knock-on effect on the level of pipelining required to meet certain performance criteria. The choice of submodules will also be an important factor. The target technology will determine the maximum achievable data rates and also the physical cost of the implementation.

Again, for a generic design, choices could be made available for a range of target implementations. Parameters could be set to switch between ASIC-specific and FPGA-specific code. Even within a certain implementation platform, there should be parameters in place to support a range of target technologies or devices, so as to make the most of their capabilities and the availability of on board processors or arithmetic units.

There may also be a need for a refined architecture solution meeting the performance criteria but at a reduced area cost. This is the case when the algorithm functionality is mapped down onto a reduced number of processors, the idea being that the level of hardware for the design could be scaled to meet the performance criteria of the application. With scalable designs comes the need for scalable control circuitry and scheduling and retiming of operations. These factors form the key mechanics of a successful generic design. Generating an architecture to meet the performance criteria of a larger design is one thing, but developing the generic scheduling and control of such a design is of a different level of complexity.

Software modeling of the algorithm is essential in the design development. Initially, the model is used to functionally verify the design and to analyze finite precision effects. It then forms the basis for further development, allowing test data to be generated and used within a testbench to validate the design. For the generic IP core, the software modeling is an important part of the design-for-reuse process. A core may be available to meet the needs of a range of applications; however, analysis is still required from the outset to determine the desired criteria for the implementation, such as SNR and data wordlengths. The software model is used to determine the needs for the system, and from this analysis a set of parameters should be derived and used to generate a suitable implementation using the IP core.

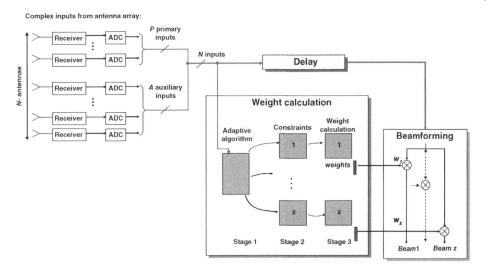

Figure 11.3 Multiple beam adaptive beamformer system

11.2.1 Adaptive Beamforming Specification

Adaptive beamforming is a general algorithm applicable in a range of applications, from medical separation of signals to military radar applications. The key factor for development of a generic design is to determine the key component within a range of adaptive beamforming applications that would be consistent to some degree and could therefore be suitable to develop as an IP core. To widen the potential of the core, it will need to be able to support a varied range of specifications dealing with issues such as the following:

Number of inputs: A varying number of auxiliary and primary inputs need to be supported (Figure 11.3). The weights are calculated for a block of the input data coming from N antennae and then applied to the same input data to generate the beamformer output for that block. For the final design, a more efficient post-processor is developed to extract the weights such as that described in Shepherd and McWhirter (1993).

Supporting a range of FPGA/ASIC technologies: By including some additional code and parameters the same core design can be re-targeted to a different technology. Doing this could enable a design to be prototyped on FPGA before targeting to ASIC.

Support for performance criteria: The variation in adaptive beamformer applications creates a wide span of desired features. For example, mobile communications power considerations and chip area could be the driving criteria, while for others a high data rate system could be the primary objective.

Scalable architecture: May need to be created to support a range of design criteria. Some key points driving the scalable architecture are desired data rate, area constraints, clock rate constraints and power constraints.

Clock rate performance: Depends on the architecture design and target technology chosen. Specifying the system requirements enables the designer to make a choice regarding the target technology and helps reach a compromise with other performance criteria such as power and area.

Wordlength: As different applications require different wordlengths, a range of wordlengths should be supported.

Level of pipelining: The desired clock rate may rely on pipelining within the design to reduce the critical path. Giving a choice of pipelining within the submodules of the design will greatly influence performance.

These values will form the basis from which to develop the adaptive beamformer solution from the generic architecture. The surrounding software models and testbenches should include the same level of scalability so as to complete the parameterization process.

11.2.2 Algorithm Development

The function of a typical adaptive beamformer is to suppress signals from every direction other than the desired "look direction" by introducing deep nulls in the beam pattern in the direction of the interference. The beamformer output is a weighted combination of signals received by a set of spatially separated antennae. An adaptive filtering algorithm calculating the filter weights is a central process of the adaptive beamforming application.

The aim of an adaptive filter is to continually optimize itself according to the environment in which it is operating. A number of mathematically and highly complex algorithms exist to calculate the filter weights according to an optimization criterion. Typically the target is to minimize an error function, which is the difference between a desired performance and the actual performance. Figure 11.4 highlights this process.

A great deal of research has been carried out into different methods for calculating the filter weights (Haykin 2002). The algorithms range in complexity and capability, and detailed analysis is required in order to determine a suitable algorithm. However there is no distinct technique for determining the optimum adaptive algorithm for a specific

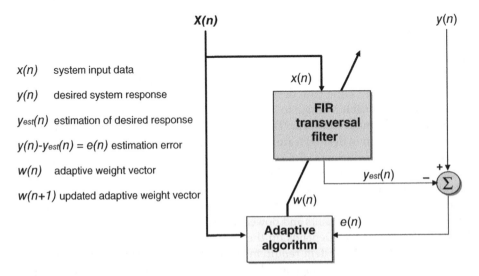

Figure 11.4 Adaptive filter system

application. The choice comes down to a balance of the range of characteristics defining the algorithms, such as:

- rate of convergence, i.e. the rate at which the adaptive algorithm reaches an optimum solution;
- steady-state error, i.e. the proximity to an optimum solution;
- ability to track statistical variations in the input data;
- computational complexity;
- ability to operate with ill-conditioned input data;
- sensitivity to variations in the wordlengths used in the implementation.

As was discussed in Chapter 2, two methods for deriving recursive algorithms for adaptive filters use Wiener filter theory and the method of least squares, resulting in the LMS and the RLS algorithms, respectively. Whilst the LMS algorithm is simpler, its limitations lie in its sensitivity to the condition number of the input data matrix as well as slow convergence rates. In contrast, the RLS algorithm is more elaborate, offering superior convergence rates and reduced sensitivity to ill-conditioned data. On the negative side, the RLS algorithm is substantially more computationally intensive than the LMS equivalent, although it is preferred here.

In particular, the QR-RLS decomposition is seen as the algorithm for adaptively calculating the filter weights (Gentleman and Kung 1982; Kung 1988; McWhirter 1983). It reduces the computation order of the calculations and removes the need for a matrix inversion, giving a more stable implementation.

11.3 Algorithm to Architecture

A key aspect of achieving a high-performance implementation is to ensure an efficient mapping of the algorithm into hardware. This involves developing a hardware architecture in which independent operations are performed in parallel so as to increase the throughput rate. In addition, pipelining may be employed within the processor blocks to achieve faster throughput rates. One architecture that uses both parallelism and pipelining is the systolic array (Kung 1988). As well as processing speed, Chapter 13 highlights its impact on power consumption. The triangular systolic array (Figure 11.5), first introduced in Chapter 2, consists of two types of cells, referred to as BCs and ICs. Figure 11.6 illustrates the process from algorithm to architecture for this implementation.

It starts with the RLS algorithm solved by QR decomposition, shown as equations. The next stage depicts the RLS algorithm solved through QR decomposition using a sequential algorithm; at each iteration a new set of values are input to the equations, thus continuously progressing towards a solution. The new data are represented by $\underline{x}^T(n)$ and $y(n)$, where x is the input data (auxiliary) matrix and y is the desired (primary) data. The term n represents the iteration of the algorithm. The QR operation can be depicted as a triangular array of operations. The data matrix is input at the top of the triangle and with each row another term is eliminated, eventually resulting in an upper triangular matrix.

The dependence graph (DG) in Figure 11.6 depicts this triangularization process. The cascaded triangular arrays within the diagram represent the iterations through time, i.e.

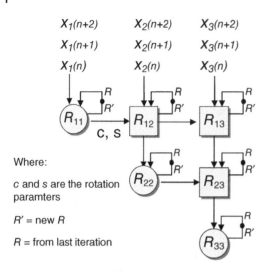

$$X_1(n+2) \quad X_2(n+2) \quad X_3(n+2)$$
$$X_1(n+1) \quad X_2(n+1) \quad X_3(n+1)$$
$$X_1(n) \quad X_2(n) \quad X_3(n)$$

Figure 11.5 Triangular systolic array for QRD RLS filtering

Where:

c and s are the rotation paramters

$R' = \text{new } R$

$R = \text{from last iteration}$

each one represents a new iteration. The arrows between the cascaded arrays highlight the dependency through time.

11.3.1 Dependence Graph

The dependencies between data can be identified in a DG. This allows the maximum level of concurrency to be identified by breaking the algorithm into nodes and arrows. The nodes outline the computations and the direction of the arrows shows the dependence of the operations. This is shown for the QR algorithm by the three-dimensional DG in Figure 11.7. The diagram shows three successive QR iterations, with arcs

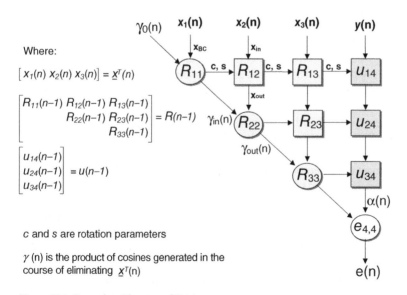

Where:

$$[x_1(n)\ x_2(n)\ x_3(n)] = \underline{x}^T(n)$$

$$\begin{bmatrix} R_{11}(n-1)\ R_{12}(n-1)\ R_{13}(n-1) \\ R_{22}(n-1)\ R_{23}(n-1) \\ R_{33}(n-1) \end{bmatrix} = R(n-1)$$

$$\begin{bmatrix} u_{14}(n-1) \\ u_{24}(n-1) \\ u_{34}(n-1) \end{bmatrix} = u(n-1)$$

c and s are rotation parameters

$\gamma(n)$ is the product of cosines generated in the course of eliminating $\underline{x}^T(n)$

Figure 11.6 From algorithm to architecture

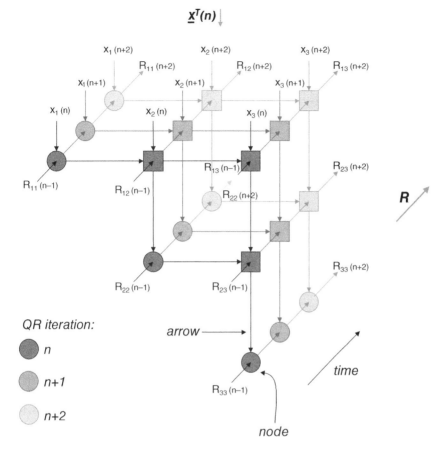

Figure 11.7 Dependence graph for QR decomposition

connecting the dependent operations. Some of the variable labels have been omitted for clarity.

In summary, the QR array performs the rotation of the input $\underline{x}^T(n)$ vector with R values held within the memory of the QR cells so that each input x value into the BCs is rotated to zero. The same rotation is continued along the line of ICs via the horizontal arrows between QR cells. From this DG, it is possible to derive a number of SFG representations. The most obvious projection, which is used here, is to project the DG along the time (i.e. R) arrows.

11.3.2 Signal Flow Graph

The transition from DG to SFG is clearly depicted in Figure 11.8. To derive the SFG from the DG, the nodes of the DG are assigned to processors, and then their operations are scheduled on these processors. One common technique for processor assignment is linear projection of all identical nodes along one straight line onto a single processor, as indicated by the projection vector **d** in Figure 11.8. Linear scheduling is then used to determine the order in which the operations are performed on the processors.

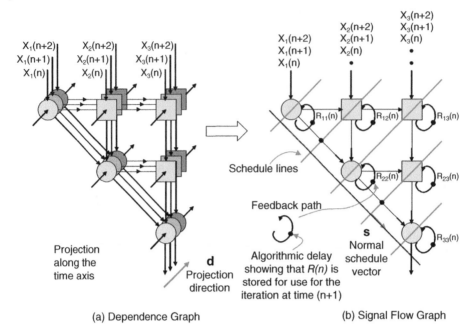

(a) Dependence Graph   (b) Signal Flow Graph

Figure 11.8 From dependence graph to signal flow graph

The schedule lines in Figure 11.8 indicate the operations that are performed in parallel at each cycle. Mathematically they are represented by a schedule vector **s** normal to the schedule lines, which points in the direction of dependence of the operations. That is, it shows the order in which each line of operations is performed.

There are two basic rules that govern the projection and scheduling, and ensure that the sequence of operations is retained. Given a DG and a projection vector **d**, the schedule is permissible if and only if:

- all the dependence arcs flow in the same direction across the schedule lines;
- the schedule lines are not parallel with the projection vector **d**.

In the QR example in Figure 11.8, each triangular array of cells within the DG represents one QR update. When cascaded, the DG represents a sequence of QR updates. By projecting along the time axis, all the QR updates may be assigned onto a triangular SFG as depicted in Figure 11.8(b). In the DG, the R values are passed through time from one QR update to another, represented by the cascaded triangular arrays. This transition is more concisely represented by the loops in Figure 11.8(b), which feed the R values back into the cells via an algorithmic delay needed to hold the values for use in the next QR update. This is referred to as a recursive loop.

The power of the SFG is that it assumes that all operations performed within the nodes take one cycle, as with the algorithmic delays, represented by small black nodes, which are a necessary part of the algorithm. The result is a more concise representation of the algorithm than the DG.

The rest of this chapter gives a detailed account of the processes involved in deriving an efficient architecture and hence hardware implementation of the SFG representation

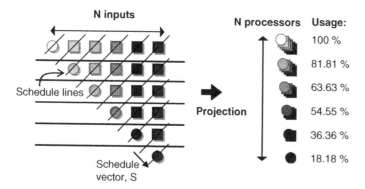

Figure 11.9 Simple linear array mapping

of the algorithm. In particular, the emphasis is on creating an intuitive design that will be parameterizable, therefore enabling a fast development for future implementations.

11.4 Efficient Architecture Design

With the complexity of the SGR QR-RLS algorithm, coupled with the number of processors increasing quadratically with the number of inputs, it is vital to generate efficient QR array architectures tailored to the applications that meet desired performance with the lowest area cost. This is achievable by mapping the triangular functionality down onto a smaller array of processors. Deriving an efficient architecture for this QR array is complicated by its triangular shape and the position of the BCs along the diagonal. A simple projection of operations from left to right onto a column of N processors leads to an architecture where the processors are required to perform both the IC and BC operations (which were described in Chapter 2). In addition, while the first processor is used 100% efficiently, this rate usage decreases down the column of processors such that the Nth processor is only used once in every N cycles. This results in an overall efficiency of about 60% as shown in Figure 11.9.

Rader (1992 1996) solved the issue of low processor usage by mirroring part B in the x-axis (see Figure 11.10) and then folding it back onto the rest of the QR array. Then, all the

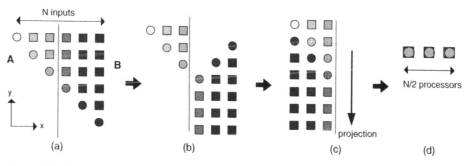

Figure 11.10 Radar mapping (Rader 1992, 1996)

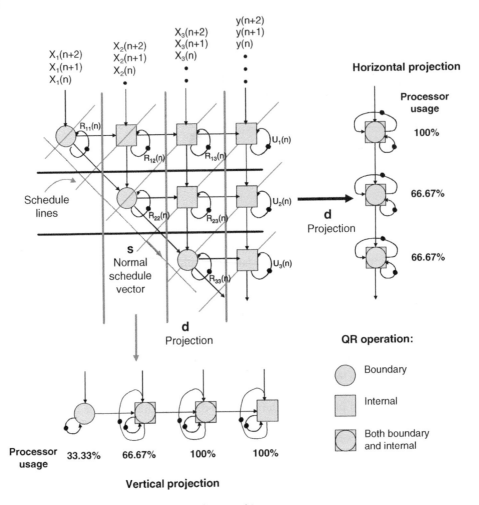

Figure 11.11 Projecting the QR array onto a linear architecture

operations were mapped down onto a linear architecture of $N/2$ processors. This works quite effectively but the BC and IC operations still need to be performed on the same processor, involving the design of a generic cell architecture or an implementation based on the CORDIC algorithm, see Hamill (1995). Another solution (Tamer and Ozkurt 2007) used a tile structure on which to map the QR cells.

Figure 11.11 gives an example of how the QR operations need to be scheduled. It shows a simplified QR array with just three auxiliary (x) inputs and one primary (y) input. The schedule lines show the sequence in which the QR operations need to be performed due to the dependence on variables passing between the cells. On each schedule line, there are a number of operations that can be performed at the same time. The normal schedule vector, **s**, then depicts the order of the operations, that is, the order of the schedule lines. Two examples are given for the projection vector, **d**. There is a horizontal projection of the QR operations onto a column of three processors. Likewise, a vertical projection is possible down to four processors. As with the example above, the resulting architectures require that both the BC and IC operations are performed on the same processor.

Another mapping (Walke 1997) solves this issue of requiring QR cells that perform both operations. The mapping assigns the triangular array of $2m^2 + 3m + 1$ cells (i.e. $N = 2m + 1$ inputs) onto a linear architecture consisting of one BC processor and m IC processors. It folds and rotates the triangular array so that all the BC operations may be assigned to one processor, while all the IC operations are implemented on a row of separate processors. All processors in the resulting linear architecture are locally interconnected and used with 100% efficiency, thus displaying the characteristics of a systolic array and hence offering all the advantages associated with these structures. This procedure is depicted in Figure 11.12 for a seven-input triangular array (for a more detailed description, see Lightbody 1999; Lightbody *et al.* 2003; Walke 1997).

For clarity, each QR operation is assigned a coordinate originating from the R (or U) term calculated by that operation, i.e. the operation $R_{1,2}$ is denoted by the coordinate 1, 2, and $U_{1,7}$ is denoted by 1, 7. To simplify the explanation, the multiplier at the bottom of the array is treated as a BC, denoted by 7, 7.

The initial aim of mapping a triangular array of cells down onto a smaller architecture is to maneuver the cells so that they form a locally interconnected regular rectangular array. This can then be partitioned evenly into sections, each to be assigned to an individual processor. This should be done in such a way as to achieve 100% cell usage and a nearest neighbor connected array. Obtaining the rectangular array is achieved through the following four stages. The initial triangular array is divided into two smaller triangles, A and B. A cut is then made after the $(m + 1)$th BC at right angles to the diagonal line of BCs (Figure 11.12(a)). Triangle A forms the bottom part of a rectangular array, with $m + 1$ columns and $m + 1$ rows.

Triangle B now needs to be manipulated so that it can form the top part of the rectangular array. This is done in two stages. By mirroring triangle B first in the x-axis, the BCs are aligned in such a way that they are parallel to the BCs in the triangle A, forming a parallelogram, as shown in Figure 11.12(b). The mirrored triangle B is then moved up along the y-axis and left along the x-axis to above A forming the rectangular array (Figure 11.12(c)). As depicted, the BC operations are aligned down two columns and so the rectangular array is still not in a suitable format for assigning operations onto a linear architecture.

The next stage aims to fold the large rectangular array in half so that the two columns of BC operations are aligned along one column. This fold interleaves the cells so that a compact rectangular processor array (Figure 11.12(d)) is produced. From this rectangular processor array, a reduced architecture can be produced by projection down the diagonal onto a linear array, with all the BC operations assigned to one BC processor and all the IC operations assigned to a row of m IC processors (Figure 11.12(e)). The resulting linear architecture is shown in more detail in Figure 11.13.

The lines drawn through each row of processors in Figure 11.12(e) (labeled 1, ... , 7), represent the set of QR operations that are performed on each cycle of the linear array. They are used to derive the schedule for architecture, as denoted more compactly by a schedule vector **s**, normal to the schedule lines. In Figure 11.13, it is assumed that registers are present on all processor outputs to maintain the data between the cycles of the schedule. Multiplexers are present at the top of the array so that system inputs to the

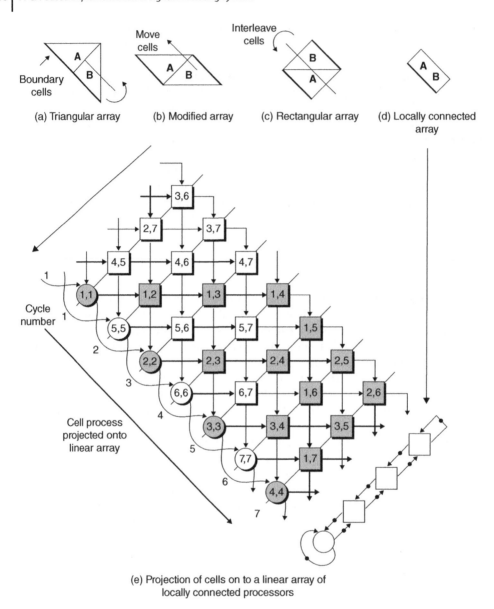

(a) Triangular array (b) Modified array (c) Rectangular array (d) Locally connected array

(e) Projection of cells on to a linear array of locally connected processors

Figure 11.12 Interleaved processor array

QR array can be supplied to the cells at the right instance in time. The linear array has only local interconnections, so all the cell inputs come from adjacent cells. The bottom multiplexers govern the different directions of dataflow that occur between rows of the original array.

The folding of the triangular QR array onto an architecture with reduced number of processors means that the R values need to be stored for more than one clock cycle. They are held locally within the recursive data paths of the QR cells, rather than external

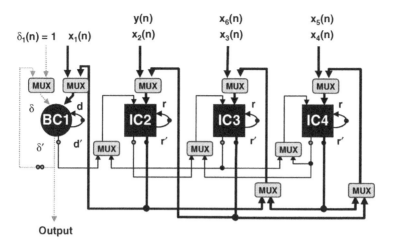

Figure 11.13 Linear architecture for a seven-input QR array

memory (i.e. the values are pipelined locally to delay them until they are needed). Some of the required delays are met by the latency of existing operations within the loop and the remainder are achieved by inserting additional registers.

11.4.1 Scheduling the QR Operations

The derivation of the architecture is only a part of the necessary development as a valid schedule needs to be determined to ensure that the data required by each set of operations are available at the time of execution. This implies that the data must flow across the schedule lines in the direction of the schedule vector. The rectangular processor array in Figure 11.12(d) contains all the operations required by the QR algorithm, showing the sequence in which they are to be implemented on the linear architecture. Therefore, this diagram can be used to show the schedule of the operations to be performed on the linear architecture.

An analysis of the scheduling and timing issues can now be refined. Looking at the first schedule line, it can be seen that operations from two different QR updates have been interleaved. The shaded cells represent the current QR update at time n and the unshaded cells represent the previous unfinished update at time $n - 1$. Effectively the QR updates have been interleaved. This is shown in more clarity in Figure 11.14. The first QR operation begins at cycle 1, then after $2m + 1$ cycles of the linear architecture the next QR operation begins. Likewise, after a further $2m + 1$ cycles the third QR operation is started. In total, it takes $4m + 1$ cycles of the linear architecture to complete one specific QR update.

From Figure 11.14, it can be seen that the x inputs into the QR cells come from either external system data, i.e. from the snapshots of data forming the input $x(n)$ matrix and $y(n)$ vector, or internally from the outputs of other processors. The external inputs are fed into the linear architecture every $2m + 1$ clock cycles.

If each QR cell takes a single clock cycle to produce an output, then there will be no violation of the schedule shown in Figure 11.12. However, additional timing issues must be taken into account as processing units in each QR cell have detailed timing

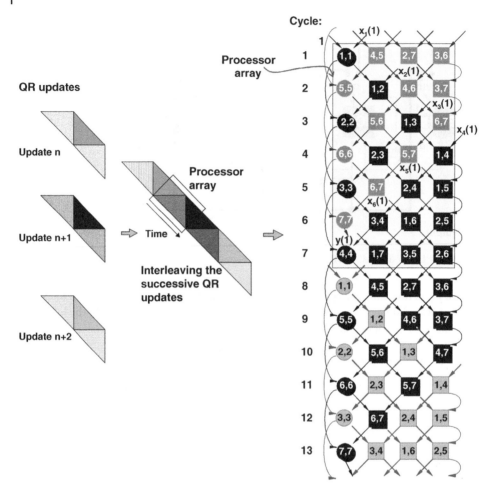

Figure 11.14 Interleaving successive QR operations. (Source: Lightbody 2003. Reproduced with permission of IEEE.)

requirements. The retiming of the operations is discussed in more detail later on in this chapter.

Note that the processor array highlighted in Figure 11.14 is equivalent to the processor array given in Figure 11.12(d). This processor array is the key starting point from which to develop a generic QR architecture.

11.5 Generic QR Architecture

The technique shown so far was applied to a QR array with only one primary input. More generally, the QR array would consist of a triangular part and a rectangular part (Figure 11.15(a)), the sizes of which are determined by the number of auxiliary and primary inputs, respectively. Typically, the number of inputs to the triangular part is at least a

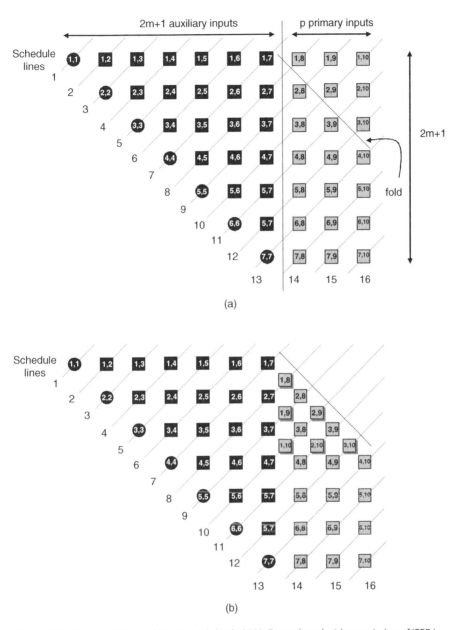

Figure 11.15 Generic QR array. (Source: Lightbody 2003. Reproduced with permission of IEEE.)

factor greater than the number of inputs to the rectangular part, with example numbers for radar being 40 inputs for the triangular part and only 2 for the rectangular part.

The mapping procedure presented in this section implements both the triangular and rectangular components of the QR array in a single architecture. As before, the BC and IC operations are kept to two distinct processors. The additional factor presented by the generic mapping technique is that a choice of linear or rectangular architectures is available. The number of IC processors may be reduced further, allowing more flexibility in the level of hardware reduction. However, at least one BC processor is required, even if the number of ICs is reduced below one row. Note that the connections have been removed from Figure 11.15 and in later following diagrams in order to reduce the complexity of the diagram and aid clarity.

11.5.1 Processor Array

In the previous section, the triangular structure of the QR array was manipulated into a rectangular processor array of locally interconnected processors, as shown in Figure 11.12(d). From this starting point, the operations can be mapped onto a reduced architecture. A simplified method for creating the processor array is demonstrated in the following example.

The processor array is obtained through two steps. Firstly, a fold is made by folding over the corner of the array after the *m*th cell from the right-hand side, as depicted in Figure 11.15. The cells from the fold are interleaved between the rows of unfolded cells as shown. The next stage is to remove the gaps within the structure by interleaving successive QR updates in the same manner as shown in Figure 11.14. The choice of position of the fold and the size of the triangular part of the array are important. By placing the fold after the *m*th cell from the right-hand side, a regular rectangular array of operations can be produced.

This is shown in greater detail in the Figure 11.16, which shows that there is a section which repeats over time and contains each of all the required QR operations. This section is referred to as the processor array. It is more clearly depicted in Figure 11.17, which shows just the repetitive section from Figure 11.16.

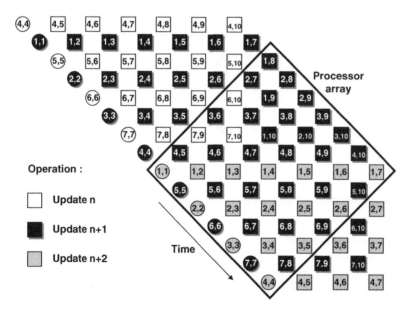

Figure 11.16 Repetitive section

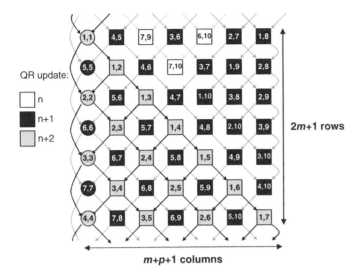

QR update:

□ n

■ n+1

▨ n+2

2m+1 rows

m+p+1 columns

Figure 11.17 Processor array. (Source: Lightbody 2003. Reproduced with permission of IEEE.)

In this example, the processor array contains QR operations built up from three successive QR updates, represented by the differently shaded cells. The interconnections are included within this diagram, showing that all cells are locally connected. The size of the processor array is determined by the original size of the triangular QR array, which in turn is governed by the number of auxiliary and primary inputs, $2m + 1$ and p, respectively. The resulting processor array has $2m + 1$ rows and $m + p + 1$ columns. As expected, the product of these two values gives the number of operations within the original QR array. From the processor array, a range of architectures with reduced number of processors can be obtained by dividing the array into partitions and then assigning each of the partitions to an individual processor. There are several possible variants of QR architecture:

Linear architecture: The rectangular array is projected down onto a linear architecture with one BC and $m + p$ ICs.

Rectangular architecture: The rectangular array is projected down onto a number of linear rows of cells. The architecture will have r rows (where $1 < r \leq 2m + 1$), and each row will have one BC and $m + p$ ICs.

Sparse linear architecture: The rectangular array is projected down onto a linear architecture with one BC and less than $m + p$ ICs.

Sparse rectangular architecture: The rectangular array is projected down onto a number of linear rows of cells. The architecture will have r rows (where $1 < r \leq 2m + 1$), and each row will have one BC and less than $m + p$ ICs.

Linear Array

The linear array is derived by assigning each column of operations to an individual processor, as shown in Figure 11.18. In total, it takes $4m + p + 1 = 16$ cycles of the linear array to complete each QR operation. In addition, there are $2m + 1 = 7$ cycles between the start of successive QR updates. This value is labeled as T_{QR}. Note that so far the

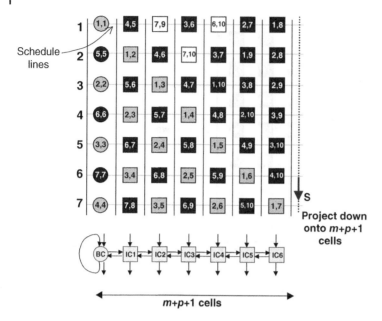

Figure 11.18 Linear array. (Source: Lightbody 2003. Reproduced with permission of IEEE.)

latency of the QR cells is considered to be one clock cycle, i.e. on each clock cycle one row of QR operations is performed on the linear architecture. Later sections will examine the effect of a multi-cycle latency, which occurs when cell processing elements with detailed timings are used in the development of the generic QR architecture.

Sparse Linear Array

A further level of hardware reduction is given in Figure 11.19, resulting in a sparse linear array. Here the number of IC processors has been halved. When multiple columns (i.e. N_{IC} columns) of IC operations are assigned to each processor then the number of iterations of the architecture is increased by this factor. Hence, for the sparse linear array, T_{QR} is expressed as the product of $2m + 1$ (used in the linear array) and N_{IC}. The schedule for the sparse linear array example is illustrated in Figure 11.20.

Rectangular Array

The processor array can be partitioned by row rather than by column so that a number of rows of QR operations are assigned to a linear array of processors. The example below shows the processor array mapped down on an array architecture. As the processor array consisted of 7 rows, 4 are assigned to one row and 3 are assigned to the other. To balance the number of rows for each linear array, a dummy row of operations is needed and is represented by the cells marked by the letter D.

On each clock cycle, the rectangular array processor executes two rows of the original processor array. Each QR iteration takes 18 cycles to be completed, two more clock cycles than for the linear array due to the dummy row of operations. However, the QR

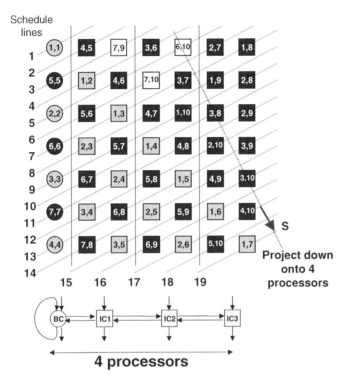

Figure 11.19 Sparse linear array

updates are started more frequently. In this case T_{QR} is 4, compared to the linear array which took 7 cycles. For the array architecture, T_{QR} is determined by

$$T_{QR} = \frac{(2m + 1) + N_D}{N_{rows}},$$

where N_{rows} is the number of lines of processors in the rectangular architecture, and N_D is the number of rows of dummy operations needed to balance the schedule. The resulting value relates to the number of cycles of the architecture required to perform all the operations within the processor array.

Sparse Rectangular Array

The sparse rectangular array assigns the operations to multiple rows of sparse linear arrays. A number of rows of the processor array are assigned to each linear array. The columns are also partitioned so that multiple columns of operations are assigned to each IC processor, as shown in Figure 11.22.

The QR update takes 34 cycles for completion and each update starts every 7 cycles, i.e. $T_{QR} = 7$. Including the term N_{IC}, the equation for T_{QR} becomes

$$T_{QR} = \frac{((2m + 1) + N_D)N_{IC}}{N_{rows}}.$$

For example, $T_{QR} = ((2 \times 3 + 1 + 0) \times 2)/2 = 7$ cycles.

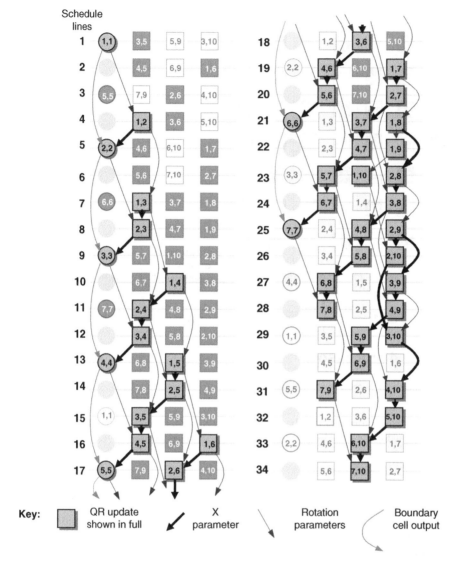

Figure 11.20 One QR update scheduled on the sparse linear array. (Source: Lightbody 2003. Reproduced with permission of IEEE.)

The discussion to date has concentrated on mapping QR arrays that have an odd number of auxiliary inputs. The technique can be applied to an array with an even number with a slight reduction in overall efficiency.

11.6   Retiming the Generic Architecture

The QR architectures discussed so far have assumed that the QR cells have a latency of one clock cycle. The mapping of the architectures is based on this factor; hence there will be no conflicts of the data inputs. However, the inclusion of actual timing details

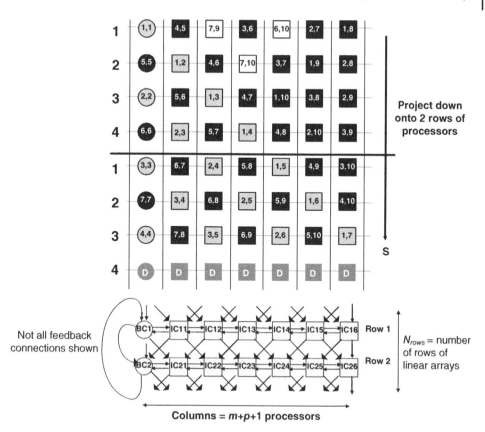

Figure 11.21 Rectangular array

within the QR cells will affect this guarantee of a valid data schedule. The arithmetic IP processors (McCanny *et al.* 1997; Northeastern University 2007), used to implement the key arithmetic functions such as multiplication, addition and division involve timing details which will impact the overall circuit timing. Embedding processor blocks with specific timing information, coupled with the impact of truncation and internal word growth, means that detailed retiming of the original SFGs of the QR cells must be performed before the processors can be used to implement the QR architecture (Trainor *et al.* 1997). The overall effect of retiming is to incur variable latencies in the output data paths of the QR cells. The effect of real timing information within the QR cells is discussed in this section.

The choice for the QR array was to use floating-point arithmetic to support the dynamic range of the variables within the algorithm. The floating-point library used supported variable wordlengths and levels of pipelining, as depicted in Figure 11.23.

In adaptive beamforming, as with many signal processing applications, complex arithmetic representations are needed as incoming signals contain a magnitude and phase component. This is implemented using one signal for the real part and another for the imaginary part, and gives the BC and IC operations shown in the SFGs depicted in Figure 11.24.

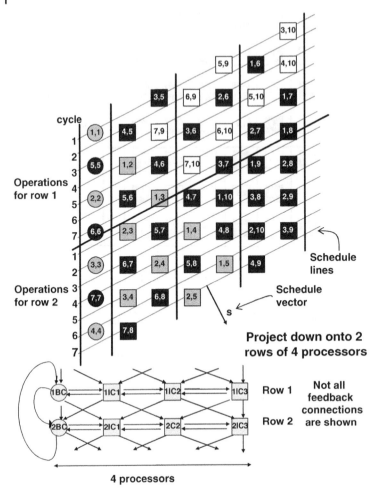

Figure 11.22 Sparse rectangular array. (Source: Lightbody 2003. Reproduced with permission of IEEE.)

The floating-point complex multiplication is built up from four real multiplications and two real additions: $(a + jb)(c + jd) = (ac - bd) + j(ad + bc)$. An optimization is available to implement the complex multiplication using three multiplications and fives additions/subtractions as illustrated in Section 6.2.2. However, given that an addition is of a similar area to multiplication within floating-point arithmetic due to the costly exponent calculation, this is not beneficial. For this reason, the four-multiplication version is used. The detail of the complex arithmetic operations is given in Figure 11.25.

The SFGs for the BCs and ICs are given in Figures 11.26 and 11.27, respectively. These diagrams show the interconnections of the arithmetic modules within the cell architectures. Most functions are self-explanatory, except for the shift-subtracter. For small values of x, the operation $\sqrt{1-x}$ can be approximated by $1 - x^2$ which may be implemented by a series of shifts denoted by $D = A - \text{Shift}(A, N)$. This operation is used to implement the forgetting factor, β, within the feedback paths of the QR cells. This value, β, is close to 1, therefore x is set to $1 - \beta$ for the function application.

Floating point block	Add	Sub Add	Shft Sub	Mult	Div	Round
Function	Addition S = A+B	Adder/sub S = A+B when Sub = 0 else S = A−B	Shift-subtractor: D = A− Shift (A, N).	Multiplier: P = X×Y	Divider: Q = N/D	Rounder
Symbol	$\begin{array}{c} A \\ \downarrow \\ B \rightarrow \oplus \\ \downarrow \\ S \end{array}$	$\begin{array}{c} A \\ \downarrow \\ B \rightarrow \oplus \\ Sub \\ \downarrow \\ S \end{array}$	$\begin{array}{c} A \\ \downarrow \\ \rightarrow \bullet \\ \downarrow \\ D \end{array}$	$\begin{array}{c} X \\ \downarrow \\ Y \rightarrow \otimes \\ \downarrow \\ P \end{array}$	$\begin{array}{c} N \\ \downarrow \\ D \rightarrow \oslash \\ \downarrow \\ Q \end{array}$	$\begin{array}{c} F_{in} \\ \downarrow \\ \boxed{R} \\ \downarrow \\ F_{out} \end{array}$
Latency	0 – 3	0 – 3	0 – 1	0 – 2	0 – Mbits+1	0 – 1
Label for Latency	P_A	P_A	P_S	P_M	P_D	P_R

Figure 11.23 Arithmetic modules. (Source: Lightbody 2003. Reproduced with permission of IEEE.)

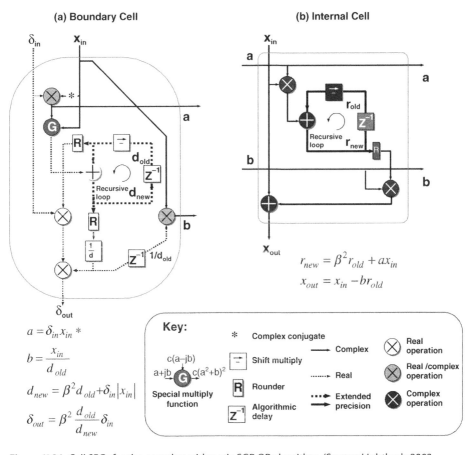

(a) Boundary Cell

(b) Internal Cell

$$r_{new} = \beta^2 r_{old} + ax_{in}$$
$$x_{out} = x_{in} - br_{old}$$

$$a = \delta_{in} x_{in} *$$

$$b = \frac{x_{in}}{d_{old}}$$

$$d_{new} = \beta^2 d_{old} + \delta_{in} |x_{in}|$$

$$\delta_{out} = \beta^2 \frac{d_{old}}{d_{new}} \delta_{in}$$

Key:

* Complex conjugate	
→ Complex	⊗ Real operation
$\overrightarrow{\Box}$ Shift multiply	⊗ Real /complex operation
$c(a{-}jb)$ $a{+}jb \rightarrow \circled{G} \rightarrow c(a^2{+}b)^2$ Special multiply function	⊗ Complex operation
$\boxed{R}$ Rounder	······▶ Real
$\boxed{Z^{-1}}$ Algorithmic delay	····▶ Extended precision

Figure 11.24 Cell SFGs for the complex arithmetic SGR QR algorithm. (Source: Lightbody 2003. Reproduced with permission of IEEE.)

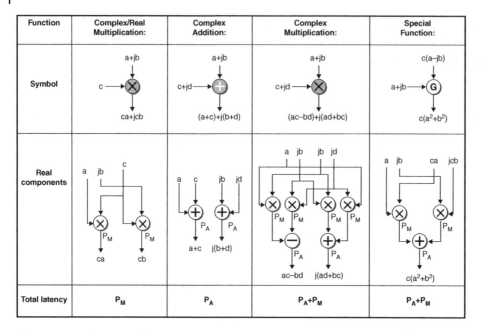

Figure 11.25 Arithmetic modules. (Source: Lightbody 2003. Reproduced with permission of IEEE.)

There are a number of feedback loops within the QR cells, as shown in Figures 11.26 and 11.27. These store the R values from one RLS iteration to the next. These loops will be a fundamental limitation to achieving a throughput rate that is close to the clock rate and, more importantly, could lead to considerable inefficiency in the circuit utilization. In other words, even when using a full QR array, the delay in calculating the new R values will limit the throughput rate.

Figures 11.26 and 11.27 show the QR cell descriptions with generic delays placed within the data paths. These are there to allow for the re-synchronization of operations due to the variable latencies within the arithmetic operators, i.e. to ensure correct timing. The generic expressions for the programmable delays are listed in Tables 11.1 and 11.2 for the BC and IC, respectively.

Secondly, to maintain a regular data schedule, the latencies of the QR cells are adjusted so that the x values and rotation parameters are output from the QR cells at the same time. The latency of the IC in producing these outputs can be expressed generically using a term L_{IC}. The latencies of the BC in producing the rotation parameters, a and b, are also set to L_{IC} to keep outputs synchronized. However, the latency of the BC in producing the δ_{out} is set to double this value, $2L_{IC}$, as this relates back to the original scheduling of the full QR array, which showed that no two successive BC operations are performed on successive cycles. By keeping the structure of the data schedule, the retiming process comes down to a simple relationship.

11.6.1 Retiming QR Architectures

This subsection continues with the discussion of retiming issues and how to include them in a generic architecture.

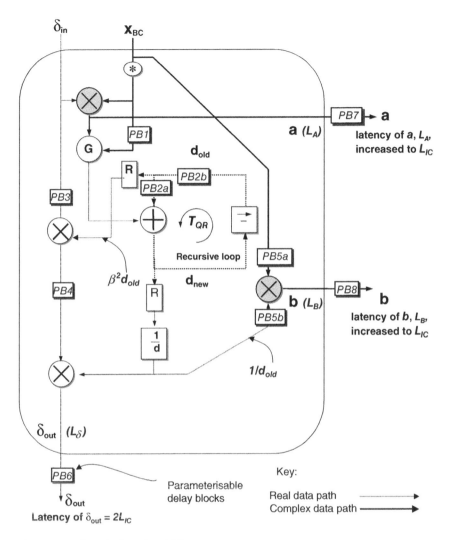

Figure 11.26 Generically retimed BC

Retiming of the Linear Array Architecture

The latency has the effect of stretching out the schedule of operations for each QR update. This means that iteration $n = 2$ begins $2m + 1$ clock cycles after the start of iteration $n = 1$. However, the introduction of processor latency stretches out the scheduling diagram such that iteration $n = 2$ begins after $(2m + 1)L_{IC}$ clock cycles. This is obviously not an optimum use of the linear architecture as it would only be used every L_{IC}th clock cycle. A factor, T_{QR}, was introduced in the previous section as the number of cycles between the start of successive QR updates, as determined by the level of hardware reduction.

It can be shown that a valid schedule which results in a 100% utilization can be achieved by setting the latency L_{IC} to a value that is relatively prime to T_{QR}. That is,

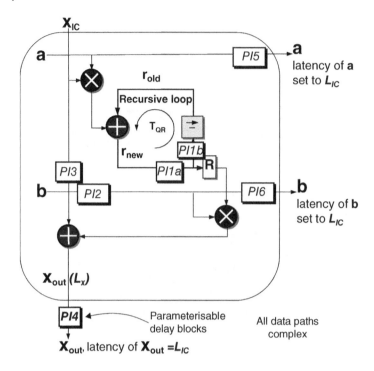

Figure 11.27 Generically retimed IC

if the two values do not share a common factor other than 1 then their lowest common multiple will be their product. Otherwise there will be data collisions at the products of L_{IC} and T_{QR} with their common multiplies. Thus if $T_{QR} = \mod c$ and $L_{IC} = \mod c$ then $T_{QR} = d \times c$ and $L_{IC} = e \times c$, giving $c = T_{QR}/d = L_{IC}/e$, where c is a common multiple of T_{QR} and L_{IC} and a positive integer other than 1, and d and e are factors of T_{QR} and

Table 11.1 BC generic timing

BC	Delay Value
B_{RL}	$2P_A + 2P_M + P_R + P_S - T_{QR}$
PB1	P_M
PB2	$T_{QR} - P_A - P_B$
PB2a	If $B_{RL} < 0$, then $-B_{RL}$, otherwise, 0
PB2b	If $B_{RL} < 0$, then $PB2 - PB2a$, otherwise $PB2$
PB3	If $B_{RL} > 0$, then B_{RL}, otherwise, 0
PB4	$2P_A + P_M + P_R + P_D - PB3$
PB5	$2P_A + 2P_M + P_R + P_D - T_{QR}$
PB5a	If $PB5 < 0$, then $PB5$, otherwise, 0
PB5b	If $PB5 > 0$, then $PB5$, otherwise, 0
PB6	$L_{IC} - L_\delta$
PB7	$L_{IC} - L_a$
PB8	$L_{IC} - L_b$

Table 11.2 IC generic timing

IC	Delay Value
I_{RL}	$2P_A + P_M + P_R - T_{QR}$
PI1	$T_{QR} - P_A - P_S$
PI1a	If $I_{RL} < 0, -I_{RL}$, otherwise, 0
PI1b	If $I_{RL} < 0, PI1 - PI1a$, otherwise, $PI1$
PI2	If $I_{RL} > 0, I_{RL}$, otherwise, $PI1$
PI3	$PI2 + P_A + P_M$
PI4	$L_{IC} - L_x$
PI5	L_{IC}
PI6	$L_{IC} - PI2$

Table 11.3 Generic expressions for the latencies of the BC and IC

Latency	Value
La	P_M
Lb	$P_M + PB5$
L_δ	$PB3 + PB4 + 2P_M$
L_X	$PI3 + P_A$

L_{IC} respectively. Hence, there would be a collision at $T_{QR} \times e = L_{IC} \times d$. This means that the products of both $T_{QR} \times e$ and $L_{IC} \times d$ must be less than $T_{QR} \times L_{IC}$. Therefore, there is a collision of data. Conversely, to obtain a collision free set of values, c is set to 1.

The time instance $T_{QR} \times L_{IC}$ does not represent a data collision as the value of T_{QR} is equal to $2m + 1$, as the QR operation that was in line to collide with a new QR operation will have just been completed. The other important factor in choosing an optimum value of T_{QR} and L_{IC} is to ensure that the processors are 100% efficient.

The simple relationship between T_{QR} and L_{IC} is a key factor in achieving a high utilization for each of the types of structure. More importantly, the relationship gives a concise mathematical expression that is needed in the automatic generation of a generic QR architecture complete with scheduling and retiming issues solved.

Figure 11.28 shows an example schedule for the seven-input linear array that was originally shown in Figure 11.12 where L_{IC} is 3 and T_{QR} is 7. The shaded cells represent the QR operations from different updates that are interleaved with each other and fill the gaps left by the highlighted QR update. The schedule is assured to be filled by the completion of the first QR update; hence, this is dependent on the latency, L_{IC}.

11.7 Parameterizable QR Architecture

The main areas of parameterization include the wordlength, the latency of arithmetic functions, and the value of T_{QR}. Different specifications may require different finite precision, therefore the wordlength is an important parameter. The QR cells have been built up using a hierarchical library of arithmetic functions, which are parameterized in terms of wordlength, with an option to include pipelining to increase the operation speed as

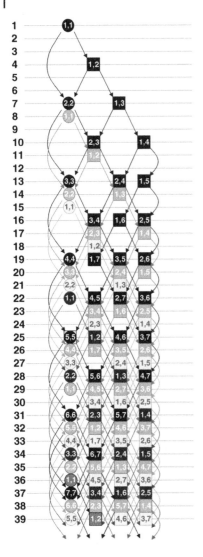

Figure 11.28 Schedule for a linear array with an IC latency of 3

required. These parameters are passed down through the hierarchy of the HDL description of the QR cells to these arithmetic functions. Another consideration is the value of T_{QR}, which determines the length of the memory needed within the recursive loops of the QR cells which hold the R and u values from one QR update to the next. Both T_{QR} and the level of pipelining within the arithmetic functions are incorporated in generic timing expressions of the SGR QR cells.

11.7.1 Choice of Architecture

Table 11.4 demonstrates the process for designing a QR architecture when given a specific sample rate and QR array size. The examples below are for a large QR array with 45 auxiliary inputs and 4 primary inputs, i.e. $m = 22$ and $p = 12$. The resulting processor array is $2m + 1 = 45$ rows by $m + p + 1 = 35$ columns. For a given sample throughput

Table 11.4 Other example architectures (clock speed = 100 MHz)

Arch.	Details	Number of processors			T_{QR}	Data rate MSPS
		BC	IC	total		
Full QR	Processor for each QR cell	45	1170	1215	4	25
Rectangular 1	Processor array assigned onto 12 linear arrays, each responsible for 4 rows	12	312	324	4	25
Rectangular 2	Processor array assigned onto 3 linear arrays, each responsible for $45/3 = 15$ rows	3	78	81	$(2m+1)/3$ (15)	6.67
Rectangular 3	Processor array assigned onto 15 linear arrays (13 ICs), each responsible for $45/3 = 15$ rows 2 columns of ICs to each	15	195	210	$(2m+1)/15$	16.67
Sparse rectangular	2 columns of ICs to each IC processor of 3 linear arrays	3	39	42	$2(2m+1)/3$ (30)	3.33
Linear	1 BC and 26 ICs	1	26	27	$2m+1$ (45)	2.22
Sparse linear	2 columns of ICs assigned to each IC processor of a linear array	1	13	14	$2(2m+1)$ (90)	1.11

rate and clock rate, we can determine the value for T_{QR}, as depicted in the table. Note that the resulting value for T_{QR} and L_{IC} must be relatively prime, but for these examples we can leave this relationship at present.

The general description for T_{QR}, as shown above, can be rearranged to give the following relationship:

$$\frac{N_{IC}}{N_{rows}} = \frac{T_{QR}}{2m+1}.$$

This result is rounded down to the nearest integer. There are three possibilities:

- If $\frac{T_{QR}}{2m+1} > 1$ then a sparse linear array is needed.

- If $\frac{T_{QR}}{2m+1} = 1$ then a linear array is needed.

- If $\frac{T_{QR}}{2m+1} < 1$ then a rectangular array is needed.

Depending on the dimensions of the resulting architecture, the designer may decide to opt for a sparse rectangular architecture.

Note that the maximum throughput that the full triangular array can meet is limited to 25 MSamples/s due to the four-cycle latency within the QR cell recursive path for the

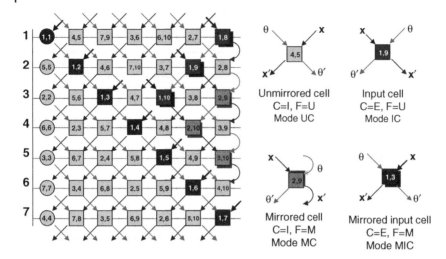

Figure 11.29 Types of cells in processor array

specific implementation listed in Lightbody et al. The first rectangular array solution is meeting the same throughput performance as the full QR array using only 408 ICs and 12 BCs, instead of the full array which requires 1530 ICs and 45 BCs.

11.7.2 Parameterizable Control

A key aspect of the design of the various architectures is the determination of the control data needed to drive the multiplexers in these structures. Due to the various mappings that have been applied, it is more relevant of think of the IC operation as having four different modes of operation: input, mirrored input, unmirrored cell and mirrored cell (Figure 11.29). The mirrored ICs are the result of the fold used to derive the rectangular processor array from the QR array and simply reflect a different dataflow. The cell orientation is governed by the multiplexers and control, and is therefore an issue concerning control signal generation.

The four modes of operation can be controlled using two control signals, C, which determines whether the x input is from the array (I) or from external data (E), and F, which distinguishes between a folded (M) and an unfolded operation (U). The latter determines the source direction of the inputs. The outputs are then from the opposite side of the cell. A mechanism for determining the control of each architecture is given next.

11.7.3 Linear Architecture

The control signals for the linear architecture were derived directly from its data schedule. The modes of operation of the cells were determined for each cycle of the schedule, as shown in Table 11.5. Figure 11.30 shows the QR cells with the applied control and multiplexers, and the control signals for a full QR operation for this example are given in Table 11.6. The control and timing of the architectures for the other variants become more complex – in particular, the effect that latency has on the control

Table 11.5 Modes of operation of the QR cells for the linear array

Cycle	BC1	IC2	IC3	IC4	IC5	IC6	IC7
1	**IC**	UC	UC	UC	UC	UC	**MIC**
2	UC	**IC**	UC	UC	UC	**MIC**	UM
3	UC	UC	**IC**	UC	**MIC**	UC	**MIC**
4	UC	UC	UC	**IC**	UM	**MIC**	UM
5	UC	UC	UC	UC	**IC**	UM	**MIC**
6	UC	UC	UC	UC	UM	**IC**	UM
7	UC	UC	UC	UC	UM	UM	**IC**

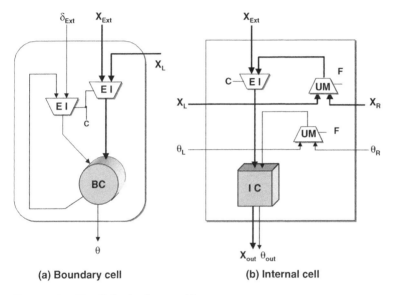

(a) Boundary cell   (b) Internal cell

Figure 11.30 QR cells for the linear architecture

Table 11.6 Linear array control for the *x*-inputs and for mirrored/not mirrored cells

Cyc.	C1	C2	C3	C4	C5	C6	C7	F1	F2	F3	F4	F5	F6
1	E	I	I	I	I	I	E	U	U	U	U	U	M
2	I	E	I	I	I	E	I	U	U	U	U	M	U
3	I	I	E	I	E	I	I	U	U	U	M	U	M
4	I	I	I	E	I	I	I	U	U	U	U	M	U
5	I	I	I	I	E	I	I	U	U	U	U	U	M
6	I	I	I	I	I	E	I	U	U	U	U	U	U
7	I	I	I	I	I	I	E	U	U	U	U	U	U
8	E	I	I	I	I	I	E	U	U	U	U	U	M
9	I	E	I	I	I	E	I	U	U	U	U	M	U
10	I	I	E	I	E	I	I	U	U	U	M	U	M
11	I	I	I	E	I	I	I	U	U	U	U	M	U
12	I	I	I	I	E	I	I	U	U	U	U	U	M
13	I	I	I	I	I	E	I	U	U	U	U	U	U

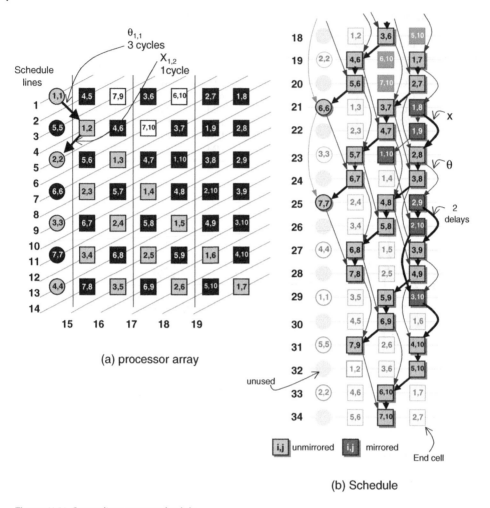

Figure 11.31 Sparse linear array schedule

sequences. In the sparse variants, extra delays need to be placed within the cells to organize the schedule, and in the rectangular variants, the cells need to be able to take x and
θ inputs from the cells above and below as well as from adjacent cells. Each of these
variants shall be looked at in turn.

11.7.4 Sparse Linear Architecture

Figure 11.31(a) shows two columns of operations being assigned onto each IC. From the
partial schedule shown in Figure 11.31(b), it can be seen that the transition of a value
from left to right within the array requires a number of delays. The transfer of θ_1 from
BC(1, 1), to the adjacent IC(1, 2) takes three cycles. However, the transfer of X_{12} from
the IC to the BC only takes one cycle.

The example in Figure 11.32 shows the partitioning of three columns of ICs. Scheduling them onto a single processor requires their sequential order to be maintained. The

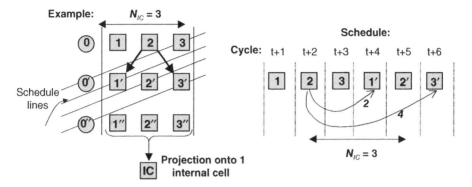

Figure 11.32 Example partitioning of three columns onto one processor. (Source: Lightbody 2003. Reproduced with permission of IEEE.)

IC operations have been numbered 1, 2, 3 for the first row, and $1'$, $2'$, $3'$ for the second row. The outputs generated from operation 2 are required for operations $1'$ and $3'$. Because all the operations are being performed on the same processor, delays are needed to hold these values until they are required by operations $1'$ and $3'$. Operation 3 is performed before operation $1'$, and operations 3, $1'$ and $2'$ are performed before operation $3'$, which relates to 2 and 4 clock cycle delays, respectively. This has been generically defined according to the number of columns of operations within the processor array assigned to each IC, N_{IC}, as shown in Figure 11.33.

Two output values, x and θ, are transferred from operation $c+1$ to c and $c+2$. The value that is fed to a specific operation depends on whether the cells perform the folded or unfolded modes of operation as summarized in Table 11.5. If the data is transferred

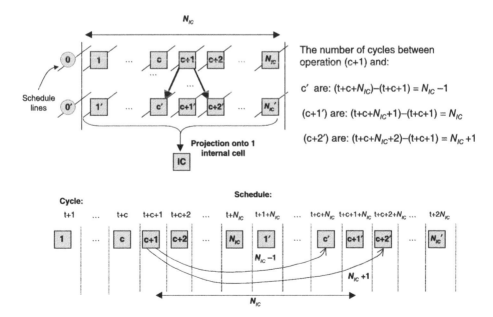

Figure 11.33 Generic partitioning of N_{IC} columns onto one processor

Table 11.7 Required delays for sparse linear array (U, not mirrored; M, mirrored)

Data transfer	Direction in terms of QR operation	Dataflow direction	Delays	Label
U → U	$(i,j) \rightarrow (i, j+1), \theta$	→	$N_{IC} + 1$	D1
	$(i,j) \rightarrow (i+1, j), x$	←	$N_{IC} - 1$	D2
M → M	$(i,j) \rightarrow (i+1, j), \theta$	←	$N_{IC} - 1$	D2
	$(i,j) \rightarrow (i, j+1), x$	→	$N_{IC} + 1$	D1
U → M(end cell)	$(i,j) \rightarrow (i, j+1), \theta$	↓	N_{IC}	D3
M → U(end cell)	$(i,j) \rightarrow (i+1, j), x$	↓	N_{IC}	D3

between the same type of cell (i.e. $U \rightarrow U$, or $M \rightarrow M$) then the delay will be either $N_{IC} - 1$ or $N_{IC} + 1$, according to Table 11.7. However, if the data transfer is between different types of cell (i.e. $U \rightarrow M$, or $M \rightarrow U$, as in the case of the end processor), then the number of delays will be N_{IC}. This is summarized in Table 11.7.

These delays are then used within the sparse linear architecture to keep the desired schedule as given in Figure 11.34. The three levels of delays are denoted by the square blocks labeled D1, D2 and D3. These delays can be redistributed to form a more efficient QR cell architectures as shown in Figure 11.35. The extra L and R control signals indicate the direction source of the inputs, with E and I control values determining whether the inputs come from an adjacent cell or from the same cell. EC refers to the end IC that differs slightly in that there are two modes of operation when the cell needs to accept inputs from its output. The control sequences for this example are given in Figure 11.36.

From Figure 11.36, it can be seen that EC is the same as R and is the inverse of L. In addition, the states alternate between E and I with every cycle, therefore, one control

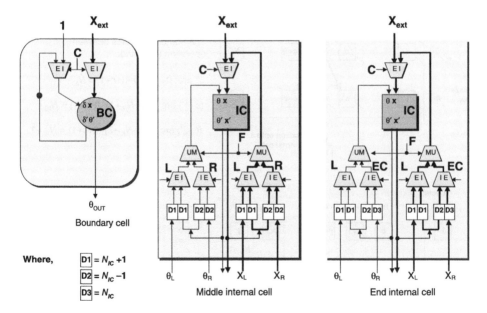

Figure 11.34 Sparse linear array cells

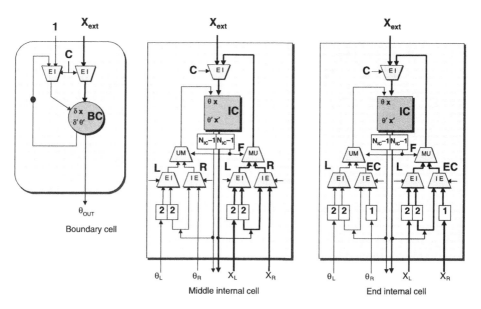

Figure 11.35 Redistributed delays for sparse linear array cells

sequence could be used to determine the control of the internal inputs. This control value has been labeled D. The control signals shall be categorized as external input control, C_i, fold control, F_i, array control, L_i and internal input control, D_i The subscripts are coordinates representing the cells to which the control signals are being fed.

One of the key issues with the sparse linear array is the effect of the latencies in the QR cells on the schedule (which previously assumed a one-cycle delay). With the linear architecture, the schedule was scaled by the latency. However, with the sparse linear array, there was a concern that the delays $N_{IC} - 1$, N_{IC}, $N_{IC} + 1$ would also need to be

Cycle	External Input	External Input Control				Fold Control			Internal Input Control		
		C1	C2	C3	C4	F2	F3	F4	L	R	EC
1	$X_1(1)$	E	I	I	I	U	U	M	E	I	E
2	$X_6(0)$	I	I	I	E	U	U	U	I	E	I
3		I	I	I	I	U	U	U	E	I	E
4	$X_2(1)$	I	E	I	I	U	U	U	I	E	I
5	$X_7(0)$	I	I	I	E	U	U	U	E	I	E
6		I	I	I	I	U	U	U	I	E	I
7	$X_3(1), X_8(0)$	I	E	I	E	U	U	M	E	I	E
8	$X_9(0)$	I	I	I	E	U	U	M	I	E	I
9	$X_{10}(0)$	I	I	E	I	U	M	U	E	I	E
10	$X_4(1)$	I	I	E	I	U	U	U	I	E	I
11		I	I	I	I	U	U	M	E	I	E
12		I	I	I	I	U	U	M	I	E	I
13	$X_5(1)$	I	I	E	I	U	U	U	E	I	E
14		I	I	I	I	U	U	U	I	E	I

Figure 11.36 Control sequence for sparse linear array

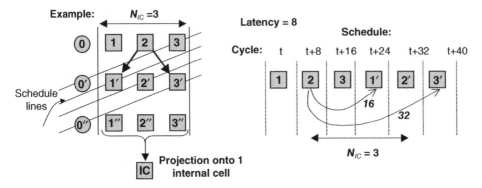

Figure 11.37 Possible effect of latency on sparse linear array schedule

scaled in order to keep the structure of the original schedule, which would cause inefficiency. This is depicted in Figure 11.37 for a latency of 8. This is not the case, as the delays $N_{IC} - 1$, N_{IC}, $N_{IC} + 1$ can be applied using the existing latency within the QR cells. The minimum allowed number of clock cycles between successive operations is the latency. By setting $N_{IC} - 1$ to this minimum value, and then setting N_{IC} to be one clock cycle more and $N_{IC} + 1$ to be two clock cycles more, a valid and efficient schedule can be achieved. This is depicted in Figure 11.38.

In the example given in Figure 11.39, the latency of the IC is 3, so this gives the minimum value for N_{IC} as 4. $N_{IC} + 1$ is therefore 5 and $N_{IC} - 1$ is 3 clock cycles. The shaded cells in Figure 11.39 show one complete QR update with interconnection included. The rest of the QR operations are shown but with limited detail to aid clarity. Since it is most probable that the latency of the IC will exceed the number of columns assigned to each processor, it figures that the delays within the linear sparse array will depend on L_{IC}, i.e. the $N_{IC} - 1$ delay will not be needed and the schedule realignment will be performed by the single- and double-cycle delays shown in Figure 11.35. The highlighted cells represent a full QR update, while the other numbered cells represent interleaved QR operations. The faded gray BCs with no numbers represent unused positions within the schedule.

11.7.5 Rectangular Architecture

The rectangular architecture consists of multiple linear array architectures that are concatenated. Therefore, the QR cells need to be configured so that they can accept inputs

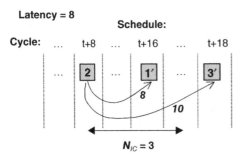

Figure 11.38 Merging the delays into the latency of the QR cells

Figure 11.39 Effect of latency on schedule for the sparse linear array ($N_{IC} = 3$)

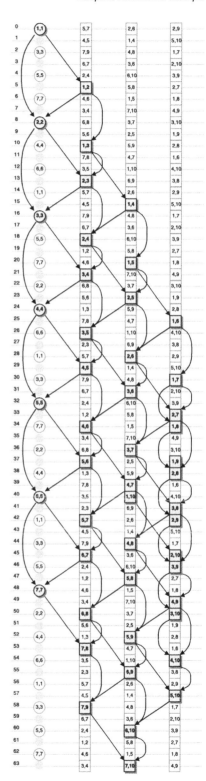

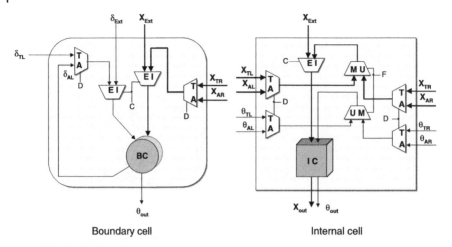

Figure 11.40 QR cells for the rectangular array

from the above linear array. In addition, the top linear array needs to be able to accept values from the bottom linear array. The QR cells are depicted in Figure 11.40. The control signals, E and I, decide on whether the X inputs are external (i.e. system inputs) or internal. The control value, T, refers to inputs from the above array and A refers to inputs from adjacent cells. When used as subscripts, TR and TL refer to values coming from the left and right cells of the array above. AR and AL refer to the values coming from the right and left adjacent cells within the same linear array.

11.7.6 Sparse Rectangular Architecture

The QR cells for the sparse rectangular array need to be able to feed inputs back to themselves, in addition to the variations already discussed with the linear and rectangular architectures. The extra control circuitry is included in the QR diagrams shown in Figure 11.41. The control and the delays required by the sparse arrays to realign the schedule are brought together into LMR multiplexer cells (Figure 11.42) that include delays needed take account of the retiming analysis demonstrated in this section.

It was discussed with the sparse linear array how certain transfer in data values required the insertion of specific delays to align the schedule. This also applies to the rectangular array and the same rules can be used.

The starting point for determining the schedule for the sparse rectangular array is the schedule for the sparse linear array. From this, the rows of operations are divided into sections, each to be performed on a specific sparse linear array. The control, therefore, is derived from the control for the linear sparse version. The next section deals with parametric ways of generating the control for the various QR architectures. In addition to the control shown so far, the next section analyzes how latency may be accounted for within the control generation.

11.7.7 Generic QR Cells

The sparse rectangular array QR cells, shown in Figure 11.41, can be used for all of the QR architecture variants, by altering the control signals and timing parameters.

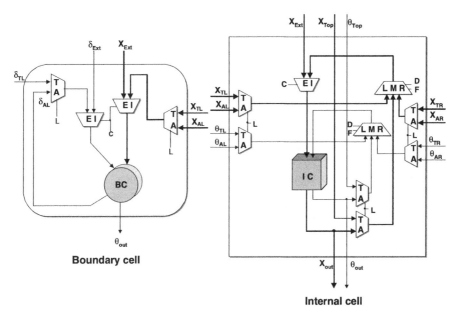

Figure 11.41 QR cells for the sparse rectangular array

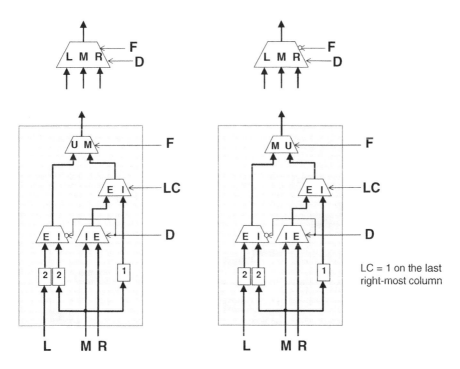

Figure 11.42 LMR control circuitry for sparse arrays

However, in the sparse variants, there are added delays embedded within the LMR control cells. These can be removed for the full linear and rectangular array versions, by allowing them to be programmable so that they may be set to zero for the non-sparse versions. The key to the flexibility in the parameterizable QR core design is the generic generation of control signals. This is discussed in the following section.

11.8 Generic Control

The previous section detailed the various architectures derived from the QR array. Some detail was given of the control signals needed to operate the circuits. This section looks at generic techniques for generating the control signals that may be applied to all the QR architecture variants. It is suggested that a software interface is used to calculate each control sequence as a bit-vector seed (of length T_{QR}) that may be fed through a linear feedback register which will allow this value to be cyclically output bit by bit to the QR cells.

The first stage in developing the control for the QR array is to look at the generic processor array which gives the control needed for the linear array. From this, the control signals may be folded and manipulated into the required sequence for the sparse linear arrays. The control for the rectangular versions may be generated quite simply from the control for the linear architectures.

11.8.1 Generic Input Control for Linear and Sparse Linear Arrays

A new external x-input is fed into a cell of the linear array on each clock cycle, starting from the leftmost cell, reaching the leftmost cell and then folding back until all the $2m + p + 1$ inputs are fed into the array for that specific QR update. This is highlighted for one set of QR inputs in Figure 11.43. The next set of inputs follow the same pattern but start after T_{QR} cycles. The result is a segment of control signals that repeat every T_{QR} cycles (which is 7 for the linear array example and 14 for the sparse linear array

Cycle	Input	Linear array							Input	Sparse linear array			
		C1	C2	C3	C4	C5	C6	C7		C1	C2	C3	C4
1	$X_1(1)$, $X_8(0)$	E	I	I	I	I	I	E	$X_1(1)$	E	I	I	I
2	$X_2(1)$, $X_9(0)$	I	E	I	I	I	E	I	$X_6(0)$	I	I	I	E
3	$X_3(1)$, $X_{10}(0)$	I	I	E	I	E	I	I		I	I	I	I
4	$X_4(1)$	I	I	I	E	I	I	I	$X_2(1)$	I	E	I	I
5	$X_5(1)$	I	I	I	I	E	I	I	$X_7(0)$	I	I	I	E
6	$X_6(1)$	I	I	I	I	I	E	I		I	I	I	I
7	$X_7(1)$	I	I	I	I	I	I	E	$X_3(1)$, $X_8(0)$	I	E	I	E
8	$X_8(1)$, $X_1(2)$	E	I	I	I	I	I	E	$X_9(0)$	I	I	I	E
9	$X_9(1)$, $X_2(2)$	I	E	I	I	I	E	I	$X_{10}(0)$	I	I	E	I
10	$X_{10}(1)$, $X_3(2)$	I	I	E	I	E	I	I	$X_4(1)$	I	I	E	I
11	$X_4(2)$	I	I	I	E	I	I	I		I	I	I	I
12	$X_5(2)$	I	I	I	I	E	I	I		I	I	I	I
13	$X_6(2)$	I	I	I	I	I	E	I	$X_5(1)$	I	I	E	I
14	$X_7(2)$	I	I	I	I	I	I	E		I	I	I	I

Figure 11.43 Control for the external inputs for the linear QR arrays

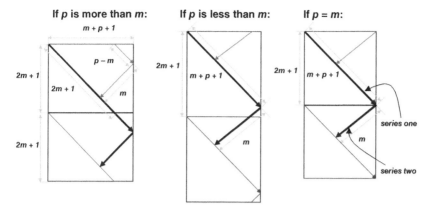

If p is more than m: **If p is less than m:**   **If p = m:**

Figure 11.44 External inputs

example). The aim is to automatically generate vectors, containing T_{QR} bits, which represent the repeating sections for each of the control signals, C1 to C7. The key point is to determine when the next set of QR inputs starts in relation to the previous set. This can be determined mathematically from the dimensions of the original QR array and the resulting processor array from which the QR architectures are derived, i.e. a QR array with $2m + 1$ auxiliary inputs and p primary inputs leads to a processor array with $2m + 1$ rows and $m + p + 1$ columns. This relationship is depicted by Figure 11.44. The heavy lines indicate the series of inputs for one QR update, and relate to the highlighted control for the external inputs for the linear array example in Figure 11.43.

Software code can be written to generate the control signals for the external inputs for the linear and sparse linear array. The inputs are broken down into two series (see Figure 11.44), one dealing with the inputs going from left to right, and the other dealing with the inputs from right to left (the change in direction being caused by the fold).

The code generates the position of the control signals within the control vector for each input into each processor. If the vector number is larger than the vector, then the vector size is subtracted from this value, leaving the modulus as the position. However, after initializing the operation of the QR array, it is necessary to delay this control signal by an appropriate value.

11.8.2 Generic Input Control for Rectangular and Sparse Rectangular Arrays

The control from the rectangular versions is then derived from these control vectors by dividing the signals vectors into parts relating to the partitions within the processor array. For example, if the control seed vectors for the linear array are eight bits wide and the rectangular array for the same system consists of two rows, then each control vector seeds would be divided into two vectors each four bits wide, one for the first rectangular array and the other for the second. The control seed for the sparse rectangular array is derived in the same manner from the control of the sparse linear array with the same value of N_{IC}. The same code may be edited to include the dummy operations that may be required for the sparse versions. Figure 11.45(a) shows an example sparse linear array mapping with $m = 4$, $p = 3$ and $N_{IC} = 2$. The control in Figure 11.45(b) can be divided

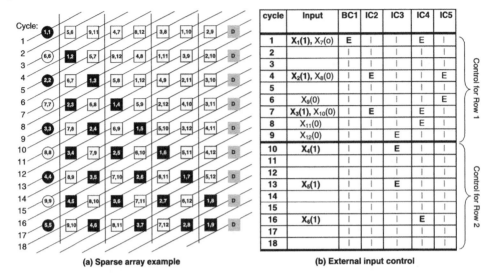

cycle	Input	BC1	IC2	IC3	IC4	IC5	
1	$X_1(1)$, $X_7(o)$	E	I	I	E	I	
2		I	I	I	I	I	
3		I	I	I	I	I	
4	$X_2(1)$, $X_8(0)$	I	E	I	I	E	
5		I	I	I	I	I	
6	$X_9(0)$	I	I	I	I	E	
7	$X_3(1)$, $X_{10}(0)$	I	E	I	E	I	
8	$X_{11}(0)$	I	I	I	E	I	
9	$X_{12}(0)$	I	I	E	I	I	
10	$X_4(1)$	I	I	E	I	I	
11		I	I	I	I	I	
12		I	I	I	I	I	
13	$X_5(1)$	I	I	E	I	I	
14		I	I	I	I	I	
15		I	I	I	I	I	
16	$X_6(1)$	I	I	I	E	I	
17		I	I	I	I	I	
18		I	I	I	I	I	

Control for Row 1 / Control for Row 2

(a) Sparse array example

(b) External input control

Figure 11.45 Partitioning of control seed

into two sections for implementing a sparse rectangular array consisting of two rows of the sparse linear array.

11.8.3 Effect of Latency on the Control Seeds

The next stage is to determine the effect that latency has on the control vectors. As discussed in this section, the sparse array needs delay values $D1$, $D2$ and $D3$, to account for assigning multiple columns, N_{IC}, of operations to each IC processor. For a system with a single cycle latency, $D1 = N_{IC} - 1$, $D2 = N_{IC}$ and $D3 = N_{IC} + 1$. However, in the real system the processors have multiple latency. It is assumed that the latency of the IC, L_{IC}, will be greater than these delays, so the delays are added onto the latency such that the appropriate delays become $D1 = L_{IC}$, $D2 = L_{IC} + 1$ and $D3 = L_{IC} + 2$. For the linear array the values $D1$, $D2$ and $D3$ are all set to L_{IC}. Then the code may be used to generate the control vectors. The only difference is when the position of the control value exceeds the width of the vector. With the single latency version, this was accounted for by subtracting the value T_{QR} from the value (where the width of the vector seed is T_{QR}).

When latency is included within the calculations, it is not sufficient to reduce the value to within the bounds of the vector width. Alternatively, the position of the control value within the vector is found by taking the modulus of T_{QR}. An analysis of the effect of latency on the control vectors is shown through an example linear array where $m = 3$, $p = 5$ and $T_{QR} = 2m + 1 = 7$.

One point to highlight is the fact that there may be several cycles of the control vector before the required input is present. For example, the vector in the above example for C4 is [I I E I I I I], but the first required input is at time 10, not 3. Therefore it is necessary to delay the start of this control signal by 7 cycles. The technique relies on the use of initialization control signals to start the cycling of the more complicated control vectors for the processors. However, the method discussed offers a parametric way of dealing with control and allows the majority of the control to be localized. In addition, the same

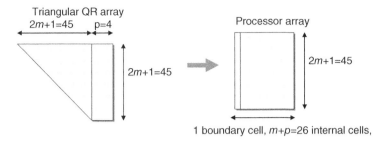

Figure 11.46 Example QR architecture derivation, $m = 22, p = 4$

principles used to develop the control signals for the timing of the external inputs may be applied for the rest of the control signals, i.e. the fold, internal input, and row control.

11.9 Beamformer Design Example

For a typical beamforming application in radar, the values of m would range from 20 to over 100. The number of primary inputs, p, would typically be from 1 to 5 for the same application. An example specification is given in Figure 11.46. One approach is to use the QR array. Assuming the fastest possible clock rate, f_{CLK}, the fundamental loop will dictate the performance and result in a design with 25% utilization. Thus the major challenge is now to select the best architecture, mostly closely matching the throughput rate with the best use of hardware. For the example here, a desired input sample rate of 15 MSPS with a maximum possible clock rate of 100 MHz is assumed.

The value for T_{QR} can be calculated using the desired sample rate, S_{QR}, and the maximum clock rate, f_{CLK}:

$$T_{QR} = \frac{f_{CLK}}{S_{QR}} = \frac{100 \times 10^6}{15 \times 10^6} = 6.67.$$

This value is the maximum number of cycles allowed between the start of successive QR updates, therefore, it needs to be rounded down to the nearest integer. The ratio N_{rows}/N_{IC} can be obtained by substituting for the known parameters into the relationship below:

$$\frac{N_{rows}}{N_{IC}} = \frac{2m + 1}{T_{QR}} = \frac{45}{6} = 7.5,$$

where $1 \leq N_{rows} \leq 2m + 1$ (i.e. 45) and $1 \leq N_{IC} \leq m + p$ (i.e. 26). Using these guidelines, an efficient architecture can be derived by setting $N_{IC} = 2$, and hence $N_{rows} = 15$. The operations are distributed over 15 sparse linear architectures, each with 1 BC and 13 ICs, as shown in Figure 11.47.

Also note that the circuit critical path within the circuit must be considered to ensure that the core can be clocked fast enough to support the desired QR operation. Here, additional pipeline stages may be added to reduce the critical path and therefore improve the clock rate. However, this has the effect of increasing the latencies and these must then be included in the architecture analysis.

Figure 11.47 Example architecture

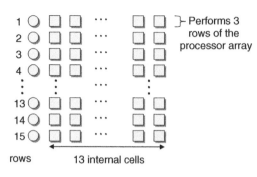

1
2
3
4
⋮

13
14
15

rows   13 internal cells

}– Performs 3
rows of the
processor array

Each row of processors is responsible for three rows of operations within the processor array, therefore, $T_{QR} = 6$, resulting in an input sample rate of 16.67 MSPS, which exceeds the required performance. The details of some example architectures for the same QR array are given in Table 11.8.

The value for T_{QR} for the full QR array implementation is determined by the latency in the recursive loop of the QR cells (consisting of a floating- point addition and a shift subtract function). For the example shown, the QR array needs to wait four clock cycles for the calculation of the value in the recursive loop, which therefore determines the sample rate of the system. This example emphasizes the poor return of performance of the full QR implementation at such a high cost of hardware. The same performance can be achieved by using the first rectangular array example with only about one quarter the number of processors.

Table 11.8 Selected architectures (clock speed = 100 MHz)

Arch.	Details	Number of processors			T_{QR}	Data rate MSPS
		BC	IC	total		
Full QR	Processor for each QR cell	45	1530	1575	4	25
Rectangular 1	Processor array assigned onto 12 linear arrays, each responsible for 4 rows	12	408	420	4	25
Rectangular 2	Processor array assigned onto 3 linear arrays, each responsible for $45/3 = 15$ rows	3	102	105	$(2m+1)/3$ (15)	6.67
Sparse rectangular	3 columns of ICs to each IC processor of 3 linear arrays	3	51	54	$2(2m+1)/3$ (30)	3.33
Linear	1 BC and 34 ICs	1	34	35	$2m+1$ (45)	2.22
Sparse linear	2 columns of ICs assigned to each IC processor of a linear array	1	17	18	$2(2m+1)$ (90)	1.11

11.10   Summary

The goal of this chapter was to document each of the stages of development for an IP core for adaptive beamforming. The main aspects covered were the design choices made with regard to:

- the decision to use design-for-reuse strategies to develop an IP core;
- determination of the algorithm;
- determination of a suitable component to design as an IP core;
- specifying the generic parameters;
- algorithm to architecture development;
- scalable architectures;
- scalable scheduling of operations and control.

Each stage listed above was detailed for the adaptive beamforming example. Background information was supplied regarding the RLS choice of algorithm decided upon for the adaptive weight calculations. The key issue with the algorithm used is its computational complexity. Techniques and background research were summarized showing the derivation of the simplified QR-RLS algorithm suitable for implementation on a triangular systolic array. Even with such reduction in the complexity there may still be a need to map the full QR array down onto a reduced architecture set.

This formed a key component of the chapter, giving a step-by-step overview of how such a process can be achieved while maintaining a generic design. Consideration was given to architecture scalability and the effects of this on operation scheduling. Further detail was given of the effects of processor latency and retiming on the overall scheduling problem, showing how such factors could be accounted for upfront. Finally, examples were given on how control circuitry could be developed so as to scale with the architecture, while maintaining performance criteria. It is envisaged that the principles covered by this chapter should be expandable to other IP core developments.

Bibliography

Athanasiadis T, Lin K, Hussain Z 2005 Space-time OFDM with adaptive beamforming for wireless multimedia applications. In *Proc. 3rd Int. Conf. on Information Technology and Applications*, pp. 381–386.

Baxter P, McWhirter J 2003 Blind signal separation of convolutive mixtures. In *Proc. IEEE Asilomar Conf. on Signals, Systems and Computers*, pp. 124–128.

Choi S, Shim D 2000 A novel adaptive beamforming algorithm for a smart antenna system in a CDMA mobile communication environment. *IEEE Trans. on Vehicular Technology*, 49(5), 1793–1806.

de Lathauwer L, de Moor B, Vandewalle J 2000 Fetal electrocardiogram extraction by blind source subspace separation. *IEEE Trans. on Biomedical Engineering*, 47(5), 567–572.

Gentleman W, Kung H 1982 Matrix triangularization by systolic arrays. In *Proc. SPIE*, 298, 19–26.

Hamill R 1995 VLSI algorithms and architectures for DSP arithmetic computations. PhD thesis, Queen's University Belfast.

Haykin S 2002 *Adaptive Filter Theory*, 4th edn. Prentice Hall, Upper Saddle River, NJ.

Hudson J 1981 *Adaptive Array Principles*. IET, Stevenage.

Kung S 1988 *VLSI Array Processors*. Prentice Hall, Englewood Cliffs, NJ.

Lightbody G 1999 High performance VLSI architectures for recursive least squares adaptive filtering. PhD thesis, Queen's University Belfast.

Lightbody G, Woods R, Walke R 2003 Design of a parameterizable silicon intellectual property core for QR-based RLS filtering. *IEEE Trans. on VLSI Systems*, 11(4), 659–678.

McCanny J, Ridge D, Hu Y, Hunter J 1997 Hierarchical VHDL libraries for DSP ASIC design. In *Proc. IEEE Int. Conf. on Acoustics, Speech, and Signal Processing*, pp. 675–678.

McWhirter J 1983 Recursive least-squares minimization using a systolic array. In *Proc. SPIE*, 431, 105–109.

Northeastern University 2007 Variable precision floating point modules. http://www.coe .neu.edu/Research/rcl/projects/floatingpoint/index.htmlprojects (accessed November 7, 2016).

Rader C 1992 MUSE –a systolic array for adaptive nulling with 64 degrees of freedom, using Givens transfomations and wafer scale integration. In *Proc. Int. Conf. on Application Specific Array Processors*, pp. 277–291.

Rader C 1996 VLSI systolic arrays for adaptive nulling. *IEEE Signal Processing Magazine*, 13(4), 29–49.

Shan T, Kailath T 1985 Adaptive beamforming for coherent signals and interference. *IEEE Trans. on Acoustics, Speech and Signal Processing*, 33(3), 527–536.

Shepherd TJ, McWhirter JG 1993 Systolic adaptive beamforming. In Haykin S, Litva J, Shepherd TJ (eds) *Array Signal Processing*, pp. 153–243. Springer, Berlin.

Tamer O, Ozkurt A 2007 Folded systolic array based MVDR beamformer. In *Proc. Int. Symp. on Signal Processing and its Applications*, pp. 1–4.

Trainor D, Woods R, McCanny J 1997 Architectural synthesis of digital signal processing algorithms using IRIS. *J. of VLSI Signal Processing*, 16(1), 41–55.

Walke R 1997 High sample rate Givens rotations for recursive least squares. PhD thesis, University of Warwick.

Wiltgen T 2007 Adaptive beamforming using ICA for target identification in noisy environments. Master's thesis, Virginia Tech.

12

FPGA Solutions for Big Data Applications

12.1 Introduction

We live in an increasingly digitized world where the amount of data being generated has grown exponentially – a world of Big Data (Manyika *et al.* 2011). The creation of large data sets has emerged as a hot topic in recent years. The availability of this valuable information presents the possibility of analyzing these large data sets to give a competitive and productivity advantage. The data come from a variety of sources, including the collection of productivity data from manufacturing shop floors, delivery times, detailed information on company sales or indeed the enormous amount of information currently being created by social media sites. It is argued that by analyzing social media trends, it should be possible to create potentially greater revenue generating products.

Big Data analytics (Zikopoulos *et al.* 2012) is the process by which value is created from these data and involves the loading, processing and analysis of large data sets. Whilst database analytics is a well-established area, the increase in data size has driven interest in using multiple distributed resources to undertake computations, commonly known as *scaling out*. This is achieved using a process known as MapReduce (Dean and Ghemawat 2004) which helps the user to distribute or *map* data across many distributed computing resources to allow the computation to be performed, and then bringing all of these computed outputs together, or *reducing*, to produce the result. Apache^{TM} Hadoop$^®$ (Apache 2015) is an open source resource for achieving this.

Given this distributed nature of processing, the authors could therefore be accused of being opportunistic in including this hot topic in a book dedicated to implementation of DSP systems using FPGA technology, but there is sound reasoning for doing so. Data analytics comprises the implementation of computationally complex algorithms, and for classes of algorithms that cannot be *scaled out* there is a need to improve processing within the single computing unit. This is known as *scaling up* and requires the realization of efficient, scalable hardware to achieve this functionality. As will be demonstrated in this chapter, the process is similar to many of those applied to signal processing examples.

There is certainly a strong case, increasingly targeted at FPGA, for developing scaled-up solutions for a number of key data mining tasks, specifically classification, regression

FPGA-based Implementation of Signal Processing Systems,
Second Edition. Roger Woods, John McAllister, Gaye Lightbody and Ying Yi.
© 2017 John Wiley & Sons, Ltd. Published 2017 by John Wiley & Sons, Ltd.

and clustering. Such algorithms specifically include decision tree classification (DTC), artificial neural networks (ANNs) and support vector machines (SVMs). There is a strong case for developing FPGA-based solutions to achieve scaling up as performance and especially power will become increasingly important. This chapter considers the implementation of such algorithms and, in particular, the k-means clustering algorithm.

Big data concepts are introduced in Section 12.2. Details are given on Big Data analytics and various forms of data mining are introduced in Section 12.3. In Section 12.4, the acceleration of Big Data analytics is discussed and the concepts of scaling up and scaling out are introduced. The case for using FPGAs to provide acceleration is made and a number of FPGA implementations reviewed, including the acceleration of the Heston model for determining share options. The computation of k-means clustering is then described in Section 12.5. The idea of using processors to make the heterogeneous FPGA-based computing platform more programmable is introduced in Section 12.6 and then applied to k-means clustering in Section 12.7. Some conclusions are given in Section 12.8.

12.2 Big Data

The number of digital information sources continues to grow as we look to digitize all sorts of information. These sources range from output from social media, storage of text information from mobile phones, personal digitized information (e.g. medical records), and information from the increasing number of security cameras. This growth of information as been labeled Big Data and is recorded in terms of exabytes (10^{18}) and zettabytes (10^{21}).

Of course, not only has the term "Big Data" emerged to define this new type of information, but marketing forces have pushed for the definition of its various forms. Big Data has been classified in terms of a number of characteristics namely, volume, velocity, variety, veracity and value, the so-called "five Vs." Some of the key features of these are outlined below.

- *Volume*, of course, refers to the amounts of data being generated. Whether this is social media data from emails, tweets etc. or data generated from sensor data (e.g. security cameras, telemetric sources), there is now a need to store these zettabytes or even brontobytes (10^{27}) of information. On Facebook alone, 10 billion messages are sent per day! Such data sets are too large to store and analyze using traditional *structured* database technology, and so there is a push to store them in *unstructured* form using distributed systems.
- *Velocity* refers to the rate at which these new data sets are generated and distributed. In our new era of instant financial transactions and smartphone connectivity, there is a need for immediate response (within seconds). This has major consequences for not only the computing infrastructure but also the communications technology to ensure fast low- latency connectivity. Lewis (2014) relates the major activity and cost involved in creating a fiber link between Chicago and New York just to shave several milliseconds off the latency for financial markets!
- *Variety* refers to the different flavors of data, whether it be social media data which may be incomplete, transitory data or even financial data which have to be secure.

Whilst structured data databases would have been used in the past for storing information, it is estimated that 80% of the world's data is now unstructured, and therefore cannot be put easily into conventional databases.

- *Veracity* refers to the trustworthiness of the data. Social media data are transitory and less reliable as they may be incorrect (and possibly deliberately so) and of poor quality, whereas security camera information is inaccurate but possibly of low quality or low information content. The challenge is then to develop algorithms to cope with the quality of data and possibly use volume as a means of improving the information content.
- *Value* is probably the most relevant data characteristic as it represents the inherent worth of the information. There is no doubt that it represents the most important aspect of Big Data as it allows us to make sense of it. However, a major challenge is to be able to extract the value from the data which is central aspect of Big Data analytics.

These are the challenges for Big Data analytics: it may be useful to have a high volume of valuable information that has strong veracity, but this is only useful if we can make sense of the information. Thus whilst Hadoop may provide an infrastructure for storing and passing information for processing, it is the implementation of very complex analytics that is critical. Note that McNulty (2014) talks about Big Data in terms of the "seven Vs," adding variability and visualization to our list above.

12.3 Big Data Analytics

The availability of such a rich form of data now presents the possibility of identifying important value. For example, insurance companies currently employ statistical models on the multiple sources of information available, including previous premium prices and even on-media and spending habits, to work out appropriate and acceptable insurance premiums! Of course, considerable potential also exists for analyzing Big Data for marketing reasons, and this represents a key driver for many of the data analytics algorithms. With 1.2 billion people using apps, blogs and forums to post, share and view content, there is a considerable body of available information.

Another very relevant area is *security*. With terrorism being a regular issue in our world today, new forms of technology are increasingly being used by terrorists to communicate with each other; these range from using mobile phones to entering information on social media sites. Extremist and terrorist groups use the internet for a wide variety of purposes, including dissemination of propaganda, recruitment, and development and execution of operational objectives.

As the volume and velocity of social media data rise exponentially, cyber threats are increasing in complexity, scale and diversity. Social media intelligence (SOCMINT) is an emerging science that aims to address this challenge through enhanced analytics that can present a step-change in a defense intelligence community's ability to instantaneously classify and interpret social media data, identify anomalies and threats, and prevent future attacks (Omand *et al.* 2012).

It is therefore clear that Big Data analytics is a growing field of study and seems to involve applying existing and new algorithms to make sense of data. It is argued that this is not simply statistics or the result of applying data mining algorithms and that the challenges of Big Data require the development of new forms of algorithms. To this end, the term "Big Data scientist" has been used to describe a specialist who is able to develop new forms of the most suitable statistical and data mining techniques to data analysis. The key is to create value from Big Data, and this presents a new computing problem as it involves analysis of data sets that are at least of the order of terabytes.

With the increase in data size, the concept of using multiple distributed resources to undertake Big Data computations is now becoming commonplace. A key development was the creation of MapReduce, which is a programming model to allow algorithms and large data sets to be distributed on a parallel, distributed computing platform and then brought back together to produce the result. The open source manifestation of MapReduce has been through Apache™ Hadoop® (Apache 2015). This is one of the earliest open source software resources for reliable, scalable, distributed computing. A number of evolutions have occurred including a structured query language (SQL) engine, *Impala* (Cloudera 2016), a generic scheduler, *Yarn*, and API file formats, namely *Crunch*.

A lot of hype surrounds Big Data, but there are also a number of major technical challenges in terms of acquiring, storing, processing and visualizing the data. From an FPGA perspective, there is a clear need to implement complex data processing algorithms in a highly parallel and pipelined manner.

12.3.1 Inductive Learning

The key aspect in many Big Data applications is that the user has to learn from data in cases where no analytical solutions exist but the data are used to construct an empirical solution (Abu-Mostafa *et al.* 2012). This is typically known as *inductive learning* and is a core area of machine learning. It involves the user spotting patterns in the information and generalizing them, hoping that they are correct but providing no guarantee that the solution will be valid. It is the source of many data mining algorithms and probably is the area where machine learning and data mining intersect. A query can then be applied to deduce a possible answer from the data (see Figure 12.1).

The key focus is to create chunks of knowledge by applying specific algorithms about some domain of interest which is presented by the data to be analyzed; this will usually be capable of providing an answer by transcending the data in such a way that the answer cannot just be provided by extracting and aggregating value from the data, i.e. creating a model that represents the data. These predictive models are the core of data mining and usually involve the application of techniques to transform the data to create the model; the model is then easier to apply as only the attributes most useful for model creation need be used and links combined to create new models to provide a better prediction.

Usually the process involves looking at a domain of the data (e.g., financial transactions, hospital bed occupancy rates) which may in many cases be incomplete and described by a set of features (Marshall *et al.* 2014). A data set is typically a subset of the domain described by a set of features; the goal is to apply a set of data mining algorithms to create one or more models from the data.

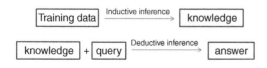

Figure 12.1 Inference of data. Source: Cichosz 2015. Reproduced with permission of John Wiley & Sons.

12.3.2 Data Mining Algorithms

The three most widely used data mining tasks, *classification*, *regression* and *clustering*, can be considered as *inductive learning* tasks (Cichosz 2015). These aim to make sense of the data:

- *Classification.* Prediction of a discrete target attribute by the assignment of the particular instance to a fixed possible set of classes, which could, for example, involve detecting odd behavior in credit card fraud.
- *Regression.* Prediction of a numerical target attributes based on some quantity of interest, for example, working out the length of stay distribution of geriatric patients admitted to one of six key acute hospitals (Marshall *et al.* 2014).
- *Clustering.* Prediction of the assignment of the instances to a set of similarly based clusters to determine the best cluster organization, i.e. similarity of data in a cluster.

12.3.3 Classification

Classification involves assigning instances X from a specific domain and is defined by a set of features which have been into a set of classes i.e. C. A simplified process is illustrated in Figure 12.2, and shows how classifier H_1 maximally separates the classes, while H_2 does not separate them. This classification process is known as a *concept c* and is defined by $X \rightarrow C$. A classification model or classifier $h : X \rightarrow C$ then produces class predictions for all instances $x \in X$ and is supposed to be a good approximation of the target concept c on the whole domain.

One way of looking at data for classification purposes is to create contingency tables or effectively histograms of the data. This is done by picking k attributes from the data set, namely $a_1, a_2, \ldots, a_k$ and then for every possible combinations of values, $a_1 = x_1, a_2 = x_2, \ldots, a_k = x_k$, recording how frequently that combination occurs. Table 12.1 shows how we can compare the school age of children against their choice of subject for a group aged between 14 and 18. Typically, on-line analytical processing (OLAP) tools can be used to view slices and aggregates of these contingency tables. Of course, these tables will have many more parameters and will comprise many more dimensions.

Classification algorithms comprise DTC, ANNs, Bayesian classifiers and SVM. DTC is carried out in two steps: a decision tree model is built up using records for which the category is known beforehand, then it is applied to other records to predict their class affiliation. They are attractive as they provide high accuracy even when the size of

Figure 12.2 Simple example of classification

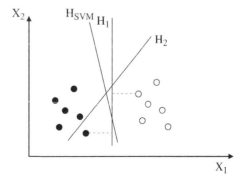

Table 12.1 Age of school children versus most popular subject

Age	Maths	English	Science	Language (α)
14	14	6	7	9
15	9	7	10	9
16	11	4	8	11
17	12	8	8	5
18	7	9	5	11

the data set increases (Narayanan *et al.* 2007). They are used for detecting spam e-mail messages and categorizing cells in MRI scans. They yield comparable or better accuracy when compared to ANNs.

ANNs comprise parallel and distributed networks of simple processing units interconnected in a layered arrangement, typically of three layers. They are based on the neural operation of the brain and are trained using some initial data. There was a lot of interest in the 1980s and 1990s in this approach, but it fell short of its potential. However, there has now been a resurgence of interest in a form of ANNs called convolutional neural networks (CNNs) (Krizevsky 2014). CNNs comprise layers of alternative local translation-invariant features, followed by switchable task-specific loss functions which are trained with stochastic gradient descent. The parameters are very large and represent a significant training challenge, but they show considerable potential. They are highly amenable to FPGA implementation due to the considerable routing resources that FPGAs offer.

Bayesian classifiers are a family of simple probabilistic classifiers based on applying Bayes' theorem which relates current probability to the prior probability. They tend to be useful in applications where the underlying probability has *independent* parameters. For example, clementine oranges may be characterized by being orange in color, squishy and 5 cm in diameter, but these features are considered to be independent of each other. So even if these parameters are interdependent, a naive Bayes classifier considers all of these properties to independently contribute to the probability that this fruit is a clementine orange!

An SVM formally constructs a hyperplane or set of hyperplanes in a high- or infinite-dimensional space, which can be used for classification, regression, or other tasks. Intuitively, a good separation is achieved by the hyperplane that has the largest distance to the nearest training data point of any class. This would be given as the hyperplane H_{SVM} in Figure 12.2 which would seem to best split the two groups by providing the best distance as well.

12.3.4 Regression

Regression is also an inductive learning task that can be thought of as *classification with continuous classes*, which means that the regression model predicts numerical values rather than discrete class labels (Cichosz 2015). It tends to be thought of in the classical statistical approach where we are trying to fit a regression model to a set of data, but if we view it as an algorithm applied in a data mining context, then we can think of regression as being able to provide numerical prediction. This leads to the development of a series of algorithms which can be used to predict the future demand for a product, volume of sales, or occupancy of beds in a hospital ward (Marshall *et al.* 2014).

In mathematical terms, for a target function f, $X \to \mathfrak{R}$ represents the true assignment of numerical values to all instances from the domain. This can be determined using a training set $T \subseteq D \subset X$ for regression which will contain some or all of the labeled instances for which the target set is available. The key is to find the relationship between the dependent and independent variables. The value of dependent variable is of most importance to researchers and depends on the value of other variables. Independent variables are used to explain the variation in the dependent variable.

Regression is classified into two types: *simple regression*, with one independent variable; and *multiple regression*, which has several independent variables. In simple regression, the aim is to create a regression equation involving several regression coefficients and then determine a best fit using the data; this involves determining the best linear relationship between the dependent and the independent variables. In multiple regression analysis, the coefficients indicate the change in dependent variables assuming the values of the other variables are constant. A number of tests of statistical significance are then applied, one of which is the F-test.

12.3.5 Clustering

A cluster is a group of objects that belong to the same class, so clustering is the process of making a group of abstract objects into classes of similar objects. It is applied to a broad range of applications such as market research, pattern recognition and image processing. Its main advantage over classification is that it is adaptable to changes and helps single out useful features that distinguish different groups.

Clustering methods can be classified as partitioning, hierarchical, density-based, grid-based, model-based and constraint-based. For a database of n objects, the partitioning method constructs $k \leq n$ partitions of the data, each of which will represent a cluster and where each group must contain an object and each object must belong to one group only. A well-known method is called k-means clustering (see Section 12.5).

12.3.6 The Right Approach

Whilst this section has highlighted a number of machine learning algorithms, it is important to identify the right estimator for the job. The key observation is that different estimators are better suited to different types of data and different problems. The scikit website (http://scikit-learn.org/stable/tutorial/machine_learning_map/index.html) provides a useful indication of how to progress from a data perspective with Python open source files.

The first stage is to determine if the user is looking to predict a category. If so, then classification or clustering will be applied; otherwise the user will look to apply regression and/or dimension reduction. If the categories and labels are known, then the user will apply classification; if the labels are not known, then the user will apply clustering to determine the organization.

If applying classification to a small data set (i.e. less than 100,000 samples), then a stochastic gradient descent learning routine is recommended; otherwise other classifiers such as SVMs are recommended (Bazi and Melgani 2006). If the data are labeled, then the choice of clustering will depend on the number samples. If the categories are known then various forms of k-means clustering are applied, otherwise it is best to apply a form of a Bayesian classifier.

Of course, this does not represent an expert view for choosing the right approach, and the main purpose of this section is just to provide an overview. We highlight some of the algorithmic characteristics which best match FPGAs; in particular, we are interested in the highly parallel computation but also the high level of interconnection of ANNs as these match well the numerous routing capabilities of modern FPGAs. This is a particularly attractive feature as this algorithmic interconnect would have to be mapped as multiple memory accesses in processors.

12.4 Acceleration

MapReduce involves scaling out the computation across multiple computers. Thus, it is possible to use resources temporarily for this purpose, and vendors such as Amazon now allow you to hire resources, with a higher premium being charged for more powerful computers in November 2016, Amazon announced a new resource called EC2 which allows access to FPGA resources (Amazon 2016). Figure 12.3 illustrates the process. For an original computation as illustrated in Figure 12.3(a), Figure 12.3(c) shows how it is possible to use Hadoop® to *scale out* the computation to improve performance. This works well for problems that can be easily parallelized and distributed, such as performing multiple searches on a distributed search engine.

12.4.1 Scaling Up or Scaling Out

It is also possible, however, to *scale up* computation as shown in Figure 12.3(b). For example, you may be performing multiple, highly complex operations on a single data set, in which case it makes more sense to scale up the resource as you then avoid the long communications delay involved in communicating between distributed computers. This is particularly relevant for highly computationally complex algorithms and also problems which cannot be easily parallelized invoking high volumes of communications which act to slow down computation. The issue of whether to scale out or scale up is a detailed decision (Appuswamy *et al.* 2013).

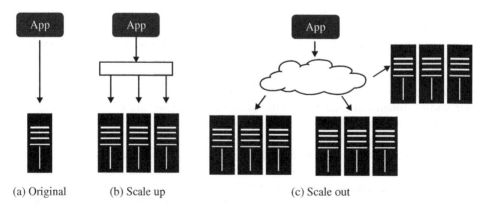

(a) Original   (b) Scale up   (c) Scale out

Figure 12.3 Scaling computing resources

12.4.2   FPGA-based System Developments

A number of major system developments have occurred which strongly indicate the potential of FPGA technology in new forms of computing architectures and therefore Big Data applications. These include the heterogeneous platform developments by Microsoft, Intel and IBM. To date, these have largely been driven by FPGA companies or academia, but there is interest among large computing companies driven mainly by the energy problems as indicated in Chapter 1 with the increased computing demand. Intel's purchase of Altera (Clark 2015) clearly indicates the emergence of FPGAs as a core component in future data centers.

Microsoft has developed flexible acceleration using FPGAs called the Catapult fabric which was used to implement a significant fraction of Bing's ranking engine. It showed an increased ranking throughput in a production search infrastructure by 95% at comparable latency to a software-only solution (Putnam *et al.* 2014). The Catapult fabric comprises a high-end Altera Stratix V D5 FPGA along with 8 GB of DRAM embedded into each server in a half-rack of 48 servers. The FPGAs are directly wired to each other in a 6×8 two-dimensional torus, allowing services to allocate scaling of the FPGA resources.

The authors implement a *query* and *document* request by getting the server to retrieve the document and its metadata and form several metastreams. A "hit vector" is generated which describes the locations of query words in each metastream; a tuple is also created for each word in the metastream that matches a query and describes the relative offset from the previous tuple, the matching query term, and a number of other properties. The frequency of use can then be computed, along with free-form expressions (FFEs) which are computed by arithmetically combining computed features. These are then used in a machine learning model which determines the document's position in the overall ranked list of documents returned to the user.

IBM and Xilinx have worked closely together to develop Memcache2, a general-purpose distributed memory caching system used to speed up dynamic database-driven searches (Blott and Vissers 2014). By developing a solution that allows tight integration between network, computer and memory and by implementing a completely separate TCP/IP stack, they are able to achieve an order-of-magnitude speed improvement over an Intel Xeon® solution. This improvement is even greater when considering power, which is a key issue in data centers.

12.4.3   FPGA Implementations

There have been a number of classification and regression implementations on FPGAs. Among these are a number of ANNs, including work implementing a general regression neural network (GRNN) used for iris plant and thyroid disease classification (Polat and Yıldırım 2009). They have developed an FPGA implementation using VHDL-based tools; it comprises summation, exponential, multiplication and division operations. However, the inability to realize the Taylor series efficiently in FPGA has meant that the implementation would not run much faster than the MATLAB model running in software on a P4 3 GHz, 256 MB RAM personal computer compared to a fixed-point Xilinx Spartan3 xc3s2000.

An FPGA-based coprocessor for SVMs (Cadambi *et al.* 2009) implemented on an off-the-shelf PCI-based FPGA card with a Xilinx Virtex-5 FPGA and 1 GB DDR2 memory gave an improvement of 20 times over a dual Opteron 2.2 GHz processor CPU with lower power dissipation. For training, it achieved end-to-end computation speeds of over 9 GMACs, rising to 14 GMACs for SVM classification using data packing. This was achieved by exploiting the low precision and highly parallel processing of FPGA technology by customizing the algorithm for low-precision arithmetic. This allowed the efficient reuse of the underlying hardware and reduction in off-chip memory accesses by packing multiple data words on the FPGA memory bus.

A number of implementations have been explored for implementing *k*-means clustering on FPGA. Lin *et al.* (2012) implemented an eight-cluster XC6VLX550T design with a clock frequency of 400 MHz; the design utilized 112 DSP blocks, 16 BRAMs, 2110 slices, 5337 LUTs and 8011 slice registers. A number of blocks were implemented for performing the distance calculations in parallel, one for each cluster. In Winterstein *et al.* (2013), *k*-means clustering was performed in FPGA without having to involve off-chip memory. A Xilinx XC4VFX12 with 5107/5549 slices, 10,216 LUTs and a maximum clock frequency of 63.07 MHz was achieved. It gave a speedup of 200 times over a MATLAB realization on a GPP Intel core 2 DUO E8400 and 3 GB RAM, and 18 times over GPU Nvidia GeForce 9600m GT graphics.

12.4.4 Heston Model Acceleration Using FPGA

The Heston model is a well-known model used in option determination in finance applications which involves calculating the risk for cases where volatility is stochastic (Heston 1993). The volatility of the underlying asset follows a Brownian motion, which in turn gives rise to a system of two stochastic differential equations:

$$dS_t = \mu S_t dt + \sqrt{V_t} S_t dW_t, \tag{12.1}$$
$$dV_t = \kappa(\theta - V_t)dt + \xi\sqrt{V_t}dW_t. \tag{12.2}$$

In these equations, S_t is the price variation, V_t is the volatility process, W_t is the correlated Brownian motion process and ξ is referred to as the volatility of the volatility. V_t is a square root mean-reverting process with a long-run mean of θ and a rate of mean reversion of κ. The mean reversion of the volatility means that the volatility is bound to revert to a certain value; so when $V_t < \theta$, the drift of the volatility encourages V_t to grow again, and conversely when $V_t > \theta$, the drift becomes negative and thus the volatility decreases.

The Heston model works well in many financial applications as the asset's log-return distribution is non-Gaussian and is characterized by big tails and peaks. It is argued that equity returns and the implied volatility are negatively correlated, which presents problems for models such as the Black–Scholes model (Black and Scholes 1973), which do not consider volatility, hence the interest in the Heston model.

A useful metric for measuring the performance of different implementations of the Heston model is the number of steps per second achieved by each technology. In this model, the cores implement a Monte Carlo simulator and the output of these is aggregated to form the final Heston output (Figure 12.4). The performance is dictated by the number of cores that can be implemented in the FPGA fabric; the achievable clock speed

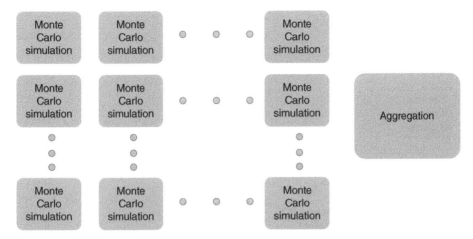

Figure 12.4 Core configuration where Monte Carlo simulations can be programmed

will vary depending on the number of cores that can be effectively placed and routed in the FPGA. A number of possible FPGA configurations have been implemented and are displayed in Table12.2. Performance is given as the throughput rate (TR) which is calculated in billions of steps per second (BSteps/s).

12.5 *k*-Means Clustering FPGA Implementation

Clustering is a process used in many machine learning and data mining applications (Jain 2010). It is an unsupervised partitioning technique which groups data sets into subsets by grouping each new data into groups with the have data points with similar features (e.g. same age groups, same image features). A flow diagram for the algorithm is shown in Figure 12.5. It is used in a range of image processing and target tracking applications (Clark and Bell 2007), when it is necessary to initially partition data before performing more detailed analytics.

The k-means algorithm requires the partitioning of a D-dimensional point set $X = \{x_j\}, j = 1, \ldots, N$, into clusters $S_i, i = 1, \ldots, k$, where k is provided as a parameter, usually

Table 12.2 32-bit Heston model implemented as both fixed- and floating-point

Data type	MCs	LUTs	Flip-flops	DSP48E	Clock (MHz)	TR (BSteps/s)
			Performance			
Fixed	64	57,757 (13%)	65,210 (8%)	320 (9%)	250	16.0
	128	64,865 (15%)	79,676 (9%)	640 (18%)	238	30.5
	256	78,061 (18%)	91,573 (11%)	640 (18%)	172	44
Floating	32	240,801 (56%)	366,642 (42%)	1280 (36%)	112	3.6

Note. MCs = Monte Carlo simulations. Percentages give the utilization of Xilinx Virtex-7 VC709 device.

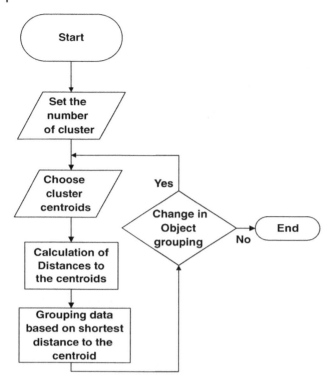

Figure 12.5 Flow chart for *k*-means algorithm

the number of required data sets, and is usually set by the user (Winterstein *et al.* 2013). The goal is to find the optimal partitioning which minimizes the objective function

$$J(\{S_i\}) = \sum_{i=1}^{k} \sum_{x_j \in S_i} \|x_j - \mu_i\|^2,$$

(12.3)

where μ_i is the geometric center (centroid) of S_i.

The ideal scenario for data organization is to group data of similar attributes closer together and farther away from data of dissimilar attributes. The *k*-means algorithm is one of the main unsupervised data mining techniques used to achieve this for large data sets (Hussain *et al.* 2011). In the *k*-means algorithm, a data set is classified into *k* centroids based on the measure of distances between each data set and the *k* centroid values (see Figure 12.6).

At the beginning, the number of centroids and their centers are chosen; each data item then belongs to the centroid with the minimum distance to it. There are many metrics for calculating distance values in the *k*-means algorithm, but the most commonly used ones are the Euclidean and Manhattan distance metrics. The Euclidean distance, D_E, is given by

$$D_E = \sqrt{\sum_{i=1}^{d} (X - C)^2}$$

(12.4)

Figure 12.6 Distance calculation

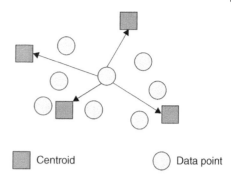

▨ Centroid	◯ Data point

where X is the data point, C is the cluster center and d is the number of dimensions of each data set. The Manhattan distance D_M is given by

$$D_M = \sum_{i=1}^{d} |X - C|. \qquad (12.5)$$

Whilst the Euclidean distance metric is more accurate (Estlick *et al.* 2001), the Manhattan distance metric is preferred as it is twice as fast as the Euclidean distance calculation and consumes less resources (Leeser *et al.* 2002).

12.5.1 Computational Complexity Analysis of k-Means Algorithm

The stages involved in k-means algorithm are: distance calculation, comparison and averaging, as shown in Figure 12.7. The centroid values are chosen from among the existing data points/pixels or by generating random values and can be viewed as negligible from a computational analysis point of view. In the distance stage, the distances from each data point to the centroids are calculated. For each data point of an RGB image, the Manhattan distance metric is given by

$$D = |X_r - C_r| + |X_g - C_g| + |X_b - C_b|. \qquad (12.6)$$

This involves 3 absolute values, 2 additions and 3 subtractions, giving rise to 8 operations. For n data points and k centroids, the number of operations involved in the distance calculation, k_D, is given by $k_D = 8nk$.

In the comparison module, the inputs are k distance values generated by each pixel. It takes $k - 1$ comparison steps to get the minimum distance. So for n data, the number of operations involved in the comparison block is given by $k_C = n(k - 1)$.

In the averaging block, data pixels in the dimension are added up and divided by the number in their dimensions in that cluster, giving an updated centroid value for the following frame. For a data pixel there are d additions, so for n data there are nd additions

Figure 12.7 Bits of data in and out of each block

and kd divisions. Hence the number of operations involved in the averaging block, K_A, is given by

$$K_A = nd + kd = d(n + k) = 3(n + k) \tag{12.7}$$

($d = 3$ for an RGB image).

The total number of operations, K_{total}, in the k-means algorithm is thus given by

$$K_{total} = 8nk + n(k - 1) + 3(n + k). \tag{12.8}$$

Figure 12.7 shows the number of operations and bits in and out of each block for 8 clusters. The number of operations in the distance calculation block is the highest and increases as the number of clusters increases. The operations in the distance calculation block are independent and as such can be put in parallel to speed up execution.

12.6 FPGA-Based Soft Processors

As some of the examples earlier in this book and even those reviewed in this chapter have indicated, FPGAs give a performance advantage when the user can develop an architecture to best match the computational requirements of the algorithm. The major problem with this approach is that generating the architecture takes a lot of design effort, as was illustrated in Chapter 8. Moreover, any simple change to the design can result in the creation of a new architecture which then incurs the full HDL-based design cycle which can be time-consuming. For this reason, there has been a lot of interest in FPGA-based software processors.

A number of FPGA-based image processors have been developed over the years, including the Xilinx MicroBlaze (Xilinx 2009) and the Altera Nios II processor (Altera 2015), both of which have used extensively. These processors can be customized to match the required applications by adding dedicated hardware for application-specific functions. The approach is supported by the FPGA company's software compilers. However, attempts to make the processor more programmable compromises the performance and have not taken advantage of recent technological FPGA developments.

A number of other processor realizations have been reported, including a vector processing approach (Russell 1978) which uses fixed, pipelined functional units (FUs) that can be interconnected; this takes advantage of the plethora of registers available in FPGAs. A soft vector processor VESPA architecture (Yiannacouras et al. 2012) employs vector chaining, control flow execution support and a banked register file to reduce execution time. Both approaches are limited to a clock rate of less than 200 MHz which is much less than the 500–700 MHz that is possible in implementing FPGA designs directly.

VENICE (Severance and Lemieux 2012) is a processor-based solution that provides support for operations on unaligned vectors, and FlexGrip (Andryc et al. 2013) is an FPGA-based multicore architecture that allows mapping of pre-compiled CUDA kernels which is scalable, programmable and flexible. However, both solutions only operate at 100 MHz. They offer flexibility, but the low frequency will result in relatively poorer implementations when compared to dedicated implementations.

Chu and McAllister (2010) have created a programmable, heterogeneous parallel soft-core processor architecture which is focused on telecommunications applications. Similarly, iDEA (Cheah *et al.* 2012) is a nine-stage, pipelined, soft-core processor which is based around the DSP48E1 and supports basic arithmetic and logical instructions by utilizing limited FPGA resources. The design runs at 407 MHz which is 1.93 times faster than Xilinx MicroBlaze and significantly faster than previous work. This improved performance provides a better proposition for achieving a soft-core-based approach for data applications. In this book, we concentrate on a processor that has been developed for image processing.

12.6.1   IPPro FPGA-Based Processor

A custom-designed DSP48-based RISC architecture, called IPPro (Siddiqui *et al.* 2014) has been developed; it uses the Xilinx DSP48E2 primitive as the ALU for faster processing and supports a wide range of instructions and various memory accesses. The following design decisions were made to optimize FPGA performance and image processing needs:

- High processing capability is required to handle the large amount of data (30–40 MB/s) needed for real-time video streaming. This is achieved by explicitly mapping the operations and logic to the underlying FPGA resource primitives and ensuring a good match. This allowed a 350–450 MIPS performance per processor to be achieved.
- Efficient memory utilization by distributing memory to hide data transfer overheads between main and local memory to keep IPPro busy in processing data. This matches the distributed nature of memory in FPGA resources. Dedicated kernel memory accelerates the linear filter operations and also reduces the code size by avoiding excessive load/store instructions and maximizing memory reusability.
- Optimized instructions/addressing modes and reduced branch penalty by decreasing the number of pipeline stages as unpredicted branches degrade performance. The creation of special instruction sets allows the acceleration of image processing operations; addressing modes to give flexibility to the programmer; and conditional execution in the form of a customizable and flexible branch controller to support mask-based conditional execution out-of-box without need of significant architectural changes.

Memory

IPPro is capable of processing 16-bit operations, and uses distributed memory to build a memory hierarchy, with register file, data memory, and kernel memory. The IPPro architecture uses a five-stage balanced, pipelined architecture as shown in Figure 12.8.

IPPro is capable of running at 337 MHz on a Xilinx SoC, in particular XC7Z020-3, using one DSP48E, one BRAM and 330 slice registers per processor. The main idea of the processor was to keep it compact, reprogrammable and scalable as much as possible to achieve high throughput rates compared to custom-made HDL designs. It contains small, fast and efficient memory to locally store data and keep ALU busy in processing data. This helps to hide data transfer overheads between the main and local memories.

It supports various instructions and memory accesses and is capable of processing signed 16-bit operations. The IPPro processor architecture uses five-stage balanced

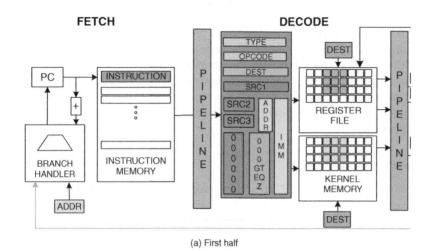

(a) First half

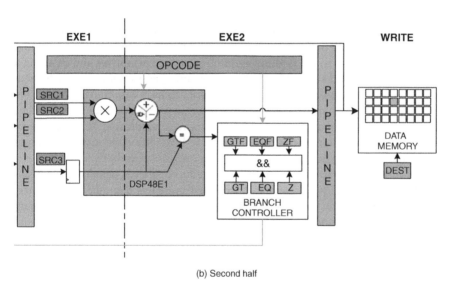

(b) Second half

Figure 12.8 IPPro architecture

pipelining and supports streaming mode operation where the input and output data are read and written back to FIFO structures, as shown in Figure 12.8.

IPPro strikes a balance between programmability and the need to maintain FPGA performance. Overall it has the following addressing modes: from local memory to local memory; from local memory to FIFO (LM–FIFO); from kernel memory to FIFO. The local memory is composed of general-purpose registers used mainly for storing operands of instructions or pixels. This memory currently contains 32 sixteen-bit registers. A FIFO is a single internal register of IPPro where the input and output streams from/to an external FIFO are stored. Kernel memory is a specialized location for coefficient storage in windowing and filtering operations with 32 sixteen-bit registers.

Data Path

The complete IPPro data path is shown in Figure 12.8. It has a loadstore five-stage balanced pipelined architecture giving a fixed latency of five clock cycles. It exploits the features of the Xilinx DSP48E1 to implement all of the supported instructions and provides a balance between hardware resource utilization, performance, throughput, latency and branch penalty. A balanced pipeline simplifies the compiler tool chain development compared to variable pipeline architecture. The deep pipeline comes at the cost of larger latency and branch penalty which adversely affects the overall performance. Various techniques predict branches, but none of them was deemed to give a shorter latency. The five pipeline stages are as follows:

1. Fetch (IF)
2. Decode (ID)
3. Execute 1 (EXE1)
4. Execute 2 (EXE2)
5. Write Back (WB).

Instruction Set

An example of the supported instructions can be seen in Table 12.3. This table shows the IPPro LM–FIFO addressing mode instructions and some miscellaneous others. The IPPro instruction set is capable of processing basic arithmetic and logical operations for different addressing modes. In addition to the unary and binary instructions, it has support for trinary expressions such as MULADD, MULSUB, MULACC.

Given the limited instruction support and requirements from the application domain, it is envisaged that coprocessor(s) could be added to provide better support for more complex processes such as division and square root. Ongoing research is being undertaken to design such a coprocessor (Kelly *et al.* 2016).

Flags and Branches

Flags are important status indicators in processors and used to handle exceptions encountered during data computation. IPPro currently supports the following data flags

Table 12.3 Instruction set

LM–FIFO		Misc	
ADD	LOR	JMP	GET
SUB	LNOR	BNEQ	PUSH
MUL	LNOT	BEQ	NOP
MULADD	LNAND	BZ	BYPASS
MULSUB	LAND	BNZ	DIV
MULACC	LSL	BS	
LXOR	LSR	BNS	
LXNR	MIN	BNGT	
	MAX	BGT	

but is flexible enough to allow new flags to be defined by modifying the branch controller shown in Figure 12.8:

1. Greater than (GTF)
2. Equal (EQF)
3. Zero (ZF)
4. Sign Flag (SF).

The flags are generated using the pattern detector function which is embedded inside the DSP48E1 block as dedicated functionality. It compares the two operands available at the input of DSP48E1 and sets the pattern detect (PD) bit in the very same clock cycle if both operands are equal. Therefore no additional clock cycle is needed to compute the flag bit which is important in the case of conditional/data dependent instructions being executed in the multicore architecture. The branch controller is flexible and scalable as it is created using combinational logic. A dataflow-based programming route has also been created (see Amiri *et al.* 2016)

12.7 System Hardware

An example of a typical system architecture is given in Figure 12.9. This gives details of the front-end processor architecture, prototyped on a Zedboard platform which comprises a Xilinx Zynq SoC which comprises on-chip dual-core ARM processors and programmable logic. The SIMD-IPPro is comprised of a number of IPPro cores connected together.

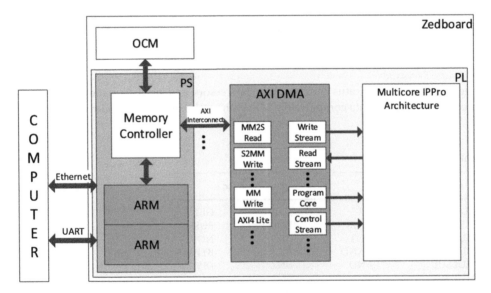

Figure 12.9 Proposed system architecture

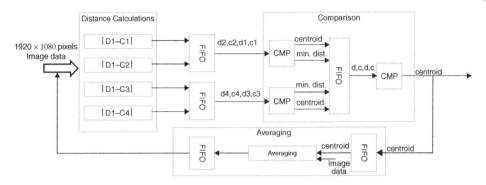

Figure 12.10 Proposed architecture for k-means image clustering algorithm on FPGA

The main aim from a design perspective for the k-means algorithm is to allocate the processing requirements (distance calculation, comparison and averaging) to the FPGA resources. The averaging is a feedback block responsible for recalculation of new centroids for new iteration. The image data is stored in off-chip memory, while the centroid values are stored in the kernel memory of the IPPro-based FPGA. Figure 12.10 shows our architecture for the k-means algorithm.

An example of IPPro code is shown in Table 12.4 for the computation of the equation

$$M10 = \sum_{i=0}^{3} (Mi * M(i+1))$$ (12.9)

12.7.1   Distance Calculation Block

The distance calculation block is the most computationally intensive block in the k-means algorithm. Since the distances between each data point and the centroids are independent, the computation was organized across a number of clusters. One IPPro is dedicated to each cluster for the distance calculation, which calculates the distance

Table 12.4 Example of IPPro code representation

S/N	IPPro code	Description
1	LD, R1, M	Load from data memory M1 to register R1
2	LD, R2, M2	Load from data memory M2 to register R2
3	LD, R3, M3	Load from data memory M3 to register R3
4	LD, R4, M4	Load from data memory M4 to register R4
5	MUL R10, R1, R2	Multiply R1 and R2 and store in register
6	MULACC R10, R3, R4	Multiply R3 and R4 and accumulate in R10
7	ST R10, M10	Store from register to data memory

between the data pixels and each centroid based on their dimensions, finds their absolute values and sums them. The outputs of the distance block are the distances and the corresponding centroid values. The distance computation takes 58 cycles and the outputs are then stored in the FIFO for use by the next block of the k-means algorithm.

12.7.2 Comparison Block

The comparison block receives the distance and the associated centroid values from the distance calculation block serially. It compares the distance values and selects the minimum value, along with the centroid value that stands for the minimum value. These outputs are part of the inputs to the next comparison block. The last comparison block produces one centroid output which replaces the image data that it represents, in order to give a clustered image in the final iteration. The centroid values as an output of the comparison stage are fed into the averaging block. It takes 81 cycles to compare eight distance values representing eight clusters.

12.7.3 Averaging

This block receives the centroid values from the comparison block and uses them to group the image data according to their order. For instance, when it receives centroid 1, and then image pixel $D1$, it puts $D1$ under the cluster 1 group. In our architecture, the averaging block avoids performing the division operation at the end of the iteration, by averaging each group's value each time there is an addition; this is done by performing a binary shift to the right by one position. This is an approximation to doing division at the end of the pixel count. The outputs of the averaging block are the new centroid values for the next iteration. It takes 76 cycles to process a pixel for eight clusters.

12.7.4 Optimizations

Some of the ways that we used to achieve the optimization are: data parallelism, cluster parallelism and code reduction. The focus of the design is to reduce the IPPro assembly code to the best possible by ensuring that the most suitable and shortest codes are used to represent the algorithmic decomposition.

The computation was profiled for different multicore arrangements and the resulting realizations then compared in terms of execution time. After a number of mappings of cores per function, it was decided that one core would be used for the distance calculation, three cores for the comparison block and one core for the averaging block. Table 12.5 shows the results for 4, 8 and 9 centroids in terms of number of cycles,

Table 12.5 Results for 4, 8 and 9 centroids image clustering

	Dist.			Comp.			Aver.		
No. of clusters	4	8	9	4	8	9	4	8	9
No. of cycles	106	106	106	52	78	104	39	75	84
Execution time (s)	0.41	0.41	0.41	0.20	0.30	0.41	0.15	0.29	0.32
Latency (μs)	0.2	0.2	0.2	0.10	0.15	0.20	0.07	0.14	0.15
Throughput (MP/s)	5	5	5	10.0	6.7	5	14	7.2	6.5

execution time, latency and throughput profiled for a high-definition image of 1920 × 1080 in a single iteration using color images when the clock frequency is 530 MHz.

12.8 Conclusions

The purpose of the chapter was to acknowledge the major developments in the area of Big Data applications. This area and the need to develop data centers to meet the increasing needs of the Big Data analytics have done more than any recent developments to see FPGAs being used in modern computing systems. The interest by Intel, IBM and Microsoft has been substantial, ranging from major joint projects with the two main FPGA companies to the purchase of Altera by Intel.

In some cases, FPGAs offer computing gains over processor-based alternatives such as CPUs, DSP microprocessors and GPUs, particularly if power consumption is taken into consideration. Indeed, this is seen as a major advantage of FPGA technologies, so if they can offer a performance advantage for even a small range of functionality, then this would be seen as beneficial. Given that some of the data mining algorithms have characteristics similar to those seen in DSP algorithms, then it would appear that FPGAs have a major role to play in future computing systems. For this reason, the authors were motivated to include a chapter in this revised edition of the book.

Certainly the abolition of high-level programming languages described in Chapter 7, and processor architectures such as the IPPro system described in this chapter, will have a major impact on how these systems will be built and programmed. In any case, the length of time need to compile high-level languages onto FPGA hardware will need to be addressed.

Bibliography

Abu-Mostafa YS, Magdon-Ismail M, Lin H-T 2012 *Learning from Data*. AMLbook.com.

Altera Corp. 2015 *Nios II (Gen2) Processor Reference Handbook*, ver 2015.04.02. Available from http://www.altera.com (accessed February, 28 2016).

Amazon, 2016 Developer Preview – EC2 Instances (F1) with Programmable Hardware. Available from https://aws.amazon.com/blogs/aws/developer-preview-ec2-instances-f1-with-programmable-hardware/ (accessed December 28, 2016)

Amiri M, Siddiqui FM, Kelly C, Woods R, Rafferty K and Bardak B 2016 FPGA-based soft-core processors for image processing applications. J. of VLSI Signal Processing, DOI: 10.1007/s11265-016-1185-7.

Andryc K, Merchant M, Tessier R 2013 FlexGrip: A soft GPGPU for FPGAs. In *Proc. Int. Conf. on Field Programmable Technology*, pp. 230–237.

Apache 2015 *Welcome to Apache*^{TM} *Hadoop*®. Available from http://hadoop.apache.org/ (accessed January 19, 2015).

Appuswamy R, Gkantsidis C, Narayanan D, Hodson O, Rowstron A 2013 Scale-up vs scale-out for Hadoop: Time to rethink? In *Proc. ACM 4th Annual Symp. on Cloud Computing*, article no. 20.

Bazi Y, Melgani F 2006 Toward an optimal SVM classification system for hyperspectral remote sensing images. *IEEE Trans. on Geoscience and Remote Sensing*, 44(11), 3374–3385.

Black F, Scholes M 1973 The pricing of options and corporate liabilities. *Journal of Political Economy*, 81 (3), pp. 637–654.

Blott M, Vissers K 2014 Dataflow architectures for 10Gbps line-rate key-value-stores. In *Proc. IEEE Hot Chips*, Palo Alto, CA.

Cadambi S, Durdanovic I, Jakkula V, Sankaradass M, Cosatto E, Chakradhar S, Graf HP 2009 A Massively Parallel FPGA-based coprocessor for support vector machines. In *Proc. IEEE Int. Symp. FPGA-based Custom Computing Machine*, pp. 115–122.

Cheah HY, Fahmy S, Maskell D 2012 iDEA: A DSP block based FPGA soft processor. In *Int. Conf. on Field Programmable Technology*, pp. 151–158.

Chu X, McAllister J 2010 FPGA based soft-core SIMD processing: A MIMO-OFDM fixed-complexity sphere decoder case study. In *Proc. IEEE Int. Conf. on Field Programmable Technology*, pp. 479–484.

Cichosz P 2015 *Data Mining Algorithms: Explained Using R*. John Wiley & Sons, Chichester.

Clark D 2015 Intel completes acquisition of Altera. *Wall Street J.*, December 28. Available from http://www. wsj.com/articles/ intel-completes-acquisition-of-altera-1451338307 (accessed March 4, 2016).

Clark D, Bell J 2007 Multi-target state estimation and track continuity for the particle PHD filter. *IEEE Trans. on Aerospace Electronics Systems*, 43(4), 1441–1453.

Cloudera 2016 *Impala SQL Language Reference*. Available from http://www.cloudera.com/documentation.html (accessed March 4, 2016).

Dean J and Ghemawat S 2004 MapReduce: Simplified data processing on large clusters. In *Proc. 6th Symp. on Operating Systems Design & Implementation*.

Estlick M, Leeser M, Theiler J, Szymanski JJ 2001 Algorithmic transformations in the implementation of k-means clustering on reconfigurable hardware. In *Proc. ACM/SIGDA 9th Int. Symp. on Field Programmable Gate Arrays*, pp. 103–110.

Heston S 1993 A closed-form solution for options with stochastic volatility. *Review of Financial Studies*, 6, 327–343.

Hussain HM, Benkrid K, Seker H, Erdogan AT 2011 FPGA implementation of k-means algorithm for bioinformatics application: An accelerated approach to clustering microarray data. In *Proc. NASA/ESA Conf. on Adaptive Hardware and Systems*, pp. 248–255.

Jain A K 2010 Data clustering: 50 years beyond K-means. *Pattern Recognition Letters*, 31(8), 651–666.

Kelly C, Siddiqui FM, Bardak B, Wu Y, Woods R 2016 FPGA based soft-core processors hardware and compiler optimizations. In *Proc. Int. Symp. of Applied Reconfigurable Computing*, pp. 78–90.

Krizevsky A 2014 One weird trick for parallelizing convolutional neural networks. arXiv:1404.5997 (accessed March 4, 2016).

Leeser ME, Belanovic P, Estlick M, Gokhale M, Szymanski JJ, Theiler JP 2002 Applying reconfigurable hardware to the analysis of multispectral and hyperspectral imagery. In *Proc. Int. Symp. on Optical Science and Technology*, pp. 100–107.

Lewis M 2014 *Flash Boys: A Wall Street Revolt*. Norton, New York.

Lin Z, Lo C, Chow P 2012 *K*-means implementation on FPGA for highdimensional data using triangle inequality. In *Proc. IEEE Int. Conf. on Field Programmable Logic*, pp. 437–442.

Manyika J, Chui M, Brown B, Bughin J, Dobbs R, Roxburgh C, Hung Byers A 2011 Big data: The next frontier for innovation, competition, and productivity. McKinsey Global Institute Report.

Marshall AH, Mitchell H, Zenga M 2014 Modelling the length of stay of geriatric patients in the Emilia Romagna hospitals using Coxian phase-type distributions with covariates. In Carpita M, Brentari E, Qannari EM (eds) *Advances in Latent Variables: Methods, Models and Applications*, pp. 1–13. Springer, Cham.

McNulty E 2014 Understanding Big Data: The seven V's. *Dataconomy*. Availavle at http://dataconomy.com/seven-vs-big-data/, (accessed March 4, 2016).

Narayanan R, Honbo D, Memik G, Choudhary A, Zambreno J 2007 An FPGA implementation of decision tree classification. In *Proc. Conf. on Design Automation and Test in Europe*, pp. 189–194.

Omand D, Bartlett J, Miller C 2012 #Intelligence. Available at http://www.demos.co.uk/publications/intelligence [Accessed 19 January 2015].

Polat Ö and Yıldırım T 2009 FPGA implementation of a general regression neural network: An embedded pattern classification system. *Digital Signal Processing*, 20(3), 881–886.

Putnam A, Caulfield AM, Chung ES, Chiou D, Constantinides K, Demme J, Esmaeilzadeh H, Fowers J, Gopal GP, Gray J, Haselman M, Hauck S, Heil S, Hormati A, Kim J-Y, Lanka S, Larus J, Peterson E, Pope S, Smith A, Thong J, Xiao PY, Burger D 2014 A reconfigurable fabric for accelerating large-scale datacenter services. In *Proc. IEEE Int. Symp. on Computer Architecture*, pp. 13–24.

Russell RM 1978 The CRAY-1 computer system. *Communications of the ACM*, 21, 63–72.

Severance A, Lemieux G 2012 VENICE: A compact vector processor for FPGA applications. In *Proc. Int. Conf. on Field-Programmable Technology*, pp. 261–268.

Siddiqui FM, Russell M, Bardak B, Woods R, Rafferty K 2014 IPPro: FPGA based image processing processor. In *Proc. IEEE Workshop on Signal Processing Systems*, pp. 1–6.

Winterstein F, Bayliss S, Constantinides G 2013 FPGA-based *K*-means clustering using tree-based data structures. In *Proc. Int. Conf. on Field Programmable Logic and Applications*, pp. 1–6.

Xilinx Inc. 2009 *MicroBlaze Processor Reference Guide Embedded Development Kit*. EDK 11.4 UG081 (v10.3). Available from http://www.xilinx.com (accessed February 28, 2016).

Yiannacouras P, Steffan JG, Rose J 2012 Portable, flexible, and scalable soft vector processors. *IEEE Trans. on VLSI Systems*, 20(8), 1429–1442.

Zikopoulos P, Eaton C, DeRoos R, Deutsch T, Lapos G 2012 *Understanding Big Data: Analytics for Enterprise Class Hadoop and Streaming Data*. McGraw-Hill, New York.

13

Low-Power FPGA Implementation

13.1 Introduction

A key trend in the introduction was technology scaling and how increasing power consumption has become a worrying aspect of modern computing design. As was highlighted in Chapter 1, this has led to serious interest by major computing companies such as Intel, IBM and Microsoft in exploiting FPGA technology. Whilst in some cases FPGA implementations may offer only moderate computational improvement over CPU/GPU implementations, the equivalent designs tend to operate at much lower clock rates and power is directly proportional to this factor.

Power consumption scales down with technology evolution, and so for much of the 1980s and 1990s the new technology evolution offered an increased number of transistors operating not only at increased speed, but also at reduced power consumption. As scaling increased, though, leakage power, caused by the increasingly imperfect performance of the gate oxide thickness, increased. As the gate leakage is inversely proportional to the gate oxide thickness, this became and continues to be an increasingly important problem.

Even though some would argue to the contrary, the switch to FPGAs could be considered to be the low-power solution for high-performance computing companies, and there are a number of important reasons to reducing FPGA power consumption. As power consumption is directly related to increased temperature, improved FPGA implementations have immediate benefits for the design of the power supply to the complete system; this can result in cheaper systems and fewer components, giving a reduction in PCB area and a reduction in thermal management costs.

System reliability is related to the issue of heat dissipation, and low values result in improved chip lifetimes. Xilinx indicates that "a decrease of 10° in device operating temperature can translate to a 2X increase in component life" (Curd 2007), thus reducing the power consumption of the FPGA implementation has clear cost and reliability implications.

Section 13.2 looks at the various sources of power and introduces the concepts of static and dynamic power consumption. Section 13.3 outlines some of the approaches applied by FPGA vendors in reducing power consumption. An introduction to power

FPGA-based Implementation of Signal Processing Systems,
Second Edition. Roger Woods, John McAllister, Gaye Lightbody and Ying Yi.
© 2017 John Wiley & Sons, Ltd. Published 2017 by John Wiley & Sons, Ltd.

Figure 13.1 Sources of leakage components
in CMOS transistor

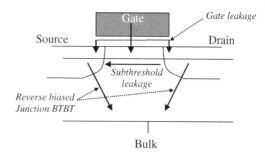

consumption minimization techniques is given in Section 13.4, and the concepts of dynamic voltage scaling and switched capacitance are introduced. The rest of the chapter is dedicated to two core topics, namely dynamic voltage scaling in Section 13.5 and switched capacitance in Section 13.6. Section 13.7 makes some final comments with regard to power consumption in FPGAs.

13.2 Sources of Power Consumption

CMOS technology comprises *static* power consumption which is that consumed when the circuit is switched on but not processing data, and *dynamic* power consumption which is when the chip is actively processing data. The static form comprises a number of components as shown in Figure 13.1: *gate leakage* is the current that flows from gate to substrate; *source-to-drain leakage*, also known as the sub-threshold current, is the current that flows in the channel from the drain to source even though the device is deemed to be off (i.e. the gate-to-source voltage, V_{GS}, is less than the threshold voltage of the transistor V_t); and *reverse biased BTBT current* is the current that flows through the source–substrate and drain–substrate junctions of the transistors when the source and drain are at higher potential than the substrate. The static power consumption is important for battery life in standby mode, and dynamic power is particularly relevant for battery life when operating.

13.2.1 Dynamic Power Consumption

For dynamic power consumption, we consider the leakage through a simple inverter as given in Figure 13.2. Assume that a pulse of data is fed into the transistor, charging up

Figure 13.2 Simple CMOS inverter

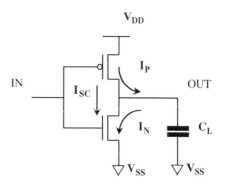

and charging down the device. Power is consumed when the gates drive their output to a new value and is dependent on the resistance values of the p and n transistors in the CMOS inverter.

Thus the current through the capacitor, $i_C(t)$, is given by (Wolf 2004)

$$i_C(t) = \frac{V_{DD} - V_{SS}}{R_p} e^{(-t/R_p C_L)}, \tag{13.1}$$

where V_{DD} represents the supply voltage, V_{SS} represents the ground voltage, R_p is the resistance of the p-type transistor, and C_L is the load capacitance. The voltage, $v_C(t)$, is given by

$$v_C(t) = V_{DD} - V_{SS}\left[1 - e^{(-t/R_p C_L)}\right], \tag{13.2}$$

and so the energy required to charge the capacitor, E_C, is

$$\begin{aligned}
E_C &= \int i_{C_L}(t) V_{C_L}(t)) dt, \\
&= C_L(V_{DD} - V_{SS})^2 \left(e^{-t/R_p C_L} - \frac{1}{2}e^{-2t/R_p C_L}\right)\Big|_0^\infty, \tag{13.3} \\
&= \frac{1}{2}C_L(V_{DD} - V_{SS})^2,
\end{aligned}$$

where V_{C_L} and i_{C_L} is the voltage and current, respectively, needed to charge the load capacitance.

The same charge will then be dissipated through the n-type transistor when the capacitance is discharging; therefore, in a cycle of operation of the transistor, the total energy consumption of the capacitance will be $\frac{1}{2}C_L(V_{DD} - V_{SS})^2$. When this is factored in with the normal operation of the design, which can be assumed to synchronous and operating at a clock frequency of f, this will define the total power consumed, namely $\frac{1}{2}C_L(V_{DD} - V_{SS})^2 f$. However, this assumes that every transistor is charging and discharging at the rate of the clock frequency, which will never happen. Therefore, a quantity denoting what proportion of transistors are changing, namely α, is introduced. For different applications, the value of α will vary as shown in Table 13.1.

This gives the expression for the dynamic power consumption of a circuit,

$$P_{dyn} = \frac{1}{2}C_L(V_{DD} - V_{SS})^2 f \alpha \tag{13.4}$$

which when V_{SS} is assumed to be 0, reduces to the more recognized expression

$$P_{dyn} = \frac{1}{2}C_L V_{DD}^2 f \alpha. \tag{13.5}$$

Table 13.1 Typical switching activity levels

Signal	Activity (α)
Clock	0.5
Random data signal	0.5
Simple logic circuits driven by random data	0.4–0.5
Finite state machines	0.08–0.18
Video signals	0.1(msb)–0.5(lsb)
Conclusion	0.05–0.5

In addition, short-circuit current can be classified as dynamic power consumption. Short-circuit currents occur when the rise/fall time at the gate input is larger than the output rise/fall time, causing imbalance and meaning that the supply voltage, V_{DD}, is short-circuited for a very short time. This will happen in particular when the transistor is driving a heavy capacitative load which, it could be argued, can be avoided in good design. To some extent, therefore, short-circuit power consumption is manageable.

13.2.2 Static Power Consumption

Technology scaling has provided the impetus for many product evolutions as it means that transistor dimensions will be adjusted as illustrated in Figure 13.3. Simply speaking, scaling by k means that the new dimensions shown in Figure 13.3(b) are given by $L' = 1/k(L)$, $W' = 1/k(W)$ and $t'_{ox} = 1/k(t_{ox})$. It is clear that this translates to a k^2 increase in the number of transistors, an increase in transistor speed and an expected decrease in transistor power as currents will also be reduced.

The expected decrease in power consumption, however, does not transpire. In order to avoid excessively high electric fields, it is necessary to scale the supply voltage, V_{DD}, which in turn requires a scaling in the threshold voltage V_t, otherwise the transistors will not turn off properly. As well as a reduction in system voltage, V_{DD}, there is a reduction in V_t which results in an increase in sub-threshold current. In order to cope with short channel effects, the oxide thickness is scaled, resulting in high tunneling through the gate insulator leading to the gate leakage. Thus, this gate leakage is inversely proportional to the gate oxide which will continue to decrease for improving technologies therefore exacerbating the problem.

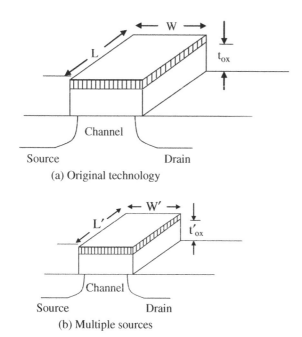

(a) Original technology

(b) Multiple sources

Figure 13.3 Impact of transistor scaling

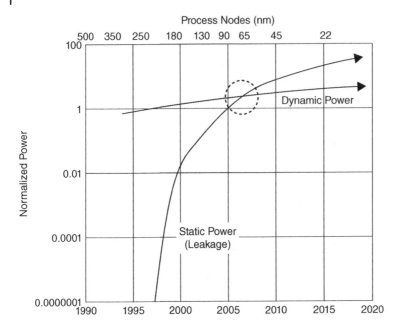

Figure 13.4 Impact of static versus dynamic power consumption with technology evolution (ITRS, 2003; Kim *et al.* 2003)

Scaled devices also require high substrate doping densities to be used near the source–substrate and drain–substrate junctions in order to reduce the depletion region. However, under high reversed bias, this results in significantly large BTBT currents through these junctions (Roy *et al.* 2003). The result is that scaling results in a dramatic increase in each of these components of leakage and with increasing junction temperatures, the impact is worsened as the leakage impact is increased (Curd 2007).

The main issue with increasing numbers of transistors is that their contribution to static power consumption is also growing. This is illustrated by the graph in Figure 13.4 which shows that a cross-over point has occurred for 90 nm and smaller technology nodes where static power began to eclipse dynamic power for many applications.

This graph has a major impact for many technologies as it now means that unlike power consumption in the previous decade where the problem was largely impacted by the operation of the device, allowing designers to reduce the impact of dynamic power consumption, it will be predicated on the normal standby mode of operation. This will have an impact on system design for fixed – but particularly for wireless – applications. A number of approaches have been actively pursued by FPGA vendors to address this.

13.3 FPGA Power Consumption

Xilinx has addressed the impact of high static power in its Virtex-5 and subsequent devices by employing a triple oxide (Curd 2007). Triple oxide is used to represent the three levels of oxide thickness used in FPGAs. A thin oxide is used for the small, fast transistors in the FPGA core, a thick oxide is used for the higher-voltage swing transistors in the I/O which do not have to be fast, and a third or middle-level oxide called a *midox* oxide is used for the configuration memory cells and interconnect pass transistors.

The midox slightly thicker gate oxide dramatically reduces leakage current when compared to the thin oxide equivalent, as the V_t reduces both the source-to-drain leakage and gate leakage. The midox transistor is used in non-critical areas in the FPGA, such as in the configuration memory used to store the user design, which do not need to be updated during device operation. In addition, it is used for the routing transistors, once again as the speed of operation is not critical. The impact is to reduce the leakage current of millions of transistors in the FPGA, thus dramatically reducing the power consumption.

In addition to the triple-oxide innovation, some architectural trends have acted to also reduce the power. This has been based largely around the shift to the six-input LUT in Virtex-5 which the company argues gives a 50% increase in logic capacity per LUT, and the shift to the larger eight-input LUT for Altera. The key effect is that more logic is mapped locally within the LUT where smaller transistors are used. Since transistor leakage is measured in current per unit width, smaller transistors will have less leakage and fewer large transistors are needed.

The Xilinx 7 series was characterized by a shift to TSMC's high-performance, low-power 28 nm process called 28HPL. It was the first to use a high-K metal gate (HKMG) process. The company argues that the shift to 20 nm in UltraScale™ should result in 40% overall device-level power savings over Xilinx 7 series FPGAs and up to 60% savings at 16 nm. This then gives an indicative reduction in static power consumption as indicated by the estimated power figures for mobile backhaul in Artix-7 (Figure 13.5).

Altera takes advantage of Intel's 14 nm Tri-Gate process. It comprises 3D tri-gate transistors and provides good dimensional scaling from the 22 nm process. The transistor "fins" are taller, thinner, and more closely spaced; this gives improved density and lower capacitance, resulting in an SRAM cell size that is almost half that for 22 nm.

With Arria 10 devices, a programmable power technology (PPT) is employed which is the company's patented approach for tuning the switching speed of logic elements in the speed-critical paths of a user design. This tuning allows the transistor's threshold voltage in a higher-speed path to be set to a lower value, increasing its switching speed. Transistors in a lower-speed path can be tuned to a higher threshold voltage, thereby reducing the static power consumption by up to 20%.

Other features include using transistors with a lower voltage threshold and small minimal channel length for high-speed operation in the DSP blocks and logic elements; a

Figure 13.5 Estimated power consumption for mobile backhaul on Artix-7

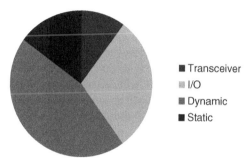

- ■ Transceiver
- ▨ I/O
- ■ Dynamic
- ■ Static

low-power transistor is then used in the less demanding areas, namely configuration RAM and memory blocks.

Whilst addressing static power consumption is therefore vital in developing a low-power FPGA solution, it is somewhat predetermined by the underlying application of many of the above technologies at fabrication; with a fixed architecture, there is little scope available to the FPGA designer. There are, however, some techniques that can act to reduce the power consumption from a dynamic power consumption designer perspective.

13.3.1 Clock Tree Isolation

One of the main contributors to static power consumption is the clock signal through its distribution network, namely a clock tree and the circuits connected to it which it will act to toggle on a regular basis, particularly if the design is synchronous. As the description in Chapter 5 indicated, most FPGAs have a number of clock signals with PPLs and individual dedicated clock tree networks which can be turned off and on as required. Static power consumption reduction is achieved by turning off parts of the clock networks on the FPGA using, for example, multiplexing and then employing clever placement and routing techniques to ensure this can be achieved effectively (Huda *et al.* 2009).

Xilinx has provided an automated capability, in its Vivado® Design Suite v2013.1 and onwards, to its standard place and route flow. It performs an analysis on all portions of the design, detecting sourcing registers that do not contribute to the result for each clock cycle. It then utilizes the abundant supply of clock enables (CEs) available in the logic to create fine-grained clock-gating. This is successful as each CE typically drives only eight registers, providing a good level of granularity to match most bus sizes. Intelligent clock-gating optimization can also be used for dedicated BRAM in simple or dual-port mode by using additional logic to control the array and avoiding unnecessary memory accesses.

13.4 Power Consumption Reduction Techniques

It is clear from equation (13.5) that a number of factors impact dynamic power consumption. The voltage, V_{DD}, will have been predetermined by the FPGA vendor (and optimized to provide low-power operation), and any scope to reduce this voltage will be made available to the user via the design software. The process for achieving this has been outlined by the vendors (Chapman and Hussein 2012).

This only leaves scope for adjusting the other parameters, namely the toggling rate presumed to be the clock frequency, f times the switching activity α and the load capacitance, C_L. However, any technique that acts to adjust the clock frequency f and/or switching activity α should be developed on the understanding that the overall clock rate for the system will generally have been determined by the application and that the switching activity will be governed again by the application domain, meaning that levels shown in Table 13.1 should be given consideration.

Generally speaking, power reduction techniques (Chandrakasan and Brodersen 1996) either act to minimize the switched capacitance (Cf) or employ techniques to increase performance by reducing the supply voltage, thereby achieving a squared reduction in power at the expense of a linear increase in area, i.e. C_L or frequency, f. In voltage minimization techniques, transformations are used to speed up the system's throughput beyond that necessary; the voltage is then reduced, slowing up performance until the required throughput rate is met but at a lower power consumption budget. This has considerable scope for SoC systems where the system can be developed knowing that the circuit will be implemented using a range of voltage values. It is a little more difficult in FPGAs, but work by Chow *et al.* (2005) and Nunez-Yanez (2015) suggests viable techniques which are described later.

There is also some scope to reduce the capacitance and switching activity, but rather than consider this separately, it is useful to think about reducing the *switched capacitance* of a circuit, i.e. the sum of all of toggling activity of each node multiplied by the capacitance of that node. This is an important measure of power consumption as opposed to just circuit capacitance alone, as a circuit can either have a large capacitive net with a low switching activity which will not contribute greatly to power consumption, or a number of low-capacitance nets with a lot of switching activity which can make a not insubstantial contribution to power consumption. The same argument applies to switching activity levels as some nets can have high switching activity but low capacitance, and so on. A large proportion of the techniques fall into this domain and so more of the discussion is centered upon this aspect.

13.5 Dynamic Voltage Scaling in FPGAs

As the name indicates, dynamic voltage scaling involves reducing the supply voltage of the circuit in such a way that it can still operate correctly. Typically, the designer will have exploited any voltage capacity by applying design techniques to slow down the circuit operation, presumably by achieving an area reduction or some other gain. Thus reducing the voltage may cause a circuit failure as the critical path timing may not be met. This is because scaling the voltage causes as impact of circuit delay, t_d, as given by the expression (Bowman *et al.* 1999)

$$t_d = \frac{kV_{DD}}{(V_{DD} - V_t)^2},$$
(13.6)

where k and α are constants with $1 < \alpha < 2$. As V_{DD} is scaled, the circuit delay increases.

Voltage scaling should be only applied to the FPGA core as it is important that the I/O pins operate to the specifications to which they have been designed. Whilst the circuit delay increases, only parts of the circuit need to operate at the shortest circuit delay, which means that there is scope for reducing the voltage for a large portion of the circuit without impacting performance. Of course, the design has to be reliable across a range of devices, and there can be a variation in delay times as well as operating temperature.

The approach shown in Figure 13.6 outlines how a transformation can be applied to reduce power consumption. Parallelism and voltage scaling can be employed to reduce the power consumption of a circuit. If the performance is met with the functional units shown in Figure 13.6(a), then parallelism can be used to give the circuit shown in Figure 13.6(b); as this circuit can now operate much faster than the original, there is scope to

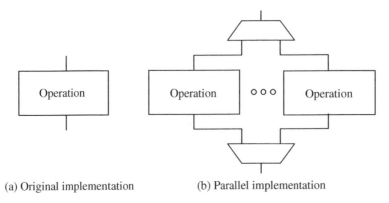

(a) Original implementation   (b) Parallel implementation

Figure 13.6 Use of parallelism (and voltage scaling) to lower power consumption

reduce power by scaling the voltage to the resulting circuit. This may seem counterproductive as the circuit has a larger area, resulting in increased capacitance and switching activity, but the power reduction due to the V_{DD}^2 scaling in addition to the scaling in frequency will more than outweigh any increase.

The impact of adaptive voltage scaling can be addressed in a number of ways, adding circuitry to detect exactly when this happens with a specific FPGA implementation and detecting the correct voltage threshold to achieve lower power operation (Chow *et al.* 2005; Ryan and Calhoun 2010) or applying design techniques to speed up circuit operation. Chow *et al.* (2005) work on the first principle of assuming the original circuit, scaling down the internal voltage supply and then checking for any possible errors on correct operation. Of course, the approach can be used for parallelization as shown in Figure 13.6.

The authors argue that on the basis that the designer can observe any two types of design errors as a result of the voltage scaling in normal operation, it is just a case of trying to work out two other types of error, namely *I/O errors* (due to the lower-voltage core circuit having to interface with the I/O which is operating at the original voltage) and *delay errors* (occurring as a result of the critical path now possibly not meeting the timing). In the case of I/O errors, the danger is that a high output signal from the core will be too small for the threshold voltage of the I/O buffer to correctly detect its value.

The lowest supply voltage is estimated at runtime and adjusted accordingly. A logic delay measurement circuit (LDCM) (Gonzalez *et al.* 1997) is used with an external monitor to adjust the FPGA internal voltage at 200 ms intervals. Typical power savings of 20–30% were achieved with a Xilinx Virtex 300E-8 device in a number of experiments. Nunez-Yanez *et al.* (2007) present a similar approach that applies dynamic voltage scaling (DVS) by adjusting first the voltage, then searching for a suitable operating frequency using the LDCM. Energy savings of up to 60% on an XC4VSX35-FF668-10C FPGA, were achieved by scaling down from 1.2 V to 0.9 V.

Nunez-Yanez (2015) extends this work by proposing adaptive voltage scaling (AVS) which uses voltage scaling together with dynamic reconfiguration and clock management. By exploiting the available application-dependent timing margins, a power reduction up to 85% from operating at 0.58 V (compared to a nominal 1 V) is achieved. He argues that the energy requirements at 0.58 V are approximately five times lower compared to the nominal voltage. This is achieved through the development of an AVS unit which monitors the delay properties and adjusts the voltage to operate at the lowest-energy point for a given frequency.

These techniques are very design-specific and involve implementation of dedicated circuitry to achieve the power reduction. They are probably relevant for designs where strict power consumption needs to be achieved and the design is less likely to change. Moreover, they may have to be checked for every FPGA component used as chip performance tends to vary.

13.6 Reduction in Switched Capacitance

The previous techniques require that the voltage is scaled (typically only the internal voltage) but do not deal with the results of the application of this scaling. However, as suggested earlier, it is also possible to reduce the switched capacitance of the circuit. A number of techniques are considered which are well understood in the literature even though, in some cases, some have not been directly applied to FPGAs.

13.6.1 Data Reordering

In DSP processor implementations described in Chapter 4, the architecture is typically composed of data and program memory connected to the processor via data buses; this is also the case in the increasing number of FPGA-based processors such as the one described in detail in Chapter 12. In these architectures, therefore, the capacitance of the buses will be fixed, but in some cases it may be possible to reorder the data computation in order to minimize the Hamming difference and thereby achieve a reduction in the switching activity on large capacitative buses. Consider the trivial example of a four-tap filter given by $y(n) = a_0 x(n) + a_1 x(n-1) + a_2 x(n-2) + a_3 x(n-3)$ where the coefficient are as follows:

$a0$	1011		$a0$	1011
		3			1
$a1$	0110		$a2$	1001
		4			3
$a2$	1001		$a3$	0100
		3			1
$a3$	0100		$a1$	0110
		4			3
$a0$	1011		$a0$	1011
	14 transitions			8 transitions

It can be seen that if the coefficients are loaded in the normal numerical order, namely a_0, a_1, a_2, a_3 and back to a_0, then this will require 14 transitions which will involve charging and discharging of the line capacitance of the interconnect, which could be considerable. By changing the order of loading, the number of transitions can be reduced as shown above. The main issue is then to resolve the *out-of-order* operation of the filter.

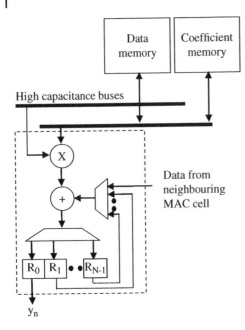

Figure 13.7 Generic MAC time domain filter implementation

In Erdogan and Arslan (2002), the authors show how this can be applied to the design of FIR filters to achieve a reported 62% reduction in power consumption. The architecture reproduced from their earlier paper (Erdogan and Arslan 2000) presents a MAC structure comprising a multiplier and adder which are fed by program and coefficient data from the memory. This type of structure is shown in Figure 13.7. The structure can be made cascadable by feeding the output of the previous section into the current block.

This structure thus allows an out-of-order operation to be performed on the data accumulating in the structure, resulting in a reduction of the switching data from the coefficient memory via the large-coefficient data bus. By exploiting the direct-form FIR filter implementation rather than possibly the transposed form, this also reduces the switching on the data bus as one data word is loaded and then reused.

13.6.2 Pipelining

An effective method to reduce power consumption is by using pipelining coupled with power-aware component placement. Pipelining, as illustrated in Figure 13.8, breaks the processing of the original circuit (Figure 13.8(a)) into short stages as illustrated in Figure 13.8(b), thereby providing a speedup but with an increase in the latency in terms of the number of clock cycles, although this does not necessarily mean a large increase in time (as the clock period has been shortened). The increase in processing speed can be used in a similar way to the use of parallelism in Figure 13.6, to allow the voltage to be reduced, thereby achieving a power reduction (Chandrakasan and Brodersen 1996).

In addition to providing the speedup, though, pipelining provides a highly useful mechanism to reduce power consumption (Keane *et al.* 1999; Raghunathan *et al.* 1999). Work by Wilton *et al.* (2004) has outlined in detail the application of pipelining to FPGA

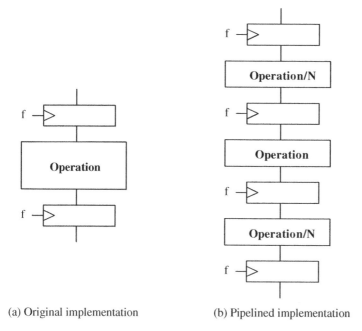

(a) Original implementation (b) Pipelined implementation

Figure 13.8 Application of pipelining

and the impact that can be achieved. Whilst this works well for SoC, it has greater potential in FPGAs, as the registers were previously available and would have been contributing to power even if unused, as they are in the FPGA in any case.

Glitches can contribute to dynamic power consumption and occur when logic signals do not arrive at time at the LUTs as they traverse different levels of logic and perhaps different routing paths. They contribute significantly to dynamic power consumption (from 20% to 70%) through the unplanned switching activity of signals (Shen *et al.* 1992). It is argued that pipelining acts to reduce the logic depth of combinatorial paths and thus decrease the probability of glitches and prevent them being propagated from one pipeline stage to the next (Boemo *et al.* 2013). Moreover, there is a reduction in the net lengths and an overall reduction in the switching activity of longer net activity; by reducing a high contributor to power and shortening high-frequency nets, the dynamic power dissipated can be significantly reduced.

This technique is particularly effective in FPGA technology because the increasing flexibility comes at a power budget cost due to the long routing tracks and programmable switches. These features provide the programmability but are laden with parasitic capacitance (Chen *et al.* 1997) as illustrated by Figure 13.9, which shows a model of a typical FPGA route. Another benefit of implementing pipelining in a FPGA is that it may be using an underutilized resource, namely the flip-flop at the output of the logic cell, thereby only providing a small area increase (Wilton *et al.* 2004).

There are also other advantages to applying pipelining in FPGA designs. The aim of the place and route tools is to achieve the best placement in order to achieve the required speed. By applying pipelining, this provides a more rigorous structure to the design and allows faster placement of the logic (McKeown and Woods 2008). It also acts to reduce

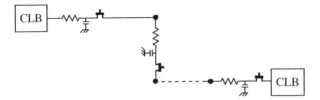

Figure 13.9 Typical FPGA interconnection route

the number of longer nets that result in the design, allowing speed targets to be more easily met.

Pipelining Examples

Work by Boemo *et al.* (2013) reports a set of experimental measurements on the impact of applying pipelining on a range of FPGA technologies for a range of integer multipliers. Some of the results are presented in Table 13.2. The power consumption was obtained by measuring the actual current from an internal core power supply on the FPGA board. The *power reduction factor* is expressed by the ratio between the power consumption of the best pipeline version and the power consumption of the original combinatorial circuit.

In the paper, they review a range of publications reporting the use of pipelining across a number of applications. They indicate that power reductions around 50% tend to be obtained. In addition, higher values of the power reduction factor tend to be achieved with larger wordlengths.

Consider the application of pipelining to the FIR filter shown in Figure 13.10. A number of filter realizations and designs were investigated, namely a 4-, 8-, 16-, and 32-tap FIR filter implemented on a Virtex-II XC-2V3000bf 957-6 FPGA. The filter was initialized by loading coefficients using an address bus, data bus and enable signal so that this was consistent for all implementations. No truncation was employed and the output word length is 24 and 27 bits for the 4-tap and 32-tap filters, respectively. Expansion is handled by including a word growth variable which is defined for each filter size to prevent truncation.

Xilinx ISE^{TM} Project Navigator (version 6.2) was used to translate, map, place and route the designs, and sub-programs of the ISE^{TM} design suite were used to compile component libraries, manually place and route and generate post place and route VHDL files for XPower. Xpower was then used to generate simulation results for Table 13.3.

Table 13.2 Pipelined, 54-bit multiplier running at 50 MHz in Spartan-6 FPGA

Pipeline stages	mA	Flip-flops	Power reduction factor
1	67.2	216	1.00
3	34.9	666	0.52
5	27.9	2346	0.42
7	—	—	—
9	23.1	3206	0.34

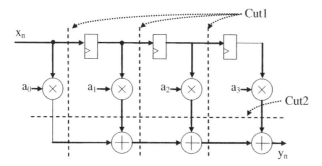

Figure 13.10 Pipelined FIR filter implementation for low power

The unpipelined version (PL0) was used to benchmark the results, with the only change being the latency imposed by pipelining stages. A speech data file was used to simulate the designs, the filter clock speed was 20 MHz and the simulation time was 200 μs. Two levels of pipelining were investigated; a layer of pipelining after the multipliers as shown by Cut2 and given as PL1 in Table 13.3; another pipeline cut in the adder (and delay) chain, Cut1, in addition to the first level of pipelining was then applied and defined as PL2 in Table 13.3, i.e. PL2 encompasses Cut1 *and* Cut2.

The results show power reductions of 59–63% for single-stage pipelined versions and 82–98% for two-stage pipelined versions, depending on filter size. Whilst this is quite dramatic and only based on simulation, it gives some idea of the power reduction possible. It is clear that pipelining becomes more effective as filter size and thus design area and interconnection lengths increase, giving greater opportunity for power-aware placement. Net lengths are shortened by placing interconnected components closely together.

This acts to reduce power consumption in two ways: firstly by decreasing the net capacitance, and secondly by reducing toggling. Partitioning the design into pipelined stages further reduces power consumption by diminishing the ripple effect of propagation delays. This can be seen in Figure 13.11, which shows post place and route capacitance in pF rounded to the nearest integer plotted against the summation of the toggling activity on nets with equivalent capacitance. These values are plotted for PL0, PL1 and PL2, showing that not only is toggle activity reduced on high-capacity nets but overall there are fewer toggles in the design when power reduction techniques are implemented.

Table 13.3 Internal signal/logic power consumption of various filters

	FIR filter tap size			
Technique	4	8	16	32
PL0	8.4	89.7	272.0	964.2
PL1	3.1 (−63%)	29.1 (−68%)	89.7 (−67%)	391.7 (−59%)
PL2	1.5 (−82%)	6.7 (−93%)	8.1 (−97%)	16.4 (−98%)

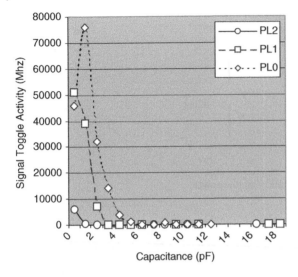

Figure 13.11 Tap signal capacitance and toggle activity

The results presented in Wilton *et al.* (2004) are equally impressive but more thorough; they are based on data from a real board setup. Results are presented for a 64-bit unsigned integer array multiplier, a triple-DES encryption circuit, an eight-tap floating-point FIR filter and a CORDIC circuit to compute the sine and cosine of an angle. Power results just for the FIR filter and the CORDIC circuit are given in the Table 13.4. They were taken from the circuits implemented on an Altera Nios Development Kit (Stratix Professional Edition) which contains a 0.13 μm CMOS Stratix EP1S40F780C5 device which gave the original FPGA power results, and the estimated power was taken from a Quartus simulator and power estimator.

The results show the impact of applying of different levels of pipelining. The authors quote savings overall of 40–82% and indicate, when they factor out quiescent power from the results, that the savings on the dynamic logic block energy can be as high as 98%. They indicate that lower-level physical design optimizations presented in the work in Lamoureux and Wilton (2003) can achieve energy savings of up 23%, highlighting the importance of applying system-level optimizations and highlighting the impact of pipelining generally.

Table 13.4 Pipelining results for 0.13 μm FPGA

Benchmark circuit	Pipeline stages	Estimated power	Original FPGA power
	2	4,420	7,866
8-tap floating point FIR filter	4	2,468	5,580
	Max.	776	3,834
	4	971	5,139
CORDIC circuit to compute	8	611	4,437
sine and cosine of angle	16	565	4,716
	Max.	567	4,140

13.6.3   Locality

It is clear from the discussion on switched capacitance and indeed from Figure 13.11 that the length of interconnection can have a major impact on power consumption. Thus, locality can be a highly attractive feature in circuit architectures for signal and data processing architectures. One class that features localized interconnection is systolic arrays (Kung and Leiserson 1979; Kung 1988); they were initially introduced to address issues of design complexity and the increasing problem of long interconnect in VLSI designs (Mead and Conway 1979).

In addition to locality, systolic array architectures also employ pipelining which makes them attractive for implementing regular computations such as matrix–matrix multiplication and LU decomposition. They benefit immensely from the highly regular nature of DSP computations and offer huge performance potential. The concept was extended to the *bit level*, resulting in bit-level systolic arrays (Woods *et al.* 2008).

The key challenge is to be able to map algorithms into these structures (Choi and Prasanna 2003; Woods *et al.* 2008). Kung (1988) classified DSP algorithms as "locally recursive," for example as with matrix multiplication, and "globally recursive." In locally recursive algorithms, data dependency is limited to adjacent elements, whereas in globally recursive algorithms inherently complex communication networks are required as some cells need to communicate with numerous others.

In McKeown and Woods (2011), some attempt is made to define the locality in terms of a parameter called the *index space separation*. Index space is defined as a lattice of points in an *n*-dimensional discrete space (Parashar and Browne 2000). Strictly, then, index space separation is defined as a measure of total distance values between indices. Defining the hierarchical space where each position in the lattice vector space is a Cartesian coordinate allows the index space to be defined in Euclidean geometry: the index space separation is the Euclidean distance between indices. The index space separation, η, between the two indices $A = (a_1, a_2, \ldots, a_n)$ and $B = (b_1, b_2, \ldots, b_n)$ is thus defined as

$$\eta = \sqrt{(a_1 + b_1)^2 + (a_2 + b_2)^2 + \ldots + (a_n + b_n)^2} = \sqrt{\sum (a_n + b_n)^2}. \qquad (13.7)$$

McKeown and Woods (2011) argue that data dependency can be measured using the index space separation as the relative distances over which data must be passed between consecutive operations. This is not particularly important for power consumption in processor-style implementations, but in FPGA designs it relates directly to the separate individual interconnections created as a result of the FPGA place and route process. This directly relates to the FPGA dynamic power consumption.

Application to FFT Implementation

The Cooley–Tukey algorithm (Cooley and Tukey 1965) is commonly used to compute the FFT (see Section 2.3.2). The problem is that its globally recursive nature manifests itself in the irregular routing where data are routinely passed to non-adjacent global PEs as shown in Figure 2.5.

By removing the in-place restriction of the Cooley–Tukey algorithm, matrix decomposition can be used to increase data locality, thereby minimizing the index space separation. By identifying periodicity and symmetry in the structured transform matrix along which the algorithm can be factorized, a decomposition is achieved with a smaller

$$X_0 = X(n_1, n_2, n_3) \qquad X_1 = X(n_1, n_2, k_3) \qquad X_2 = X(n_1, k_1, k_2) \qquad X_3 = X(k_1, k_2, k_3)$$

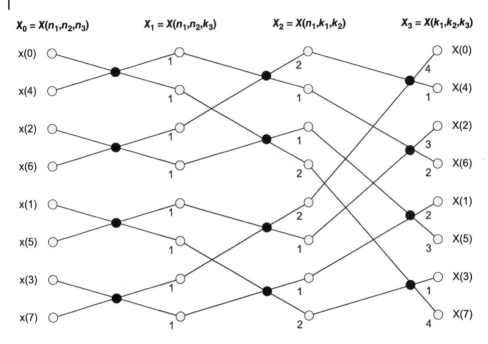

Figure 13.12 Eight-point radix-2 modified flow graph

radix which requires only N/r_m radix-r_m elements at stage m, with the elementary radix operations mapped as a small locally interconnected array.

This gives the modified version shown in Figure 13.12, where the index space separation is reduced by 28.6% from 56 to 40 for the eight-point version. The numbers on the figure represent the index space separation and sum to 40; examination of Figure 2.5 reveals that it is 56 there. It is shown that this can be generalized to a reduction from $O(N(N-1))$ to $O(N(N + \log_2 N - 1)/2)$ in radix-2 for all N, representing a reduction of 40% or more for point sizes greater than 32. The range of values is given in Table 13.5.

Table 13.5 Radix-2 FFT index space separation

N	Figure 13.12	Figure 13.13	Reduction (%)
2	2	2	0
4	12	10	16.7
8	56	40	28.6
16	240	152	36.7
32	992	576	41.9
64	4032	2208	45.2
128	16,256	8,576	47.2
256	65,280	33,664	48.4
512	261,632	133,120	49.1
1024	1,047,552	528,896	49.5
2048	4,192,256	2,107,392	49.7
4096	16,773,120	8,411,136	49.9

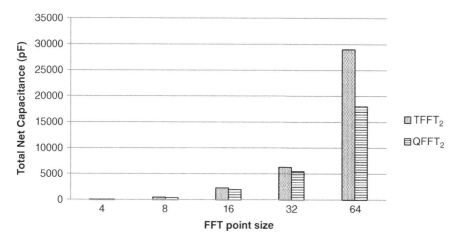

Figure 13.13 Interconnect capacitance versus point size for FFT designs in Xilinx Virtex-7 FPGA technology

The index schedule, where $n_1, n_2, n_3 = 0, 1$ and $k_1, k_2, k_3 = 0, 1$ for iterations I_1, I_2, I_3, I_4, for the in-place mapping schedule, is given by

$$x(n) = I_1(n_1 + 2n_2 + 4n_3)$$
$$X(k) = I_4(4k_3 + 2k_2 + 1k_1), \tag{13.8}$$

and the out-of-place mapping schedule, given by

$$x(n) = I_1(n_1 + 2n_2 + 4n_3)$$
$$X(k) = I_4(1k_1 + 2k_2 + 4k_3), \tag{13.9}$$

is shown in Figure 13.12.

Validation of the research is given in McKeown and Woods (2008). The TFFT design was described in Figure 2.5, and Figure 13.12 represents the QFFT design. The designs were synthesized using Version 11 of the Xilinx ISE tools, giving the results in Figure 13.13. The chart shows the saving in accumulated routing capacitance for QFFT design as the FFT point sizes increases. In addition, Figure 13.14 gives the detailed distribution of the number of capacitance values against sizes for a specific point size, namely a 64-point FFT. It clearly shows how the lower index space separation results in placed and routed designs with lower capacitance. Power savings of 36–37% were achieved.

13.6.4 Data Mapping

FPGA vendors have developed optimized dedicated processing elements such as the DSP48E blocks in the Xilinx FPGA families and DSP blocks in the Altera family. The data size has been predetermined to be 9–18 bits in most cases, and so it is a case of FPGA vendors trying to predetermine the largest finite word size that will work well for a range of applications. Their aim is to provide sufficient dynamic range to meet all operating conditions.

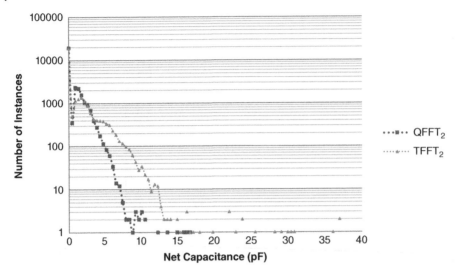

Figure 13.14 Interconnect capacitance for a 64-point FFT design implemented using Xilinx Virtex-7 FPGA technology

There are some applications, however, such as in radar applications that do not even require this dynamic range but require the processing of data at multi-GSPS. One example is in electronic warfare (EW), where radar is used to detect and classify airborne, naval and land-based threats; flexibility is needed to meet changeable tactical and mission requirements to ensure high intercept probability. Power is critical and systems typically limit the maximum functionality, resulting in a narrow and inflexible design space (McKeown and Woods 2013).

In these cases, there is an opportunity to process more than one data stream in the dedicated resources in the dynamic range, typically 18 bits in Altera and Xilinx FPGA technology. For example, the Altera DSP block supports numerous wordlengths (9 bits, etc.). It would seem a little contrived but, as the next section indicates, there is a low-precision mode which operates on the large portion of the data and a high-precision mode when an event is detected.

FFT-Based Digital Receiver

The EW receiver demonstrator (Figure 13.15) contains four distinct functions, including ADCs, conversion into the frequency domain using the FFT and frequency domain analysis with selection and detection of signals of interest. For this application, power consumption is the dominating factor and FFT is the key factor to such an extent that system functionality is designed around the power requirements of the FFT cores.

It is critical in EW environments to entirely encapsulate a pulse in as short a window as possible for time of arrival (TOA) and sensitivity reasons. This creates a clear need

Figure 13.15 Digital receiver architecture for radar system

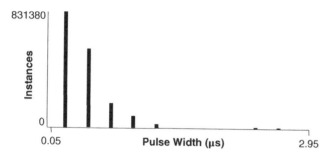

Figure 13.16 Pulse width instances: mixture of nautical and airborne radars

to run lots of FFTs of different point sizes and also to run overlapping FFT functions in real time. The time spent at each point size depends on the precise mission criteria and operating environment and ultimately determines the FPGA resources. The pulse width shown in Figure 13.16 gives the aggregated number of measured instances for various pulse widths in a typical extremely active radar environment. The majority of pulse widths are relatively short, suggesting high numbers of shorter- length FFTs with a need to perform overlapping FFTs.

To be effective, pulse envelopes must be captured in a high-resolution digital representation which requires a large dynamic range to encapsulate pulse envelope peaks during worst-case operating conditions, leading to underutilization during significant time periods. Figure 13.17 outlines the typical dynamic range of aggregated instances and shows that only 8% of samples exceed half of the dynamic range, meaning that there is 50% underutilization for 92% of the time. This is typical of this application domain as, although the radar environment is extremely active, the vast majority of peak pulses are in the sub-microsecond range.

There is a conflict between making the window large enough to minimize the probability of a pulse being split across two windows, and yet still small enough to ensure that the TOA is contained in the window. This is overcome by using the processing resources in a much more dynamic fashion for variable-length algorithms. This is achieved by exploiting the commonality in the fundamental operations of DSP algorithmic variations, mapping them to a static architecture of underlying processing elements, and using a flexible routing structure to implement the required functionality.

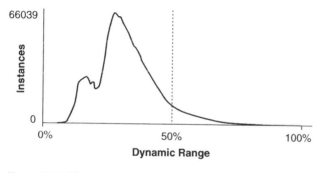

Figure 13.17 DR instances: nautical and airborne radars

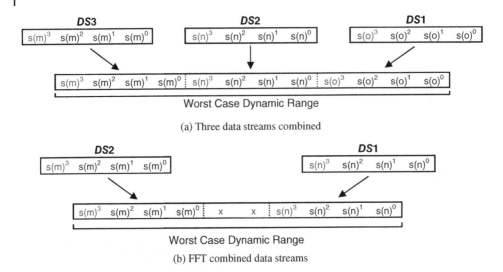

Figure 13.18 Combining data streams

Depending on the exact correlation of the standard deviation between worst-case and normal operation dynamic ranges, two or more data streams can be computed by incorporating them into a single word stream. This is shown for three input streams (DS1, DS2 and DS3) in Figure 13.18(a) and two internal streams which allows for word growth (Figure 13.18(b)). This is done by dynamically grouping processing threads to meet real-time operating requirements.

The difficulty is that while multiple data streams can be routed through some elements such as delays and multiplexers directly, others like adders and multipliers require special consideration to prevent overflow due to word growth during operations. Fixed hardware in the FPGA fabric means carry-chain modification is not possible, but an alternative approach is presented which overcomes these limitations by cleverly exploiting the existing hardware at the bit level to achieve computational separation as outlined in McKeown and Woods (2013).

13.7 Final Comments

Since the first edition of this book, power consumption has become an even more critical aspect of FPGAs. The lower power consumption of FPGA has made them highly attractive compared to their computing alternatives such as CPUs and GPUs and has largely been instrumental in the interest of such organizations such as IBM, Intel, Amazon and Microsoft in adopting the technology. Of course, FPGAs have not been immune to the problems that have driven this interest and have had to address particularly static power consumption by adopting technology, circuit and architectural solutions as covered in this chapter.

Unlike many computing alternatives, it is possible to derive the FPGA architecture to best match the algorithmic requirements and thus optimize the dynamic power consumption. This can be achieved to some extent by employing approaches to speed up the throughput rate and then reducing the voltage to achieve reduction in power consumption, but this has limited potential as the FPGA technology has already been optimized for low-voltage operation.

A number of other approaches exist for reducing the switch capacitance. One of the most natural is to employ pipelining, as the availability of registers in the FPGA fabric makes this an obvious approach. Introducing pipeline registers has a number of attractive attributes from a power consumption perspective; it acts to reduce glitching and also the interconnection capacitance. The increase in throughput rate also enables a reduction in the amount of computation needed to provide the solution and employ hardware sharing. The chapter has also outlined some DSP-specific optimizations which have been applied in a real FFT-based application.

Bibliography

Boemo E, Oliver JP, Caffarena G 2013 Tracking the pipelining-power rule along the FPGA technical literature. In *FPGAWorld '13*, Stockholm, Sweden.

Bowman KA, Austin LB, Eble JC, Tang X, Meindl JD 1999 A physical alpha-power-law MOSFET model. *IEEE J. of Solid-State Circuits*, 32, 1410–1414.

Chandrakasan A, Brodersen R 1996 *Low Power Digital Design*, Kluwer, Dordrecht.

Chapman K, Hussein J 2012 Lowering power using the voltage identification bit. Application Note: Virtex-7 FPGAs, XAPP555 (v1.1) (accessed February 28, 2016).

Chen C-S, Hwang TT, Liu CL 1997 Low power FPGA design – a re-engineering approach. In *Proc. of Design Automation Conf.*, pp. 656–661.

Choi S, Prasanna VK 2003 Time and energy efficient matrix factorization using FPGAs. In *Proc. Int. Conf. on Field Programmable Logic and Applications*, pp. 507–519.

Chow CT, Tsui LSM, Leong PHW, Luk W, Wilton S 2005 Dynamic voltage scaling for commercial FPGAs. In *Proc. Int. Conf. on Field-Programmable Technology*, pp. 215–222.

Cooley JW, Tukey JW 1965 An algorithm for the machine calculation of complex Fourier series. *Mathematics of Computation*, 19, 297–301.

Curd D 2007 Power consumption in 65 nm FPGAs. White Paper: Virtex-5 FPGAs. Available from www.xilinx.com (accessed February 28, 2016).

Erdogan AT, Arslan T 2000 High throughput FIR filter design for low power SoC applications. In *Proc. 13th Annual IEEE Conf. on ASIC/SOC*, pp. 374–378.

Erdogan AT, Arslan T 2002 Implementation of FIR filtering structures on single multiplier DSPs. *IEEE Trans. on Circuits and Systems II*, 49(3), 223–229.

Gonzalez R, Gordon B, Horowitz M 1997 Supply and threshold scaling for low power CMOS. *IEEE J. of Solid-State Circuits*, 32(8), 1210–1216.

Huda S, Mallick M, Anderson JH 2009 Clock gating architectures for FPGA power reduction. In *Proc. Int. Conf. on Field Programmable Logic and Applications*, pp. 112–118.

ITRS 2003 International Roadmap for Semiconductors. Available from http://public.itrs.net (accessed February 28, 2016).

Keane G, Spanier JR, Woods R 1999 Low-power design of signal processing systems using characterization of silicon IP cores. In *Proc of 33rd Asilomar Conference on Signals, Systems and Computers*, pp. 767–771.

Kim NS, Austin T, Baauw D, Mudge T, Flautner K, Hu JS, Irwin MJ, Kandemir M, Narayanan V 2003 Leakage current: Moore's law meets static power. *IEEE Computer*, 36(12), 68–75.

Kung HT, Leiserson CE 1979 Systolic arrays (for VLSI). In *Proc. on Sparse Matrix*, pp. 256–282.

Kung SY 1988 *VLSI Array Processors*. Prentice Hall, Englewood Cliffs, NJ.

Lamoureux J, Wilton SJE 2003 On the interaction between power-aware FPGA CAD algorithms. In *Proc. IEEE/ACM Int. Conf. on Computer Aided Design*.

McKeown S, Woods R 2008 Algorithmic factorisation for low power FPGA implementation through increased data locality. In *Proc. of IEEE Int. Symp. VLSI Design, Automation and Test*, pp. 271–274.

McKeown S, Woods R 2011 Low power FPGA implementation of fast DSP algorithms: Characterisation and manipulation of data locality. *IET Proc. on Computer and Design Techniques*, 5(2), 136–144.

McKeown S, Woods R 2013 Power efficient FPGA implementations of transform algorithms for radar-based digital receiver applications. *IEEE Trans. on Industrial Electronics*, 9(3), 1591–1600.

Mead C, Conway L 1979 *Introduction to VLSI Systems*. Addison-Wesley, Reading, MA.

Nunez-Yanez J 2015 Adaptive voltage scaling with in-situ detectors in commercial FPGAs. *IEEE Trans. on Computers*, 641, 45–53.

Nunez-Yanez J, Chouliaras V, Gaisler J 2007 Dynamic voltage scaling in a FPGA-based system-on-chip. In *Proc. Int. Conf. on Field Programmable Logic and Applications*, pp. 459–462.

Parashar M, Browne JC 2000 Systems engineering for high performance computing software: The HDDA/DAGH infrastructure for implementation of parallel structured adaptive mesh refinement. In Baden SB, Chrisochoides NP, Gannon DB, Norman ML (eds) *Structured Adaptive Mesh Refinement (SAMR) Grid Methods*, IMA Volumes in Mathematics and its Applications, 117, pp. 1–18. Springer, New York

Raghunathan A, Dey S, Jia NK 1999 Register transfer level power optimization with emphasis on glitch analysis and reduction. *IEEE Trans. Computer Aided Design*, 18(8), 114–131.

Roy K, Mukhopadhyay S, Meimand H 2003 Leakage current mechanisms and leakage reduction technqiues in deep-submicron CMOS circuits. *Proc. IEEE*, 91(2), 305–327.

Ryan J, Calhoun B 2010 A sub-threshold FPGA with low-swing dual-VDD interconnect in 90nm CMOS. In *Proc. IEEE Custom Integrated Circuits Conf.*, pp. 1–4.

Shen A, Kaviani A, Bathala K 1992 On average power dissipation and random pattern testability of CMOS combinational logic networks. In *IEEE Int. Conf. on Computer Aided Design*, pp. 402–407.

Wilton SJE, Luk W, Ang SS 2004 The impact of pipelining on energy per operation in field-programmable gate arrays. *Proc. of Int. Conf. on Field Programmable Logic and Applications*, pp. 719—728.

Wolf W 2004 *FPGA-Based System Design*. Prentice Hall, Upper Saddle River, NJ.

Woods RF, McCanny JV, McWhirter JG 2008 From bit level systolic arrays to HDTV processor chips. *J. of VLSI Signal Processing*, 53(1–2), 35–49.

14

Conclusions

14.1 Introduction

The aim of this book has been to cover many of the techniques needed to create FPGA solutions for signal and data processing systems. Interest in FPGAs for such systems has grown since the first edition as vendors have targeted this market and introduced innovations into their technology offerings to make them more attractive. This has included dedicated DSP blocks typically comprising 18-bit MAC blocks and increased memory units but also increased parallelism in terms of resources and the ability to employ pipelining to improve speed of operation. To support this activity, FPGA companies have created a range of IP cores (for customers to use in their designs) and design tools to create such implementations.

The main attraction of using FPGAs is that the resulting designs provide a very high quality of performance, particularly if $MSPS/mm^2/W$ is considered. This is possible as the designer is able to create an architecture which is a good match to the algorithmic needs, rather than striving to map the requirements onto a fixed architecture as is the case in microcontrollers, DSP microprocessors or GPUs, even though vendors increasingly offer multicore platforms. The main design challenge is to create a suitable architecture for the algorithmic requirements.

The creation of this suitable FPGA architecture comes from employing the right level of parallelism and pipelining to match initially the throughput rate and then area and power consumption requirements. The key approach taken in this book has been to derive an efficient circuit architecture which successfully utilizes the underlying resources of the FPGA to best match the computational and communication requirements of the applications. This was demonstrated using simple design examples such as FIR, IIR and lattice filer structures in Chapter 8 as well as more complex examples such as the fixed beamformer in Chapter 10 and the adaptive beamformer in Chapter 11.

The purpose of this chapter is to give some attention to emerging issues and provide some insight into future challenges for FPGAs. In Section 14.2, attention is given to the changes in design methods as FPGA architectures have emerged. The rest of the chapter considers a range of issues likely to be important in the future. Firstly, more consideration is given in Section 14.3 to the use of FPGAs in Big Data applications and

FPGA-based Implementation of Signal Processing Systems,
Second Edition. Roger Woods, John McAllister, Gaye Lightbody and Ying Yi.
© 2017 John Wiley & Sons, Ltd. Published 2017 by John Wiley & Sons, Ltd.

the implications for FPGAs in high-performance computing. Connected to this is the need to allow FPGAs to be more effectively integrated in future computing systems, and this is considered in Section 14.4. In Sections 14.5 and 14.6, the key issues of floating-point arithmetic and memory architectures are then covered.

14.2 Evolution in FPGA Design Approaches

The emergence of a dedicated DSP block as outlined in Chapter 5 has now simplified the mapping of design functions. This is particularly relevant for fixed-point DSP systems with wordlengths from 8 to 12 bits (a core target market for FPGAs) as these designs readily map into the 18-bit DSP blocks if wordlength growth has been addressed. The use of pipelining is easily achievable through use of programmable pipeline registers in the DSP blocks and the plethora of scalable resisters in the main programmable logic fabric.

Moreover, since the first edition of this book, there have been a number of innovations in FPGA design, primarily focused around design tools. The ability to trade off levels of parallelism and pipelining has been encapsulated to some extent within FPGA vendor synthesis tools such as the Xilinx Vivado, where the starting position is a C description; this is clearly aimed at engineering companies as C is key design language. The Altera perspective has been to start with an OpenCL description and then use the conventional FPGA place and route implementation tools to produce the final bit files for programming the FPGA.

There has been a major growth in the availability of soft IP cores such as parameterized HDL cores for a range of telecommunications, signal and image processing applications, as well as dedicated memory interface circuitry and soft processor cores. The availability of commercial cores through the Design & Reuse website (http://www.design-reuse.com/), which has 16,000 IP from 450 vendors, and open source cores from the OpenCores website (opencores.org) provides a strong "plug-and-play" design ethos to system design; this will be an increasingly important aspect as system complexities grow.

The evolution of FPGAs to SoC platforms has transformed the design problem from one only concerned with HDL-based implementation to a highly parallel, programmable logic fabric to a hardware/software system design challenge involving the incorporation of IP cores. Of course, the shift toward hardware/software FPGAs is not new, as in the early 2000s the Xilinx Virtex-II incorporated a PowerPC processor into the FPGA die, but this was not truly supported as a hardware/software environment. This has now meant an explosion in requirements for design teams with skills in embedded system programming, memory partitioning and incorporation of system components and accompanying system timing issues.

14.3 Big Data and the Shift toward Computing

An interesting development highlighted in Chapters 1 and 12 is the interest taken by major computing companies such as Microsoft, Intel and IBM in FPGAs. Whilst FPGAs have been around for many decades, it has only been in the last couple of years that these major companies have shown an interest in employing this technology. This has been driven by energy concerns and this computing infrastructure driven by the emergence of data science and accompanying data centers to provide the infrastructure for undertaking such data inquiries.

The new computing infrastructures are challenging as they need to access large and varied sources of data distributed across websites and then create in-memory databases in a dynamic fashion to create microservers that will perform both transactional and analytical processing. The former is being undertaken on general- purpose computers using dedicated hardware to undertake the latter form of processing. One example is the Nanostreams project (www.nanostreams.eu) which is looking to create an application-specific heterogeneous analytics-on-chip (AoC) engine to perform such processing. It includes an AoC accelerator being developed by Analytics Engines Ltd. that is based on FPGAs in the form of a programmable, customizable processing core called Nanocore which acts to give improvements in performance and energy-efficiency over GPUs.

Microsoft has undertaken research into novel, FPGA-based data center architectures and created a reconfigurable fabric called Catapult. They argue that "datacenter providers are faced with a conundrum: they need continued improvements in performance and efficiency, but cannot obtain those improvements from general-purpose systems" (Putnam *et al.* 2014). The system comprises a bed of 1632 servers equipped with FPGAs giving gains in search throughput and latency for Bing, giving 95% greater ranking throughput in a production search infrastructure at comparable latency to a software-only solution (Putnam *et al.* 2014).

The IBM work with Xilinx has focused on accelerating Memcache2, a general-purpose distributed memory caching system used to speed up dynamic database-driven searches (Blott and Vissers 2014). Intel's purchase of Altera (Clark 2015) clearly indicates a clear aim of employing FPGAs in heterogeneous computing for data centers.

Unlike processors, FPGAs offer the capability of turning off and on resources, thus allowing scaling. This provides a more direct relationship between power consumed and the processing employed. Therefore, we can create an implementation where the dynamic power can be adjusted to match the processing granularity needed by the user. This can be further refined by adjusting the voltage and employing clock gating to reduce the overall static power consumed. As outlined in Chapter 13, this further improvement may be possible but at the cost of increased design effort.

14.4 Programming Flow for FPGAs

A topic closely associated with the further adoption of FPGAs in computing applications is the design flow. Whilst progress has been made in increasing the level of abstraction, thus removing the need to get software designers to learn specialized HDLs for programming FPGAs and allowing them to employ C descriptions in Vivado and even OpenCL, the compile times will still seem strangely long to programmers.

These tools now allow developers to write their programs in high-level languages (well, high-level for an engineer!) and then use aspects of the tools to gauge speed requirements against FPGA resources. However, compile times of typically hours will seem alien to programmers and will be prohibitive to further adoption of the technology. Considerable efforts are being made to overcome this limitation (including efforts by this book's authors) by building multi-processors which can then be programmed in software.

This seems counterproductive for several reasons. Firstly, the FPGA vendors have already introduced the MicroBlaze (Xilinx) and NIOS (Altera) processors. Secondly, the advantage of the FPGA is in creating application- specific implementations, thus overcoming the fixed multi-processor architecture which is exactly what is being proposed here. So this is the conundrum: how to gain the advantages of FPGA performance without having to undertake much of the work highlighted in this book.

14.5 Support for Floating-Point Arithmetic

A conscious decision to first introduce scalable adder structures in early FPGAs and then dedicated multiplicative complexity in latter versions, such as the Stratix® III family from Altera and the Virtex^{TM}-5 FPGA family from Xilinx, has greatly influenced the use of FPGAs for DSP systems. Along with the availability of distributed memory, this has driven further interest in using FPGAs for computing due to the extremely high computation rates required.

If FPGAs are to be applied to computing or even supercomputing applications, then support for floating-point is needed. Chapter 5 outlined the progress that FPGA vendors have made in incorporating floating-point arithmetic in FPGAs, particularly in the Altera Arria® 10 family. This FPGA contains multiple IEEE 754 single-precision multipliers and IEEE 754 single-precision adders in each DSP block. This provides support for a variety of addition, multiplication and MAC floating-point operations, useful for a variety of vector operations.

The Arria® 10 FPGA family gives a peak performance of 3340 GMACs and 1366 GFLOPS. The technology has been incorporated into the new 510T from Nallatech which is termed an "extreme compute acceleration" technology targeted at data centers. It is an FPGA PCIe Gen3 card comprising two Altera Arria 10 1150 GX FPGAs providing up to 3 TFLOPS with 4 GB DDR3 per FPGA. The card offers hybrid memory cube memory architectures using a high-speed process technology through-silicon via bonded memory die. Altera offers a single Arria FPGA card called a DK-SOC-10AS066S-ES development kit. Other FPGA-based platform vendors include Alphadata, Accelize and Picocomputing (now Micron).

14.6 Memory Architectures

The support for parallel and pipelining operations was highlighted as the major attraction of FPGAs when considered for implementing DSP systems. However, one factor that has received some attention throughout this book is the availability of a wide range of different sizes of parallel memory, whether in the form of distributed RAM blocks, simple LUTs or a single register.

As highlighted by Wulf and McKee (1995), the memory wall gives a depressing view for fixed computer architectures as the ratio of the memory access time to the processor cycle time increases. Whilst some approaches try to address this via technology and increased use of multi-level caches (Baer and Wang 1988), FPGAs get around the problem by naturally developing a highly parallel solution with a distributed memory architecture. This happens through deliberate derivation of a distributed memory architecture or as a result of an algorithmic optimization, as for example in the application of pipelining which, in effect, results in the creation of distributed memory. This approach is particularly suited to many signal and data systems due to data independence and high computation rates.

This means that there needs to be a focus on ensuring memory utilization rather than computation. This was seen in the Imagine processor (Kapasi *et al.* 2002) where the memory architecture was developed for the class of algorithms needed, and in some of the FPGA examples in Chapters 6, where different memory, i.e. LUTs in the forms of SRLs, was selected in preference to flip-flops to provide more efficient implementation delay chains, either because of lack of flip-flop resources or more relevant selection of resources. However, this has tended to be a good design decision or optimization using the routines available in design tools, rather than a conscious need to develop memory-orientated architectures. Work by Fischaber *et al.* (2010) has suggested how design of memory can be directed from the dataflow level.

Bibliography

Baer J-L, Wang W-H. 1988. On the inclusion properties for multi-level cache hierarchies. In *Proc. 15th Ann. Int. Symp. on Computer Architecture*, 73–80.

Blott M, Vissers K 2014 Dataflow architectures for 10Gbps line-rate key-value-stores. In *Proc. IEEE Hot Chips*, Palo Alto, CA.

Clark D 2015 Intel completes acquisition of Altera. *Wall Street J.*, December 28.

Fischaber S, Woods R, McAllister J 2010 SoC memory hierarchy derivation from dataflow graphs. *Journal of Signal Processing Systems*, 60(3), 345–361.

Kapasi UJ, Dally WJ, Rixner S, Owens JD, Khailany B 2002 The Imagine stream processor. In *Proc. IEEE Int. Conf on Computer Design: VLSI in Computers and Processors*, pp. 282–288.

Putnam A, Caulfield AM, Chung ES, Chiou D, Constantinides K, Demme J, Esmaeilzadeh H, Fowers J, Gopal GP, Gray J, Haselman M, Hauck S, Heil S, Hormati A, Kim J-Y, Lanka S, Larus J, Peterson E, Pope S, Smith A, Thong J, Xiao PY, Burger D 2014 A reconfigurable fabric for accelerating large-scale datacenter services. In *Proc. IEEE Int. Symp. on Computer Architecture*, pp. 13–24.

Wulf WA, McKee SA 1995 Hitting the memory wall: implications of the obvious. In *SIGARCH Comput. Archit. News*, 23(1), 20–24.

Index

FPGA-based Implementation of Signal Processing Systems,
Second Edition. Roger Woods, John McAllister, Gaye Lightbody and Ying Yi.
© 2017 John Wiley & Sons, Ltd. Published 2017 by John Wiley & Sons, Ltd.

Printed and bound by CPI Group (UK) Ltd, Croydon, CR0 4YY

27/10/2024

14580360-0002